Electronic Circuits: Fundamentals and Applications

Third Edition

Michael Tooley BA
Formerly Vice Principal
Brooklands College of Further and Higher Education

ELSEVIER

AMSTERDAM • BOSTON • HEIDELBERG • LONDON • NEW YORK •OXFORD
PARIS • SAN DIEGO • SAN FRANCISCO • SINGAPORE • SYDNEY • TOKYO

Newnes is an imprint of Elsevier

Newnes

Newnes is an imprint of Elsevier
Linacre House, Jordan Hill, Oxford OX2 8DP, UK
30 Corporate Drive, Suite 400, Burlington, MA 01803, USA

First edition 2006
Reprinting 2007

British Library Cataloguing in Publication Data
A catalogue record for this book is available from the British Library

Library of Congress Cataloging-in-Publication Data
A catalog record for this book is available from the Library of Congress

ISBN–13: 978-0-7506-6923-8
ISBN–10: 0-7506-6923-3

For information on all Newnes publications
visit our website at www.newnespress.com

Printed and bound in *The Netherlands*

07 08 09 10 10 9 8 7 6 5 4 3 2

Working together to grow
libraries in developing countries

www.elsevier.com | www.bookaid.org | www.sabre.org

ELSEVIER   BOOK AID International   Sabre Foundation

Electronic Circuits: Fundamentals and Applications

Please return on

2012 | 15

efore the date below.

Contents

Preface

This is the book that I wish I had when I first started exploring electronics nearly half a century ago. In those days, transistors were only just making their debut and integrated circuits were completely unknown. Of course, since then much has changed but, despite all of the changes, the world of electronics remains a fascinating one. And, unlike most other advanced technological disciplines, electronics is still something that you can 'do' at home with limited resources and with a minimal outlay. A soldering iron, a multi-meter, and a handful of components are all that you need to get started. Except, of course, for some ideas to get you started—and that's exactly where this book comes in!

The book has been designed to help you understand how electronic circuits work. It will provide you with the basic underpinning knowledge necessary to appreciate the operation of a wide range of electronic circuits including amplifiers, logic circuits, power supplies and oscillators.

The book is ideal for people who are studying electronics for the first time *at any level* including a wide range of school and college courses. It is equally well suited to those who may be returning to study or who may be studying independently as well as those who may need a quick *refresher*. The book has 19 chapters, each dealing with a particular topic, and eight appendices containing useful information. The approach is topic-based rather than syllabus-based and each major topic looks at a particular application of electronics. The relevant theory is introduced on a progressive basis and delivered in manageable chunks.

In order to give you an appreciation of the solution of simple numerical problems related to the operation of basic circuits, worked examples have been liberally included within the text. In addition, a number of problems can be found at the end of each chapter and solutions are provided at the end of the book. You can use these end-of-chapter problems to check your understanding and also to give you some experience of the 'short answer' questions used in most in-course assessments. For good measure, we have included 70 revision problems in Appendix 2. At the end of the book you will find 21 sample coursework assignments. These should give you plenty of 'food for thought' as well as offering you some scope for further experimentation. It is not envisaged that you should complete all of these assignments and a carefully chosen selection will normally suffice. If you are following a formal course, your teacher or lecturer will explain how these should be tackled and how they can contribute to your course assessment. While the book assumes no previous knowledge of electronics you need to be able to manipulate basic formulae and understand some simple trigonometry in order to follow the numerical examples. A study of mathematics to GCSE level (or equivalent) will normally be adequate to satisfy this requirement. However, for those who may need a refresher or have had previous problems with mathematics, Appendix 6 will provide you with the underpinning mathematical knowledge required.

In the later chapters of the book, a number of representative circuits (with component values) have been included together with sufficient information to allow you to adapt and modify the circuits for your own use. These circuits can be used to form the basis of your own practical investigations or they can be combined together in more complex circuits.

Finally, you can learn a great deal from building, testing and modifying simple circuits. To do this you will need access to a few basic tools and some minimal test equipment. Your first purchase should be a simple multi-range meter, either digital or analogue. This instrument will allow you to measure the voltages and currents present so that you can compare them with the predicted values. If you are attending a formal course of instruction and have access to an electronics laboratory, do make full use of it!

A note for teachers and lecturers

The book is ideal for students following formal courses (e.g. GCSE, AS, A-level, BTEC, City and Guilds, etc.) in schools, sixth-form colleges, and further/higher education colleges. It is equally well suited for use as a text that can support distance or flexible learning and for those who may need a 'refresher' before studying electronics at a higher level.

While the book assumes little previous knowledge students need to be able to manipulate basic formulae and understand some simple trigonometry to follow the numerical examples. A study of mathematics to GCSE level (or beyond) will normally be adequate to satisfy this requirement.

However, an appendix has been added specifically to support students who may have difficulty with mathematics. Students will require a scientific calculator in order to tackle the end-of-chapter problems as well as the revision problems that appear at the end of the book.

We have also included 21 sample coursework assignments. These are open-ended and can be modified or extended to suit the requirements of the particular awarding body. The assignments have been divided into those that are broadly at Level 2 and those that are at Level 3. In order to give reasonable coverage of the subject, students should normally be expected to complete between four and five of these assignments. Teachers can differentiate students' work by mixing assignments from the two levels. In order to challenge students, minimal information should be given to students at the start of each assignment. The aim should be that of giving students 'food for thought' and encouraging them to develop their own solutions and interpretation of the topic.

Where this text is to be used to support formal teaching it is suggested that the chapters should be followed broadly in the order that they appear with the notable exception of Chapter 14. Topics from this chapter should be introduced at an early stage in order to support formal lab work. Assuming a notional delivery time of 4.5 hours per week, the material contained in this book (together with supporting laboratory exercises and assignments) will require approximately two academic terms (i.e. 24 weeks) to deliver in which the total of 90 hours of study time should be divided equally into theory (supported by problem solving) and practical (laboratory and assignment work). The recommended four or five assignments will require about 25 to 30 hours of student work to complete. Finally, when constructing a teaching programme it is, of course, essential to check that you fully comply with the requirements of the awarding body concerning assessment and that the syllabus coverage is adequate.

Mike Tooley
January 2006

A word about safety

When working on electronic circuits, personal safety (both yours and of those around you) should be paramount in everything that you do. Hazards can exist within many circuits—even those that, on the face of it, may appear to be totally safe. Inadvertent misconnection of a supply, incorrect earthing, reverse connection of a high-value electrolytic capacitor, and incorrect component substitution can all result in serious hazards to personal safety as a consequence of fire, explosion or the generation of toxic fumes.

Potential hazards can be easily recognized and it is well worth making yourself familiar with them but perhaps the most important point to make is that electricity acts very quickly and you should always think carefully before working on circuits where mains or high voltages (i.e. those over 50 V, or so) are present. Failure to observe this simple precaution can result in the very real risk of electric shock.

Voltages in many items of electronic equipment, including all items which derive their power from the a.c. mains supply, are at a level which can cause sufficient current flow in the body to disrupt normal operation of the heart. The threshold will be even lower for anyone with a defective heart. Bodily contact with mains or high-voltage circuits can thus be lethal. The most critical path for electric current within the body (i.e. the one that is most likely to stop the heart) is that which exists from one hand to the other. The hand-to-foot path is also dangerous but somewhat less dangerous than the hand-to-hand path.

So, before you start to work on an item of electronic equipment, it is essential not only to switch off but to disconnect the equipment at the mains by removing the mains plug. If you have to make measurements or carry out adjustments on a piece of working (or 'live') equipment, a useful precaution is that of using one hand only to perform the adjustment or to make the measurement. Your 'spare' hand should be placed safely away from contact with anything metal (including the chassis of the equipment which may, or may not, be earthed).

The severity of electric shock depends upon several factors including the magnitude of the current, whether it is alternating or direct current, and its precise path through the body. The magnitude of the current depends upon the voltage which is applied and the resistance of the body. The electrical energy developed in the body will depend upon the time for which the current flows. The duration of contact is also crucial in determining the eventual physiological effects of the shock. As a rough guide, and assuming that the voltage applied is from the 250 V 50 Hz a.c. mains supply, the following effects are typical:

Current	Physiological effect
less than 1 mA	Not usually noticeable
1 mA to 2 mA	Threshold of perception (a slight tingle may be felt)
2 mA to 4 mA	Mild shock (effects of current flow are felt)
4 mA to 10 mA	Serious shock (shock is felt as pain)
10 mA to 20 mA	Motor nerve paralysis may occur (unable to let go)
20 mA to 50 mA	Respiratory control inhibited (breathing may stop)
more than 50 mA	Ventricular fibrillation of heart muscle (heart failure)

It is important to note that the figures are quoted as a guide—there have been cases of lethal shocks resulting from contact with much lower voltages and at relatively small values of current. The upshot of all this is simply that **any potential in excess of 50 V should be considered dangerous.** Lesser potentials may, under unusual circumstances, also be dangerous. As such, it is wise to **get into the habit of treating all electrical and electronic circuits with great care.**

1

Electrical fundamentals

This chapter has been designed to provide you with the background knowledge required to help you understand the concepts introduced in the later chapters. If you have studied electrical science, electrical principles, or electronics beyond school level then you will already be familiar with many of these concepts. If, on the other hand, you are returning to study or are a newcomer to electronics or electrical technology this chapter will help you get up to speed.

Fundamental units

You will already know that the units that we now use to describe such things as length, mass and time are standardized within the International System of Units. This SI system is based upon the seven **fundamental units** (see Table 1.1).

Derived units

All other units are derived from these seven fundamental units. These **derived units** generally have their own names and those commonly encountered in electrical circuits are summarized in Table 1.2 together with the corresponding physical quantities.

Table 1.1 SI units

Quantity	Unit	Abbreviation
Current	ampere	A
Length	metre	m
Luminous intensity	candela	cd
Mass	kilogram	kg
Temperature	Kelvin	K
Time	second	s
Matter	mol	mol

(Note that 0 K is equal to −273°C and an **interval** of 1 K is the same as an **interval of 1°C**.)

If you find the exponent notation shown in Table 1.2 a little confusing, just remember that V^{-1} is simply $1/V$, s^{-1} is $1/s$, m^{-2} is $1/m^{-2}$, and so on.

Example 1.1

The unit of flux density (the Tesla) is defined as the magnetic flux per unit area. Express this in terms of the fundamental units.

Solution

The SI unit of flux is the Weber (Wb). Area is directly proportional to length squared and, expressed in terms of the fundamental SI units, this is square metres (m^2). Dividing the flux (Wb) by the area (m^2) gives Wb/m^2 or $Wb\ m^{-2}$. Hence, in terms of the fundamental SI units, the Tesla is expressed in $Wb\ m^{-2}$.

Table 1.2 Electrical quantities

Quantity	Derived unit	Abbreviation	Equivalent (in terms of fundamental units)
Capacitance	Farad	F	$A\ s\ V^{-1}$
Charge	Coulomb	C	$A\ s$
Energy	Joule	J	$N\ m$
Force	Newton	N	$kg\ m\ s^{-1}$
Frequency	Hertz	Hz	s^{-1}
Illuminance	Lux	lx	$lm\ m^{-2}$
Inductance	Henry	H	$V\ s\ A^{-1}$
Luminous flux	Lumen	lm	$cd\ sr$
Magnetic flux	Weber	Wb	$V\ s$
Potential	Volt	V	$W\ A^{-1}$
Power	Watt	W	$J\ s^{-1}$
Resistance	Ohm	Ω	$V\ A^{-1}$

Example 1.2

The unit of electrical potential, the Volt (V), is defined as the difference in potential between two points in a conductor which, when carrying a current of one Amp (A), dissipates a power of one Watt (W). Express the Volt (V) in terms of Joules (J) and Coulombs (C).

Solution

In terms of the derived units:

$$\text{Volts} = \frac{\text{Watts}}{\text{Amperes}} = \frac{\text{Joules/seconds}}{\text{Amperes}}$$

$$= \frac{\text{Joules}}{\text{Amperes} \times \text{seconds}} = \frac{\text{Joules}}{\text{Coulombs}}$$

Note that: Watts = Joules/seconds and also that Amperes × seconds = Coulombs

Alternatively, in terms of the symbols used to denote the units:

$$V = \frac{W}{A} = \frac{J/s}{A} = \frac{J}{As} = \frac{J}{C} = JC^{-1}$$

Hence one Volt is equivalent to one Joule per Coulomb.

Measuring angles

You might think it strange to be concerned with angles in electrical circuits. The reason is simply that, in analogue and a.c. circuits, signals are based on repetitive waves (often sinusoidal in shape). We can refer to a point on such a wave in one of two basic ways, either in terms of the time from the start of the cycle or in terms of the angle (a cycle starts at 0° and finishes as 360° (see Fig. 1.1)). In practice, it is often more convenient to use angles rather than time, however, the two methods of measurement are interchangeable and it's important to be able to work in either of these units.

In electrical circuits, angles are measured in either degrees or radians (both of which are strictly dimensionless units). You will doubtless already be familiar with angular measure in degrees where one complete circular revolution is equivalent to an angular change of 360°. The alternative method of

measuring angles, the **radian,** is defined somewhat differently. It is the angle subtended at the centre of a circle by an arc having length which is equal to the radius of the circle (see Fig. 1.2).

You may sometimes find that you need to convert from radians to degrees, and vice versa. A complete circular revolution is equivalent to a rotation of 360° or 2π radians (note that π is approximately equal to 3.142). Thus one radian is equivalent to 360/2π degrees (or approximately 57.3°). Try to remember the following rules that will help you to convert angles expressed in degrees to radians and vice versa:

- **From degrees to radians, divide by 57.3.**

- **From radians to degrees, multiply by 57.3.**

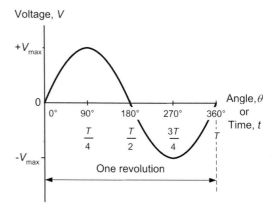

Figure 1.1   One cycle of a sine wave voltage

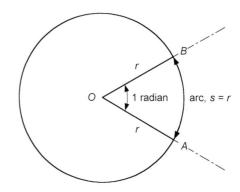

Figure 1.2   Definition of the radian

Example 1.3

Express a quarter of a cycle revolution in terms of:

(a) degrees;
(b) radians.

Solution

(a) There are 360° in one complete cycle (i.e. one full revolution. Hence there are (360/4)° or 90° in one quarter of a cycle.
(b) There are 2π radians in one complete cycle. Thus there are $2\pi/4$ or $\pi/2$ radians in one quarter of a cycle.

Example 1.4

Express an angle of 215° in radians.

Solution

To convert from degrees to radians, divide by 57.3. So 215° is equivalent to 215/57.3 = 3.75 radians.

Example 1.5

Express an angle of 2.5 radians in degrees.

Solution

To convert from radians to degrees, multiply by 57.3. Hence 2.5 radians is equivalent to 2.5 × 57.3 = 143.25°.

Electrical units and symbols

Table 1.3 shows the units and symbols that are commonly encountered in electrical circuits. It is important to get to know these units and also be able to recognize their abbreviations and symbols. You will meet all of these units later in this chapter.

Multiples and sub-multiples

Unfortunately, many of the derived units are either too large or too small for convenient everyday use but we can make life a little easier by using a standard range of multiples and sub-multiples (see Table 1.4).

Table 1.3   Electrical units

Unit	Abbrev.	Symbol	Notes
Ampere	A	I	Unit of electric **current** (a current of 1 A flows when a charge of 1 C is transported in a time interval of 1 s)
Coulomb	C	Q	Unit of electric **charge** or quantity of electricity
Farad	F	C	Unit of **capacitance** (a capacitor has a capacitance of 1 F when a potential of 1 V across its plates produced a charge of 1 C)
Henry	H	L	Unit of **inductance** (an inductor has an inductance of 1 H when an applied current changing at 1 A/s produces a potential difference of 1 V across its terminals)
Hertz	Hz	f	Unit of **frequency** (a signal has a frequency of 1 Hz if one cycle occurs in an interval of 1 s)
Joule	J	W	Unit of **energy**
Ohm	Ω	R	Unit of **resistance**
Second	s	t	Unit of **time**
Siemen	S	G	Unit of **conductance** (the reciprocal of resistance)
Tesla	T	B	Unit of **magnetic flux density** (a flux density of 1 T is produced when a flux of 1 Wb is present over an area of 1 square metre)
Volt	V	V	Unit of electric **potential** (e.m.f. or p.d.)
Watt	W	P	Unit of **power** (equivalent to 1 J of energy consumed in 1 s)
Weber	Wb	ϕ	Unit of **magnetic flux**

Table 1.4   Multiples and sub-multiples

Prefix	Abbreviation	Multiplier
tera	T	10^{12} (= 1 000 000 000 000)
giga	G	10^{9} (= 1 000 000 000)
mega	M	10^{6} (= 1 000 000)
kilo	k	10^{3} (= 1 000)
(none)	(none)	10^{0} (= 1)
centi	c	10^{-2} (= 0.01)
milli	m	10^{-3} (= 0.001)
micro	μ	10^{-6} (= 0.000 001)
nano	n	10^{-9} (= 0.000 000 001)
pico	p	10^{-12} (= 0.000 000 000 001)

Example 1.6

An indicator lamp requires a current of 0.075 A. Express this in mA.

Solution

You can express the current in mA (rather than in A) by simply moving the decimal point three places to the right. Hence 0.075 A is the same as 75 mA.

Example 1.7

A medium-wave radio transmitter operates on a frequency of 1,495 kHz. Express its frequency in MHz.

Solution

To express the frequency in MHz rather than kHz we need to move the decimal point three places to the left. Hence 1,495 kHz is equivalent to 1.495 MHz.

Example 1.8

A the value of a 27,000 pF in μF.

Solution

To express the value in μF rather than pF we need to move the decimal point six places to the left. Hence 27,000 pF is equivalent to 0.027 μF (note that we have had to introduce an extra zero before the 2 and after the decimal point).

Exponent notation

Exponent notation (or **scientific notation**) is useful when dealing with either very small or very large quantities. It's well worth getting to grips with this notation as it will allow you to simplify quantities before using them in formulae.

Exponents are based on **powers of ten.** To express a number in exponent notation the number is split into two parts. The first part is usually a number in the range 0.1 to 100 while the second part is a multiplier expressed as a power of ten.

For example, 251.7 can be expressed as 2.517×100, i.e. 2.517×10^{2}. It can also be expressed as $0.2517 \times 1,000$, i.e. 0.2517×10^{3}. In both cases the exponent is the same as the number of noughts in the multiplier (i.e. 2 in the first case and 3 in the second case). To summarize:

$$251.7 = 2.517 \times 10^{2} = 0.2517 \times 10^{3}$$

As a further example, 0.01825 can be expressed as 1.825/100, i.e. 1.825×10^{-2}. It can also be expressed as 18.25/1,000, i.e. 18.25×10^{-3}. Again, the exponent is the same as the number of noughts but the minus sign is used to denote a fractional multiplier. To summarize:

$$0.01825 = 1.825 \times 10^{-2} = 18.25 \times 10^{-3}$$

Example 1.9

A current of 7.25 mA flows in a circuit. Express this current in Amperes using exponent notation.

Solution

$1 \text{ mA} = 1 \times 10^{-3} \text{ A}$ thus $7.25 \text{ mA} = 7.25 \times 10^{-3} \text{ A}$

Example 1.10

A voltage of 3.75×10^{-6} V appears at the input of an amplifier. Express this voltage in (a) V and (b) mV, using exponent notation.

Solution

(a) 1×10^{-6} V = 1 μV so 3.75×10^{-6} V = 3.75 μV
(b) There are 1,000 μV in 1 mV so we must divide the previous result by 1,000 in order to express the voltage in mV. So 3.75 μV = 0.00375 mV.

Multiplication and division using exponents

Exponent notation really comes into its own when values have to be multiplied or divided. When multiplying two values expressed using exponents, you simply need to add the exponents. Here's an example:

$$(2 \times 10^2) \times (3 \times 10^6) = (2 \times 3) \times 10^{(2+6)} = 6 \times 10^8$$

Similarly, when dividing two values which are expressed using exponents, you only need to subtract the exponents. As an example:

$$(4 \times 10^6) \div (2 \times 10^4) = 4/2 \times 10^{(6-4)} = 2 \times 10^2$$

In either case it's important to remember to specify the units, multiples and sub-multiples in which you are working (e.g. A, kΩ, mV, μF, etc).

Example 1.11

A current of 3 mA flows in a resistance of 33 kΩ. Determine the voltage dropped across the resistor.

Solution

Voltage is equal to current multiplied by resistance (see page 7). Thus:

$$V = I \times R = 3 \text{ mA} \times 33 \text{ k}\Omega$$

Expressing this using exponent notation gives:

$$V = (3 \times 10^{-3}) \times (33 \times 10^3) \text{ V}$$

Separating the exponents gives:

$$V = 3 \times 33 \times 10^{-3} \times 10^3 \text{ V}$$

Thus $V = 99 \times 10^{(-3+3)} = 99 \times 10^0 = 99 \times 1 = 99 \text{ V}$

Example 1.12

A current of 45 μA flows in a circuit. What charge is transferred in a time interval of 20 ms?

Solution

Charge is equal to current multiplied by time (see the definition of the ampere on page 3). Thus:

$$Q = I\,t = 45 \text{ }\mu\text{A} \times 20 \text{ ms}$$

Expressing this in exponent notation gives:

$$Q = (45 \times 10^{-6}) \times (20 \times 10^{-3}) \text{ Coulomb}$$

Separating the exponents gives:

$$Q = 45 \times 20 \times 10^{-6} \times 10^{-3} \text{ Coulomb}$$

Thus $Q = 900 \times 10^{(-6-3)} = 900 \times 10^{-9} = 900 \text{ nC}$

Example 1.13

A power of 300 mW is dissipated in a circuit when a voltage of 1,500 V is applied. Determine the current supplied to the circuit.

Solution

Current is equal to power divided by voltage (see page 9). Thus:

$$I = P / V = 300 \text{ mW} / 1,500 \text{ V} \quad \text{Amperes}$$

Expressing this in exponent notation gives:

$$I = (300 \times 10^{-3})/(1.5 \times 10^3) \text{ A}$$

Separating the exponents gives:

$$I = (300/1.5) \times (10^{-3}/10^3) \text{ A}$$

$$I = 300/1.5 \times 10^{-3} \times 10^3 \text{ A}$$

Thus, $I = 200 \times 10^{(-3-3)} = 200 \times 10^{-6} = 200 \text{ }\mu\text{A}$

Conductors and insulators

Electric current is the name given to the flow of **electrons** (or negative charge carriers). Electrons orbit around the nucleus of atoms just as the earth orbits around the sun (see Fig. 1.3). Electrons are held in one or more **shells**, constrained to their orbital paths by virtue of a force of attraction towards the nucleus which contains an equal number of **protons** (positive charge carriers). Since like charges repel and unlike charges attract, negatively charged electrons are attracted to the positively charged nucleus. A similar principle can be demonstrated by observing the attraction between two permanent magnets; the two North poles of the magnets will repel each other, while a North and South pole will attract. In the same way, the unlike charges of the negative electron and the positive proton experience a force of mutual attraction.

The outer shell electrons of a **conductor** can be reasonably easily interchanged between adjacent atoms within the **lattice** of atoms of which the substance is composed. This makes it possible for the material to conduct electricity. Typical examples of conductors are metals such as copper, silver, iron and aluminium. By contrast, the outer shell electrons of an **insulator** are firmly bound to their parent atoms and virtually no interchange of electrons is possible. Typical examples of insulators are plastics, rubber and ceramic materials.

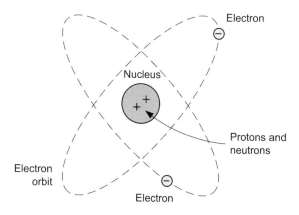

Figure 1.3   A single atom of helium (H_e) showing its two electrons in orbit around its nucleus

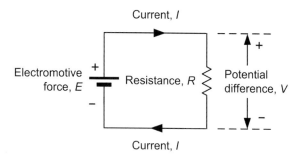

Figure 1.4 Simple circuit to illustrate the relationship between voltage (V), current (I) and resistance (R). Note that the direction of conventional current flow is from positive to negative

Voltage and resistance

The ability of an energy source (e.g. a battery) to produce a current within a conductor may be expressed in terms of **electromotive force** (e.m.f.). Whenever an e.m.f. is applied to a circuit **a potential difference** (p.d.) exists. Both e.m.f. and p.d. are measured in volts (V). In many practical circuits there is only one e.m.f. present (the battery or supply) whereas a p.d. will be developed across each component present in the circuit.

The **conventional flow** of current in a circuit is from the point of more positive potential to the point of greatest negative potential (note that electrons move in the *opposite* direction!). **Direct current** results from the application of a direct e.m.f. (derived from batteries or a d.c. power supply). An essential characteristic of these supplies is that the applied e.m.f. does not change its polarity (even though its value might be subject to some fluctuation).

For any conductor, the current flowing is directly proportional to the e.m.f. applied. The current flowing will also be dependent on the physical dimensions (length and cross-sectional area) and material of which the conductor is composed.

The amount of current that will flow in a conductor when a given e.m.f. is applied is inversely proportional to its **resistance.** Resistance, therefore, may be thought of as an opposition to current flow; the higher the resistance the lower the current that will flow (assuming that the applied e.m.f. remains constant).

Ohm's Law

Provided that temperature does not vary, the ratio of p.d. across the ends of a conductor to the current flowing in the conductor is a constant. This relationship is known as Ohm's Law and it leads to the relationship:

V / I = a constant = R

where V is the potential difference (or voltage drop) in Volts (V), I is the current in Amperes (A), and R is the resistance in Ohms (Ω) (see Fig. 1.4).

The formula may be arranged to make V, I or R the subject, as follows:

$V = I \times R$, $I = V/R$ and $R = V/I$

The triangle shown in Fig. 1.5 should help you remember these three important relationships. However, it's worth noting that, when performing calculations of currents, voltages and resistances in practical circuits it is seldom necessary to work with an accuracy of better than ±1% simply because component tolerances are usually greater than this. Furthermore, in calculations involving Ohm's Law, it can sometimes be convenient to work in units of kΩ and mA (or MΩ and μA) in which case potential differences will be expressed directly in V.

Example 1.14

A 12 Ω resistor is connected to a 6V battery. What current will flow in the resistor?

Solution

Here we must use $I = V/R$ (where $V = 6$ V and $R = 12$ Ω):

$I = V/R = 6$ V $/ 12$ Ω $= 0.5$ A (or 500 mA)

Hence a current of 500 mA will flow in the resistor.

Example 1.15

A current of 100 mA flows in a 56 Ω resistor. What voltage drop (potential difference) will be developed across the resistor?

Solution

Here we must use $V = I \times R$ and ensure that we work in units of Volts (V), Amperes (A) and Ohms (Ω).

$V = I \times R = 0.1$ A $\times 56$ Ω $= 5.6$ V

(Note that 100 mA is the same as 0.1 A.)

This calculation shows that a p.d. of 5.6 V will be developed across the resistor.

Example 1.16

A voltage drop of 15 V appears across a resistor in which a current of 1 mA flows. What is the value of the resistance?

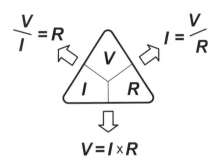

Figure 1.5 Triangle showing the relationship between V, I and R

Solution

$R = V/I = 15$ V $/ 0.001$ A $= 15,000$ Ω $= 15$ kΩ

Note that it is often more convenient to work in units of mA and V which will produce an answer directly in kΩ, i.e.

$R = V/I = 15$ V$/1$ mA $= 15$ kΩ

Resistance and resistivity

The resistance of a metallic conductor is directly proportional to its length and inversely proportional to its area. The resistance is also directly proportional to its **resistivity** (or **specific resistance**). Resistivity is defined as the resistance measured between the opposite faces of a cube having sides of 1 cm.

The resistance, R, of a conductor is thus given by the formula:

$R = \rho \times l/A$

where R is the resistance (ft), ρ is the resistivity (Ωm), l is the length (m), and A is the area (m^2).

Table 1.5 shows the electrical properties of some common metals.

Example 1.17

A coil consists of an 8 m length of annealed copper wire having a cross-sectional area of 1 mm^2. Determine the resistance of the coil.

Table 1.5   Properties of some common metals

Metal	Resistivity (at 20°C) (Ωm)	Relative conductivity (copper = 1)	Temperature coefficient of resistance (per °C)
Silver	1.626×10^{-8}	1.06	0.0041
Copper (annealed)	1.724×10^{-8}	1.00	0.0039
Copper (hard drawn)	1.777×10^{-8}	0.97	0.0039
Aluminium	2.803×10^{-8}	0.61	0.0040
Mild steel	1.38×10^{-7}	0.12	0.0045
Lead	2.14×10^{-7}	0.08	0.0040
Nickel	8.0×10^{-8}	0.22	0.0062

Solution

We will use the formula, $R = \rho\, l / A$.

The value of ρ for annealed copper given in Table 1.5 is 1.724×10^{-8} Ωm. The length of the wire is 4 m while the area is 1 mm^2 or 1×10^{-6} m^2 (note that it is important to be consistent in using units of metres for length and square metres for area).

Hence the resistance of the coil will be given by:

$$R = \frac{1.724 \times 10^{-8} \times 8}{1 \times 10^{-6}} = 13.724 \times 10^{(-8+6)}$$

Thus $R = 13.792 \times 10^{-2}$ or $0.13792\ \Omega$

Example 1.18

A wire having a resistivity of 1.724×10^{-8} Ωm, length 20 m and cross-sectional area 1 mm^2 carries a current of 5 A. Determine the voltage drop between the ends of the wire.

Solution

First we must find the resistance of the wire (as in Example 1.17):

$$R = \frac{\rho l}{A} = \frac{1.6 \times 10^{-8} \times 20}{1 \times 10^{-6}} = 32 \times 10^{-2} = 0.32\ \Omega$$

The voltage drop can now be calculated using Ohm's Law:

$V = I \times R = 5A \times 0.32\ \Omega = 1.6$ V

This calculation shows that a potential of 1.6 V will be dropped between the ends of the wire.

Energy and power

At first you may be a little confused about the difference between energy and power. Put simply, energy is the ability to do work while power is the rate at which work is done. In electrical circuits, energy is supplied by batteries or generators. It may also be stored in components such as capacitors and inductors. Electrical energy is converted into various other forms of energy by components such as resistors (producing heat), loudspeakers (producing sound energy) and light emitting diodes (producing light).

The unit of energy is the Joule (J). Power is the rate of use of energy and it is measured in Watts (W). A power of 1W results from energy being used at the rate of 1 J per second. Thus:

$P = W / t$

where P is the power in Watts (W), W is the energy in Joules (J), and t is the time in seconds (s).

The power in a circuit is equivalent to the product of voltage and current. Hence:

$P = I \times V$

where P is the power in Watts (W), I is the current in Amperes (A), and V is the voltage in Volts (V).

The formula may be arranged to make P, I or V the subject, as follows:

$$P = I \times P, \quad I = P/V \quad \text{and} \quad V = P/I$$

The triangle shown in Fig. 1.6 should help you remember these relationships.

The relationship, $P = I \times V$, may be combined with that which results from Ohm's Law $(V = I \times R)$ to produce two further relationships. First, substituting for V gives:

$$P = I \times (I \times R) = I^2 R$$

Secondly, substituting for I gives:

$$P = (V/R) \times V = V^2/R$$

Example 1.19

A current of 1.5 A is drawn from a 3 V battery. What power is supplied?

Solution

Here we must use $P = I \times V$ (where $I = 1.5$ A and $V = 3$ V).

$$P = I \times V = 1.5 \text{ A} \times 3 \text{ V} = 4.5 \text{ W}$$

Hence a power of 4.5 W is supplied.

Example 1.20

A voltage drop of 4 V appears across a resistor of 100 Ω. What power is dissipated in the resistor?

Solution

Here we use $P = V^2/R$ (where $V = 4$ V and $R = 100$ Ω).

$$P = V^2/R = (4 \text{ V} \times 4 \text{ V})/100 \text{ Ω} = 0.16 \text{ W}$$

Hence the resistor dissipates a power of 0.16 W (or 160 mW).

Example 1.21

A current of 20 mA flows in a 1 kΩ resistor. What power is dissipated in the resistor?

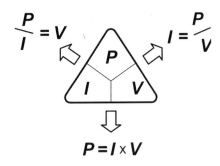

Figure 1.6 Triangle showing the relationship between P, I and V

Solution

Here we use $P = I^2 \times R$ but, to make life a little easier, we will work in mA and kΩ (in which case the answer will be in mW).

$$P = I^2 \times R = (20 \text{ mA} \times 20 \text{ mA}) \times 1 \text{ kΩ} = 400 \text{ mW}$$

Thus a power of 400 mW is dissipated in the 1kΩ resistor.

Electrostatics

If a conductor has a deficit of electrons, it will exhibit a net positive charge. If, on the other hand, it has a surplus of electrons, it will exhibit a net negative charge. An imbalance in charge can be produced by friction (removing or depositing electrons using materials such as silk and fur, respectively) or induction (by attracting or repelling electrons using a second body which is, respectively, positively or negatively charged).

Force between charges

Coulomb's Law states that, if charged bodies exist at two points, the force of attraction (if the charges are of opposite polarity) or repulsion (if the charges have the same polarity) will be proportional to the product of the magnitude of the charges divided by the square of their distance apart. Thus:

$$F = \frac{kQ_1Q_2}{r^2}$$

where Q_1 and Q_2 are the charges present at the two points (in Coulombs), r the distance separating the two points (in metres), F is the force (in Newtons), and k is a constant depending upon the medium in which the charges exist.

In vacuum or 'free space',

$$k = \frac{1}{4\pi\varepsilon_0}$$

where ε_0 is the **permittivity of free space** (8.854×10^{-12} C/Nm^2).

Combining the two previous equations gives:

$$F = \frac{kQ_1Q_2}{4\pi \times 8.854 \times 10^{-12} r^2} \quad \text{Newtons}$$

Electric fields

The force exerted on a charged particle is a manifestation of the existence of an electric field. The electric field defines the direction and magnitude of a force on a charged object. The field itself is invisible to the human eye but can be drawn by constructing lines which indicate the motion of a free positive charge within the field; the number of field lines in a particular region being used to indicate the relative strength of the field at the point in question.

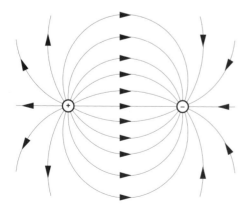

Figure 1.7   Electric field between two unlike electric charges

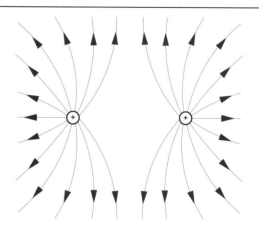

Figure 1.8   Electric field between two like electric charges (in this case both positive)

Figures 1.7 and 1.8 show the electric fields between charges of the same and opposite polarity while Fig. 1.9 shows the field which exists between two charged parallel plates. You will see more of this particular arrangement when we introduce capacitors in Chapter 2.

Electric field strength

The strength of an electric field (E) is proportional to the applied potential difference and inversely proportional to the distance between the two conductors. The electric field strength is given by:

$$E = V / d$$

where E is the electric field strength (V/m), V is the applied potential difference (V) and d is the distance (m).

Example 1.22

Two parallel conductors are separated by a distance of 25 mm. Determine the electric field strength if they are fed from a 600 V d.c. supply.

Solution

The electric field strength will be given by:

$$E = V / d = 600 / 25 \times 10^{-3} = 24 \text{ kV/m}$$

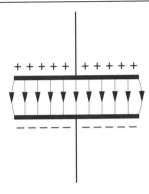

Figure 1.9 Electric field between two parallel plates

Permittivity

The amount of charge produced on the two plates shown in Fig. 1.9 for a given applied voltage will depend not only on the physical dimensions but also on the insulating dielectric material that appears between the plates. Such materials need to have a very high value of resistivity (they must not conduct charge) coupled with an ability to withstand high voltages without breaking down.

A more practical arrangement is shown in Fig. 1.10. In this arrangement the ratio of charge, Q, to potential difference, V, is given by the relationship:

$$\frac{Q}{V} = \frac{\varepsilon A}{d}$$

where A is the surface area of the plates (in m), d is the separation (in m), and ε is a constant for the dielectric material known as the **absolute permittivity** of the material (sometimes also referred to as the **dielectric constant**).

The absolute permittivity of a dielectric material is the product of the permittivity of free space (ε_0) and the **relative permittivity** (ε_r) of the material. Thus:

$$\varepsilon = \varepsilon_0 \times \varepsilon \qquad \text{and} \qquad \frac{Q}{V} = \frac{\varepsilon_0 \varepsilon_r A}{d}$$

The **dielectric strength** of an insulating dielectric is the maximum electric field strength that can safely be applied to it before breakdown (conduction) occurs. Table 1.4 shows values of relative permittivity and dielectric strength for some common dielectric materials.

Table 1.4 Properties of some common insulating dielectric materials

Dielectric material	Relative permittivity (free space = 1)	Dielectric strength (kV/mm)
Vacuum, or free space	1	∞
Air	1	3
Polythene	2.3	50
Paper	2.5 to 3.5	14
Polystyrene	2.5	25
Mica	4 to 7	160
Pyrex glass	4.5	13
Glass ceramic	5.9	40
Polyester	3.0 to 3.4	18
Porcelain	6.5	4
Titanium dioxide	100	6
Ceramics	5 to 1,000	2 to 10

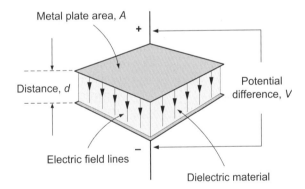

Figure 1.10 Parallel plates with an insulating dielectric material

Electromagnetism

When a current flows through a conductor a magnetic field is produced in the vicinity of the conductor. The magnetic field is invisible but its presence can be detected using a compass needle (which will deflect from its normal North–South position). If two current-carrying conductors are placed in the vicinity of one another, the fields will interact with one another and the conductors will experience a force of attraction or repulsion (depending upon the relative direction of the two currents).

Force between two current-carrying conductors

The mutual force which exists between two parallel current-carrying conductors will be proportional to the product of the currents in the two conductors and the length of the conductors but inversely proportional to their separation. Thus:

$$F = \frac{k I_1 I_2 l}{d}$$

where I_1 and I_2 are the currents in the two conductors (in Amps), l is the parallel length of the conductors (in metres), d is the distance separating the two conductors (in metres), F is the force (in Newtons), and k is a constant depending upon the medium in which the charges exist.

In vacuum or 'free space',

$$k = \frac{\mu_0}{2\pi}$$

where μ_0 is a constant known as the **permeability of free space** ($4\pi \times 10^{-7}$ or 12.57×10^{-7} H/m).

Combining the two previous equations gives:

$$F = \frac{\mu_0 I_1 I_2 l}{2\pi d}$$

or

$$F = \frac{4\pi \times 10^{-7} I_1 I_2 l}{2\pi d}$$

or

$$F = \frac{2 \times 10^{-7} I_1 I_2 l}{d} \quad \text{Newtons}$$

Magnetic fields

The field surrounding a straight current-carrying conductor is shown in Fig. 1.11. The magnetic field defines the direction of motion of a free North pole within the field. In the case of Fig. 1.11, the lines of flux are concentric and the direction of the field determined by the direction of current flow) is given by the right-hand rule.

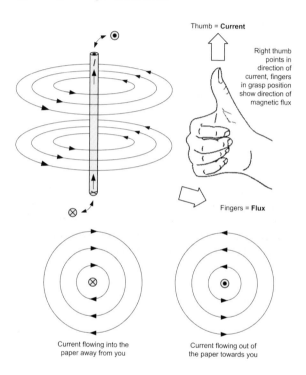

Current flowing into the paper away from you

Current flowing out of the paper towards you

Figure 1.11   Magnetic field surrounding a straight conductor

Magnetic field strength

The strength of a magnetic field is a measure of the density of the flux at any particular point. In the case of Fig. 1.11, the field strength will be proportional to the applied current and inversely proportional to the perpendicular distance from the conductor. Thus:

$$B = \frac{kI}{d}$$

where B is the magnetic flux density (in Tesla), I is

the current (in amperes), d is the distance from the conductor (in metres), and k is a constant.

Assuming that the medium is vacuum or 'free space', the density of the magnetic flux will be given by:

$$B = \frac{\mu_0 I}{2\pi d}$$

where B is the **flux density** (in Tesla), μ_0 is the permeability of 'free space' ($4\pi \times 10^{-7}$ or 12.57×10^{-7}), I is the current (in Amperes), and d is the distance from the centre of the conductor (in metres).

The flux density is also equal to the total flux divided by the area of the field. Thus:

$$B = \Phi / A$$

where Φ is the flux (in Webers) and A is the area of the field (in square metres).

In order to increase the strength of the field, a conductor may be shaped into a loop (Fig. 1.12) or coiled to form a solenoid (Fig. 1.13). Note, in the latter case, how the field pattern is exactly the same as that which surrounds a bar magnet. We will see

Example 1.23

Determine the flux density produced at a distance of 50 mm from a straight wire carrying a current of 20 A.

Solution

Applying the formula $B = \mu_0 I / 2\pi d$ gives:

$$B = \frac{12.57 \times 10^{-7} \times 20}{2 \times 3.142 \times 50 \times 10^{-3}} = \frac{251.4 \times 10^{-7}}{314.2 \times 10^{-3}}$$

from which:

$$B = 0.8 \times 10^{-4}\,\text{Tesla}$$

Thus $B = 80 \times 10^{-6}\,\text{T}$ or $B = 80\,\mu\text{T}$.

Example 1.24

A flux density of 2.5 mT is developed in free space over an area of 20 cm^2. Determine the total flux.

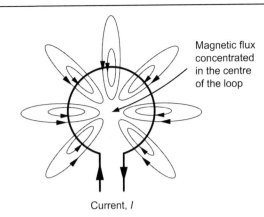

Figure 1.12   Forming a conductor into a loop increases the strength of the magnetic field in the centre of the loop

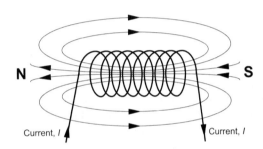

(a) Magnetic field around a solenoid

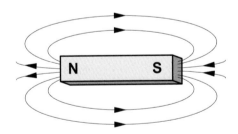

(b) Magnetic field around a permanent magnet

Figure 1.13   The magnetic field surrounding a solenoid coil resembles that of a permanent magnet

Solution

Re-arranging the formula $B = \Phi / A$ to make Φ the subject gives $\Phi = B \times A$ thus:

$$\Phi = (2.5 \times 10^{-3}) \times (20 \times 10^{-4}) = 50 \times 10^{-7}\ \text{Webers}$$

from which $B = 5\mu\text{Wb}$

Magnetic circuits

Materials such as iron and steel possess considerably enhanced magnetic properties. Hence they are employed in applications where it is necessary to increase the flux density produced by an electric current. In effect, magnetic materials allow us to channel the electric flux into a 'magnetic circuit', as shown in Fig. 1.14.

In the circuit of Fig. 1.14(b) the **reluctance** of the magnetic core is analogous to the resistance present in the electric circuit shown in Fig. 1.14(a). We can make the following comparisons between the two types of circuit (see Table 1.7).

In practice, not all of the magnetic flux produced in a magnetic circuit will be concentrated within the core and some 'leakage flux' will appear in the surrounding free space (as shown in Fig. 1.15). Similarly, if a gap appears within the magnetic circuit, the flux will tend to spread out as shown in Fig. 1.16. This effect is known as 'fringing'.

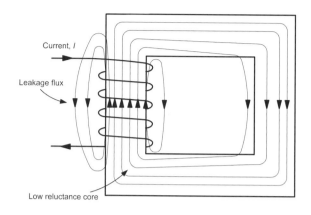

Figure 1.15   Leakage flux in a magnetic circuit

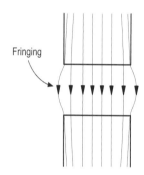

Figure 1.16   Fringing of the magnetic flux at an air gap in a magnetic circuit

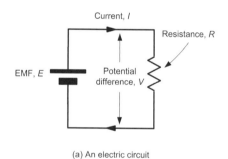

(a) An electric circuit

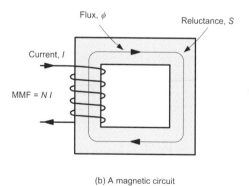

(b) A magnetic circuit

Figure 1.14   Comparison of electric and magnetic circuits

Table 1.7   Comparison of electric and magnetic circuits

Electric circuit Figure 1.14(a)	Magnetic circuit Figure 1.14(a)
Electromotive force, e.m.f. = V	Magnetomotive force, m.m.f. = $N \times I$
Resistance = R	Reluctance = S
Current = I	Flux = Φ
e.m.f. = current × resistance	m.m.f. = flux × reluctance
$V = I \times R$	$N I = S \Phi$

Reluctance and permeability

The reluctance of a magnetic path is directly proportional to its length and inversely proportional to its area. The reluctance is also inversely proportional to the **absolute permeability** of the magnetic material. Thus:

$$S = \frac{l}{\mu A}$$

where S is the reluctance of the magnetic path, l is the length of the path (in metres), A is the cross-sectional area of the path (in square metres), and μ is the absolute permeability of the magnetic material.

The absolute permeability of a magnetic material is the product of the permeability of free space (μ_0) and the **relative permeability** of the magnetic medium (μ_0). Thus

$$\mu = \mu_0 \times \mu \quad \text{and} \quad S = \frac{l}{\mu_0 \mu_r A}$$

The permeability of a magnetic medium is a measure of its ability to support magnetic flux and it is equal to the ratio of flux density (B) to **magnetizing force** (H). Thus:

$$\mu = \frac{B}{H}$$

where B is the flux density (in tesla) and H is the magnetizing force (in ampere/metre). The magnetizing force (H) is proportional to the product of the number of turns and current but inversely proportional to the length of the magnetic path.

$$H = \frac{NI}{l}$$

where H is the magnetizing force (in ampere/metre), N is the number of turns, I is the current (in amperes), and l is the length of the magnetic path (in metres).

B–H curves

Figure 1.17 shows four typical B–H (flux density plotted against permeability) curves for some common magnetic materials. If you look carefully at these curves you will notice that they flatten off due to magnetic **saturation** and that the slope of the curve (indicating the value of μ corresponding to a particular value of H) falls as the magnetizing force increases. This is important since it dictates the acceptable working range for a particular magnetic material when used in a magnetic circuit.

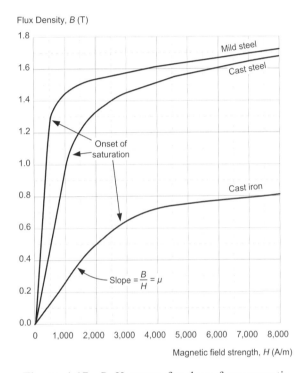

Figure 1.17 *B–H* curves for three ferromagnetic materials

Example 1.25

Estimate the relative permeability of cast steel (see Fig. 1.18) at (a) a flux density of 0.6 T and (b) a flux density of 1.6 T.

Solution

From Fig. 1.18, the slope of the graph at any point gives the value of μ at that point. We can easily find the slope by constructing a tangent at the point in question and then finding the ratio of vertical change to horizontal change.

(a)   The slope of the graph at 0.6 T is 0.6/800

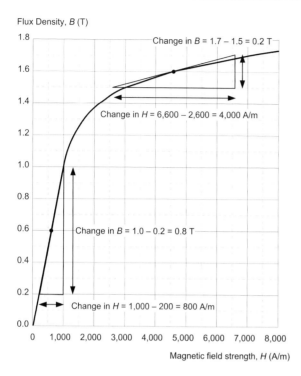

Flux Density, *B* (T)

Change in *B* = 1.7 − 1.5 = 0.2 T

Change in *H* = 6,600 − 2,600 = 4,000 A/m

Change in *B* = 1.0 − 0.2 = 0.8 T

Change in *H* = 1,000 − 200 = 800 A/m

Magnetic field strength, *H* (A/m)

Figure 1.18   *B–H* curve for a sample of cast steel

$$= 0.75 \times 10^{-3}$$

Since $\mu = \mu_0 \times \mu_r$, $\mu_r = \mu / \mu_0 = 0.75 \times 10^{-3} / 12.57 \times 10^{-7}$, thus $\mu_r = 597$ at 0.6 T.

(b)   The slope of the graph at 1.6 T is 0.2/4,000
$$= 0.05 \times 10^{-3}$$

Since $\mu = \mu_0 \times \mu_r$, $\mu_r = \mu / \mu_0 = 0.05 \times 10^{-3} / 12.57 \times 10^{-7}$, thus $\mu_r = 39.8$ at 1.6 T.

NB: This example clearly shows the effect of saturation on the permeability of a magnetic material!

Example 1.26

A coil of 800 turns is wound on a closed mild steel core having a length 600 mm and cross-sectional area 500 mm². Determine the current required to establish a flux of 0.8 mWb in the core.

Solution

Now $B = \Phi / A = (0.8 \times 10^{-3}) / (500 \times 10^{-6}) = 1.6$ T

From Fig. 1.17, a flux density of 1.6 T will occur in

mild steel when H = 3,500 A/m. The current can now be determined by re-arranging $H = N\,I\,/\,l$ as follows:

$$I = \frac{H \times l}{N} = \frac{3,500 \times 0.6}{800} = 2.625 \text{ A}$$

Circuit diagrams

Finally, and just in case you haven't seen them before, we will end this chapter with a brief word about circuit diagrams. We are introducing the topic here because it's quite important to be able to read and understand simple electronic circuit diagrams before you can make sense of some of the components and circuits that you will meet later on.

Circuit diagrams use standard symbols and conventions to represent the components and wiring used in an electronic circuit. Visually, they bear very little relationship to the physical layout of a circuit but, instead, they provide us with a 'theoretical' view of the circuit. In this section we show you how to find your way round simple circuit diagrams.

To be able to understand a circuit diagram you first need to be familiar with the symbols that are used to represent the components and devices. A selection of some of the most commonly used symbols are shown in Fig. 1.24. It's important to be aware that there are a few (thankfully quite small) differences between the symbols used in circuit diagrams of American and European origin.

As a general rule, the input to a circuit should be shown on the left of a circuit diagram and the output shown on the right. The supply (usually the most positive voltage) is normally shown at the top of the diagram and the common, 0V, or ground connection is normally shown at the bottom. This rule is not always obeyed, particularly for complex diagrams where many signals and supply voltages may be present.

Note also that, in order to simplify a circuit diagram (and avoid having too many lines connected to the same point) multiple connections to common, 0V, or ground may be shown using the appropriate symbol (see Fig. 1.24). The same applies to supply connections that may be repeated (appropriately labelled) at various points in the diagram.

A very simple circuit diagram (a simple resistance tester) is shown in Fig. 1.20. This circuit may be a little daunting if you haven't met a circuit like it before but you can still glean a great deal of information from the diagram even if you don't know what the individual components do.

The circuit uses two batteries, B1 (a 9 V multi-cell battery) and B2 (a 1.5 V single-cell battery). The two batteries are selected by means of a double-pole, double-throw (DPDT) switch. This allows the circuit to operate from either the 9 V battery (B1) as shown in Fig. 1.20(a) or from the 1.5 V battery (B2) as shown in Fig. 1.20(b) depending on the setting of S1.

A variable resistor, VR1, is used to adjust the current supplied by whichever of the two batteries is currently selected. This current flows first through VR1, then through the milliammeter, and finally through the unknown resistor, R_X. Notice how the meter terminals are labelled showing their polarity (the current flows *into* the positive terminal and *out of* the negative terminal).

The circuit shown in Fig. 1.20(c) uses a different type of switch but provides exactly the same function. In this circuit a single-pole, double-throw (SPDT) switch is used and the negative connections to the two batteries are 'commoned' (i.e. connected directly together).

Finally, Fig. 1.20(d) shows how the circuit can be re-drawn using a common 'chassis' connection to provide the negative connection between R_X and the two batteries. Electrically this circuit is identical to the one shown in Fig. 1.20(c).

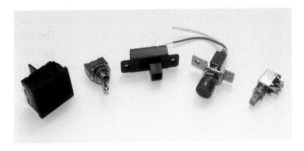

Figure 1.19 Various types of switch. From left to right: a mains rocker switch, an SPDT miniature toggle (changeover) switch, a DPDT slide switch, an SPDT push-button (wired for use as an SPST push-button), a miniature PCB mounting DPDT push-button (with a latching action).

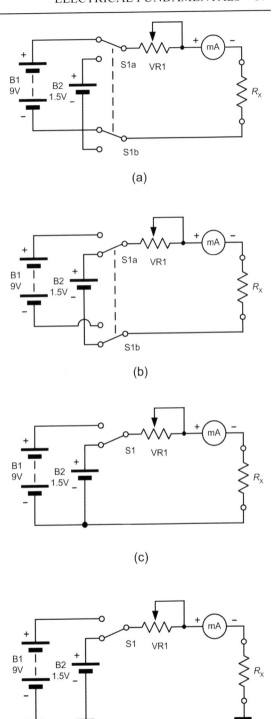

Figure 1.20 A simple circuit diagram

Practical investigation

Objective

To investigate the relationship between the resistance in a circuit and the current flowing in it.

Components and test equipment

Breadboard, digital or analogue meter with d.c. current ranges, 9 V d.c. power source (either a 9V battery or an a.c. mains adapter with a 9 V 400 mA output), test leads, resistors of 100 Ω, 220 Ω, 330 Ω, 470 Ω, 680 Ω and 1k Ω, connecting wire.

Procedure

Connect the circuit as shown in Fig. 1.21 and Fig. 1.22. Before switching on the d.c. supply or connecting the battery, check that the meter is set to the 200 mA d.c. current range. Switch on (or connect the battery), switch the multimeter on and read the current. Note down the current in the table below and repeat for resistance values of 220 Ω, 330 Ω, 470 Ω, 680 Ω and 1k Ω, switching off or disconnecting the battery between each measurement. Plot corresponding values of current (on the vertical axis) against resistance (on the horizontal axis) using the graph sheet shown in Fig. 1.23.

Measurements

Resistance (Ω)	Current (mA)
100	
220	
330	
470	
680	
1k	

Conclusion

Comment on the shape of the graph. Is this what you would expect and does it confirm that the current flowing in the circuit is inversely proportional to the resistance in the circuit? Finally, use Ohm's Law to calculate the value of each resistor and compare this with the marked value

(but before doing this, you might find it useful to make an accurate measurement of the d.c. supply or battery voltage).

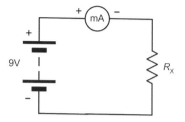

Figure 1.21 Circuit diagram

Figure 1.22 Typical wiring

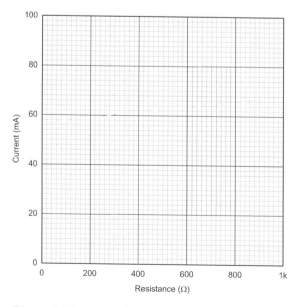

Figure 1.23   Graph layout for plotting the results

Important formulae introduced in this chapter

Voltage, current and resistance (Ohm's Law):
(page 6)

$V = I R$

Resistance and resistivity:
(page 7)

$R = \rho \, l \, / A$

Charge, current and time:
(page 5)

$Q = I t$

Power, current and voltage:
(page 8)

$P = I V$

Power, voltage and resistance:
(page 9)

$P = V^2 / R$

Power, current and resistance:
(page 9)

$P = I^2 R$

Reluctance and permeability:
(page 15)

$S = l \, / \mu \, A$

Flux and flux density:
(page 13)

$B = \Phi / A$

Current and magnetic field intensity:
(page 15)

$H = N I \, / l$

Flux, current and reluctance:
(page 14)

$N I = S \, \Phi$

Symbols introduced in this chapter

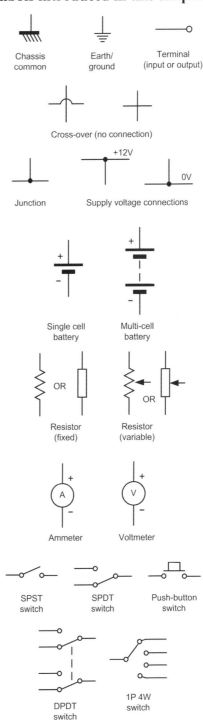

Figure 1.24   Circuit symbols introduced in this chapter

Problems

1.1   Which of the following are not fundamental units; Amperes, metres, Coulombs, Joules, Hertz, kilogram?

1.2   A commonly used unit of consumer energy is the kilowatt hour (kWh). Express this in Joules (J).

1.3   Express an angle of $30°$ in radians.

1.4   Express an angle of 0.2 radians in degrees.

1.5   A resistor has a value of 39,570 Ω. Express this in kilohms (kΩ).

1.6   An inductor has a value of 680 mH. Express this in henries (H).

1.7   A capacitor has a value of 0.00245 µF. Express this in nanofarads (nF).

1.8   A current of 190 µA is applied to a circuit. Express this in milliamperes (mA).

1.9   A signal of 0.475 mV appears at the input of an amplifier. Express this in volts using exponent notation.

1.10   A cable has an insulation resistance of 16.5 MΩ. Express this resistance in ohms using exponent notation.

1.11   Perform the following arithmetic using exponents:

(a) $(1.2 \times 10^3) \times (4 \times 10^3)$
(b) $(3.6 \times 10^6) \times (2 \times 10^{-3})$
(c) $(4.8 \times 10^9) \div (1.2 \times 10^6)$
(d) $(9.9 \times 10^{-6}) \div (19.8 \times 10^{-3})$
(e) $(4 \times 10^3) \times (7.5 \times 10^5) \times (2.5 \times 10^{-9})$

1.12   Which one of the following metals is the best conductor of electricity: aluminium, copper, silver, or mild steel? Why?

1.13   A resistor of 270 Ω is connected across a 9 V d.c. supply. What current will flow?

1.14   A current of 56 µA flows in a 120 kΩ resistor. What voltage drop will appear across the resistor?

1.15   A voltage drop of 13.2 V appears across a resistor when a current of 4 mA flows in it. What is the value of the resistor?

1.16   A power supply is rated at 15 V, 1 A. What value of load resistor would be required to test the power supply at its full rated output?

1.17   A wirewound resistor is made from a 4 m length of aluminium wire ($\rho = 2.18 \times 10^{-8}$ Ωm). Determine the resistance of the wire if it has a cross-sectional area of 0.2 mm^2.

1.18   A current of 25 mA flows in a 47 Ω resistor. What power is dissipated in the resistor?

1.19   A 9 V battery supplies a circuit with a current of 75 mA. What power is consumed by the circuit?

1.20   A resistor of 150 Ω is rated at 0.5 W. What is the maximum current that can be applied to the resistor without exceeding its rating?

1.21   Determine the electric field strength that appears in the space between two parallel plates separated by an air gap of 4 mm if a potential of 2.5 kV exists between them.

1.22   Determine the current that must be applied to a straight wire conductor in order to produce a flux density of 200 µT at a distance of 12 mm in free space.

1.23   A flux density of 1.2 mT is developed in free space over an area of 50 cm^2. Determine the total flux present.

1.24   A ferrite rod has a length of 250 mm and a diameter of 10 mm. Determine the reluctance if the rod has a relative permeability of 2,500.

1.25   A coil of 400 turns is wound on a closed mild steel core having a length 400 mm and cross-sectional area 480 mm^2. Determine the current required to establish a flux of 0.6 mWb in the core.

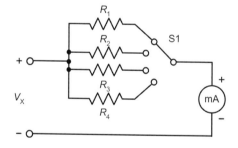

Figure 1.25   See Questions 1.26 and 1.27

1.26   Identify the type of switch shown in Fig. 1.25.

1.27   Figure 1.25 shows a simple voltmeter. If the milliammeter reads 1 mA full-scale and has negligible resistance, determine the values for R_1 to R_4 that will provide voltage ranges of 1V, 3 V, 10 V and 30 V full-scale.

Answers to these problems appear on page 374.

2

Passive components

This chapter introduces several of the most common types of electronic component, including resistors, capacitors and inductors. These are often referred to as **passive components** as they cannot, by themselves, generate voltage or current. An understanding of the characteristics and application of passive components is an essential prerequisite to understanding the operation of the circuits used in amplifiers, oscillators, filters and power supplies.

Resistors

The notion of resistance as opposition to current was discussed in the previous chapter. Conventional forms of resistor obey a straight line law when voltage is plotted against current (see Fig. 2.1) and this allows us to use resistors as a means of converting current into a corresponding voltage drop, and vice versa (note that doubling the applied current will produce double the voltage drop, and so on). Therefore resistors provide us with a means of controlling the currents and voltages present in electronic circuits. They can also act as **loads** to simulate the presence of a circuit during testing (e.g. a suitably rated resistor

can be used to replace a loudspeaker when an audio amplifier is being tested).

The specifications for a resistor usually include the value of resistance expressed in ohms (Ω), kilohms (kΩ) or megohms (MΩ), the accuracy or tolerance (quoted as the maximum permissible percentage deviation from the marked value), and the power rating (which must be equal to, or greater than, the maximum expected power dissipation).

Other practical considerations when selecting resistors for use in a particular application include temperature coefficient, noise performance, stability and ambient temperature range. Table 2.1 summarizes the properties of five of the most common types of resistor. Figure 2.2 shows a typical selection of fixed resistors with values from 15 Ω to 4.7 kΩ.

Preferred values

The value marked on the body of a resistor is not its *exact* resistance. Some minor variation in resistance value is inevitable due to production tolerance. For example, a resistor marked 100 Ω and produced within a tolerance of $\pm10\%$ will have a value which falls within the range 90 Ω to 110 Ω. A similar component with a tolerance of $\pm1\%$ would have a value that falls within the range 99 Ω to 101 Ω. Thus, where accuracy is important it is essential to use close tolerance components.

Resistors are available in several series of fixed decade values, the number of values provided with each series being governed by the tolerance involved. In order to cover the full range of resistance values using resistors having a $\pm20\%$ tolerance it will be necessary to provide six basic values (known as the **E6 series**. More values will be required in the series which offers a tolerance of $\pm10\%$ and consequently the **E12 series** provides twelve basic values. The **E24 series** for resistors of $\pm5\%$ tolerance provides no fewer than 24 basic

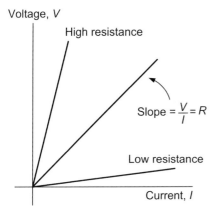

Figure 2.1 Voltage plotted against current for three different values of resistor

Table 2.1   Characteristics of common types of resistor

Property	Resistor type					
	Carbon film	Metal film	Metal oxide	Ceramic wirewound	Vitreous wirewound	Metal clad
Resistance range (Ω)	10 to 10 M	1 to 1 M	10 to 10 M	0.47 to 22 k	0.1 to 22 k	0.05 to 10 k
Typical tolerance (%)	±5	±1	±2	±5	±5	±5
Power rating (W)	0.25 to 2	0.125 to 0.5	0.25 to 0.5	4 to 17	2 to 4	10 to 300
Temperature coefficient (ppm/°C)	−250	+50 to +100	+250	+250	+75	+50
Stability	Fair	Excellent	Excellent	Good	Good	Good
Noise performance	Fair	Excellent	Excellent	n.a.	n.a.	n.a.
Ambient temperature range (°C)	−45 to +125	−45 to +125	−45 to +125	−45 to +125	−45 to +125	−55 to +200
Typical applications	General purpose	Amplifiers, test equipment, etc., requiring low-noise high-tolerance components		Power supplies, loads, medium and high-power applications		Very high power applications

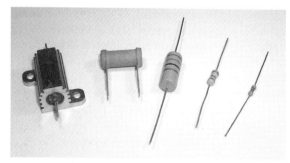

Figure 2.2   A selection of resistors including high-power metal clad, ceramic wirewound, carbon and metal film types with values ranging from 15 Ω to 4.7 kΩ

values and, as with the E6 and E12 series, decade multiples (i.e. ×1, ×10, ×100, ×1 k, ×10 k, ×100 k and × 1 M) of the basic series. Figure 2.3 shows the relationship between the E6, E12 and E24 series.

Power ratings

Resistor power ratings are related to operating temperatures and resistors should be derated at high temperatures. Where reliability is important resistors should be operated at well below their nominal maximum power dissipation.

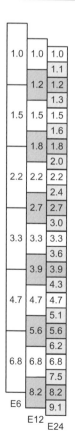

Figure 2.3   The E6, E12 and E24 series

Example 2.1

A resistor has a marked value of 220 Ω. Determine the tolerance of the resistor if it has a measured value of 207 Ω.

Solution

The difference between the marked and measured values of resistance (the error) is (220 Ω − 207 Ω) = 13 Ω. The tolerance is given by:

$$\text{Tolerance} = \frac{\text{error}}{\text{marked value}} \times 100\%$$

The tolerance is thus (13 / 220) × 100 = 5.9%.

Example 2.2

A 9 V power supply is to be tested with a 39 Ω load resistor. If the resistor has a tolerance of 10% find:

(a)   the nominal current taken from the supply;
(b)   the maximum and minimum values of supply current at either end of the tolerance range for the resistor.

Solution

(a)   If a resistor of *exactly* 39 Ω is used the current will be:

$$I = V / R = 9\text{ V} / 39\ \Omega = 231\text{ mA}$$

(b)   The lowest value of resistance would be (39 Ω − 3.9 Ω) = 35.1 Ω. In which case the current would be:

$$I = V / R = 9\text{ V} / 35.1\ \Omega = 256.4\text{ mA}$$

At the other extreme, the highest value would be (39 Ω + 3.9 Ω) = 42.9 Ω.

In this case the current would be:

$$I = V / R = 9\text{ V} / 42.9\ \Omega = 209.8\text{ mA}$$

The maximum and minimum values of supply current will thus be 256.4 mA and 209.8 mA respectively.

Example 2.3

A current of 100 mA (±20%) is to be drawn from a 28 V d.c. supply. What value and type of resistor should be used in this application?

Solution

The value of resistance required must first be calculated using Ohm's Law:

$$R = V / I = 28\text{ V} / 100\text{ mA} = 280\ \Omega$$

The nearest preferred value from the E12 series is 270 Ω (which will actually produce a current of 103.7 mA (i.e. within ±4%> of the desired value). If a resistor of ±10% tolerance is used, current will be within the range 94 mA to 115 mA (well within the ±20% accuracy specified).

The power dissipated in the resistor (calculated using $P = I \times V$) will be 2.9 W and thus a component rated at 3 W (or more) will be required. This would normally be a vitreous enamel coated wirewound resistor (see Table 2.1).

Resistor markings

Carbon and metal oxide resistors are normally marked with colour codes which indicate their value and tolerance. Two methods of colour coding are in common use; one involves four coloured bands (see Fig. 2.4) while the other uses five colour bands (see Fig. 2.5).

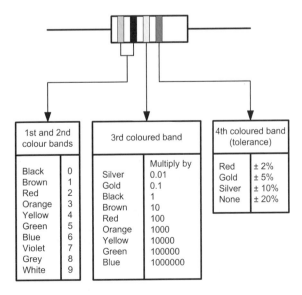

1st and 2nd colour bands		3rd coloured band		4th coloured band (tolerance)	
Black	0		Multiply by	Red	± 2%
Brown	1	Silver	0.01	Gold	± 5%
Red	2	Gold	0.1	Silver	± 10%
Orange	3	Black	1	None	± 20%
Yellow	4	Brown	10		
Green	5	Red	100		
Blue	6	Orange	1000		
Violet	7	Yellow	10000		
Grey	8	Green	100000		
White	9	Blue	1000000		

Figure 2.4   Four band resistor colour code

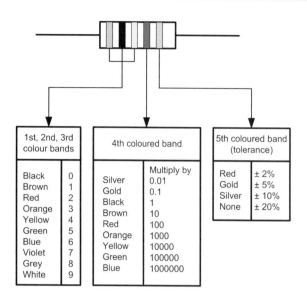

Figure 2.5   Five band resistor colour code

Example 2.4

A resistor is marked with the following coloured stripes: brown, black, red, silver. What is its value and tolerance?

Solution

See Fig. 2.6.

Example 2.5

A resistor is marked with the following coloured stripes: red, violet, orange, gold. What is its value and tolerance?

Solution

See Fig. 2.7.

Example 2.6

A resistor is marked with the following coloured stripes: green, blue, black, gold. What is its value and tolerance?

Solution

See Fig. 2.8.

Example 2.7

A resistor is marked with the following coloured stripes: red, green, black, black, brown. What is its value and tolerance?

Solution

See Fig. 2.9.

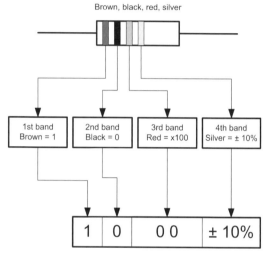

Figure 2.6   See Example 2.4

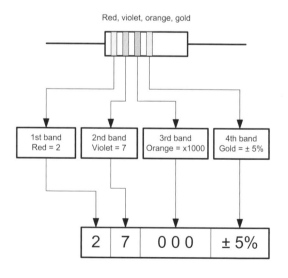

Figure 2.7   See Example 2.5

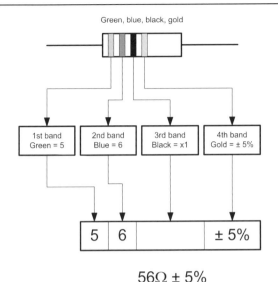

Green, blue, black, gold

| 1st band Green = 5 | 2nd band Blue = 6 | 3rd band Black = x1 | 4th band Gold = ± 5% |

| 5 | 6 | | ± 5% |

$56\Omega \pm 5\%$

Figure 2.8   See Example 2.6

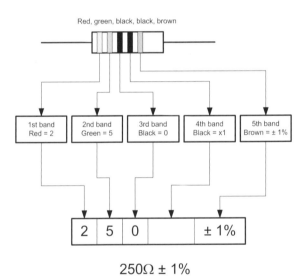

Red, green, black, black, brown

| 1st band Red = 2 | 2nd band Green = 5 | 3rd band Black = 0 | 4th band Black = x1 | 5th band Brown = ± 1% |

| 2 | 5 | 0 | | ± 1% |

$250\Omega \pm 1\%$

Figure 2.9   See Example 2.7

Example 2.8

A 2.2 kΩ of ±2% tolerance is required. What four band colour code does this correspond to?

Solution

Red (2), red (2), red (2 zeros), red (2% tolerance). Thus all four bands should be red.

BS 1852 coding

Some types of resistor have markings based on a system of coding defined in BS 1852. This system involves marking the position of the decimal point with a letter to indicate the multiplier concerned as shown in Table 2.2. A further letter is then appended to indicate the tolerance as shown in Table 2.3.

Table 2.2   BS 1852 resistor multiplier markings

Letter	Multiplier
R	1
K	1,000
M	1,000,000

Table 2.3   BS 1852 resistor tolerance markings

Letter	Multiplier
F	±1%
G	±2%
J	±5%
K	±10%
M	±20%

Example 2.9

A resistor is marked coded with the legend 4R7K. What is its value and tolerance?

Solution

4.7 Ω ±10%

Example 2.10

A resistor is marked coded with the legend 330RG. What is its value and tolerance?

Solution

330 Ω ±2%

Example 2.11

A resistor is marked coded with the legend R22M. What is its value and tolerance?

Solution

$0.22 \, \Omega \pm 20\%$

Series and parallel combinations of resistors

In order to obtain a particular value of resistance, fixed resistors may be arranged in either series or parallel as shown in Figs 2.10 and 2.11.

The effective resistance of each of the series circuits shown in Fig. 2.10 is simply equal to the sum of the individual resistances. So, for the circuit shown in Fig. 2.10(a):

$$R = R_1 + R_2$$

while for Fig. 2.10(b)

$$R = R_1 + R_2 + R_3$$

Turning to the parallel resistors shown in Fig. 2.11, the reciprocal of the effective resistance of each circuit is equal to the sum of the reciprocals of the individual resistances. Hence, for Fig. 2.11(a):

$$\frac{1}{R} = \frac{1}{R_1} + \frac{1}{R_2}$$

while for Fig. 2.12(b)

$$\frac{1}{R} = \frac{1}{R_1} + \frac{1}{R_2} + \frac{1}{R_3}$$

In the former case, the formula can be more conveniently re-arranged as follows:

$$R = \frac{R_1 \times R_2}{R_1 + R_2}$$

You can remember this as the *product* of the two resistance values *divided by* the *sum* of the two resistance values.

Example 2.12

Resistors of 22 Ω, 47 Ω and 33 Ω are connected (a)

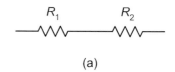

(a)

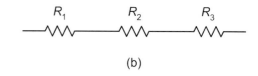

(b)

Figure 2.10   Resistors in series

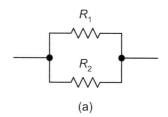

(a)

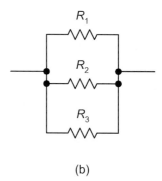

(b)

Figure 2.11   Resistors in parallel

in series and (b) in parallel. Determine the effective resistance in each case.

Solution

(a)   In the series circuit $R = R_1 + R_2 + R_3$, thus
$R = 22 \, \Omega + 47 \, \Omega + 33 \, \Omega = 102 \, \Omega$

(b)   In the parallel circuit:

$$\frac{1}{R} = \frac{1}{R_1} + \frac{1}{R_2} + \frac{1}{R_3}$$

thus

$$\frac{1}{R} = \frac{1}{22\ \Omega} + \frac{1}{47\ \Omega} + \frac{1}{33\ \Omega}$$

or

$$\frac{1}{R} = 0.045 + 0.021 + 0.03$$

from which

$$\frac{1}{R} = 0.096 = 10.42\ \Omega$$

Example 2.13

Determine the effective resistance of the circuit shown in Fig. 2.12.

Solution

The circuit can be progressively simplified as shown in Fig. 2.13. The stages in this simplification are:

(a) R_3 and R_4 are in series and they can replaced by a single resistance (R_A) of (12 Ω + 27 Ω) = 39 Ω.
(b) R_A appears in parallel with R_2. These two resistors can be replaced by a single resistance (R_B) of (39 Ω × 47 Ω)/(39 Ω + 47 Ω) = 21.3 Ω.
(c) R_B appears in series with R_1. These two resistors can be replaced by a single resistance (R) of (21.3 Ω + 4.7 Ω) = 26 Ω.

Example 2.14

A resistance of 50 Ω rated at 2 W is required. What parallel combination of preferred value resistors will satisfy this requirement? What power rating should each resistor have?

Solution

Two 100 Ω resistors may be wired in parallel to provide a resistance of 50 Ω as shown below:

$$R = \frac{R_1 \times R_2}{R_1 + R_2} = \frac{100 \times 100}{100 + 100} = \frac{10,000}{200} = 50\ \Omega$$

Note, from this, that when two resistors of the same value are connected in parallel the resulting resistance will be half that of a single resistor.

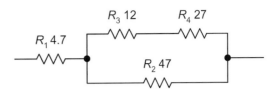

Figure 2.12 See Example 2.13

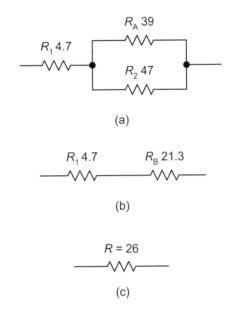

Figure 2.13 See Example 2.13

Having shown that two 100 Ω resistors connected in parallel will provide us with a resistance of 50 Ω we now need to consider the power rating. Since the resistors are identical, the applied power will be shared equally between them. Hence each resistor should have a power rating of 1 W.

Resistance and temperature

Figure 2.14 shows how the resistance of a metal conductor (e.g. copper) varies with temperature. Since the resistance of the material increases with temperature, this characteristic is said to exhibit a **positive temperature coefficient (PTC)**. Not all materials have a PTC characteristic. The resistance of a carbon conductor falls with temperature and it is therefore said to exhibit a **negative temperature coefficient (NTC)**.

The resistance of a conductor at a temperature, t, is given by the equation:

$$R_t = R_0(1 + \alpha t + \beta t^2 + \gamma t^3 \ldots)$$

where α, β, γ, etc. are constants and R_0 is the resistance at 0°C.

The coefficients, β, γ, etc. are quite small and since we are normally only dealing with a relatively restricted temperature range (e.g. 0°C to 100°C) we can usually approximate the characteristic shown in Fig. 2.14 to the straight line law shown in Fig. 2.15. In this case, the equation simplifies to:

$$R_t = R_0(1 + \alpha t)$$

where α is known as the **temperature coefficient** of resistance. Table 2.4 shows some typical values for α (note that α is expressed in $\Omega/\Omega/°C$ or just /°C).

Example 2.15

A resistor has a temperature coefficient of 0.001/°C. If the resistor has a resistance of 1.5 kΩ at 0°C, determine its resistance at 80°C.

Solution

Now

$$R_t = R_0(1 + \alpha t)$$

thus

$$R_t = 1.5\text{k}\Omega \times (1 + (0.001 \times 80))$$

Hence

$$R_t = 1.5 \times 1.08 = 1.62 \text{ k}\Omega$$

Example 2.16

A resistor has a temperature coefficient of 0.0005/°C. If the resistor has a resistance of 680 Ω at 20°C, what will its resistance be at 80°C?

Solution

First we must find the resistance at 0°C. Rearranging the formula for R_t gives:

$$R_0 = \frac{R_t}{1 + \alpha t} = \frac{680}{1 + (0.0005 \times 20)} = \frac{680}{1 + 0.01}$$

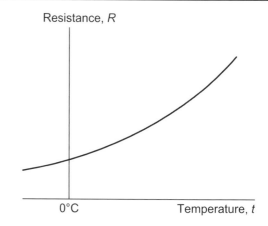

Figure 2.14 Variation of resistance with temperature for a metal conductor

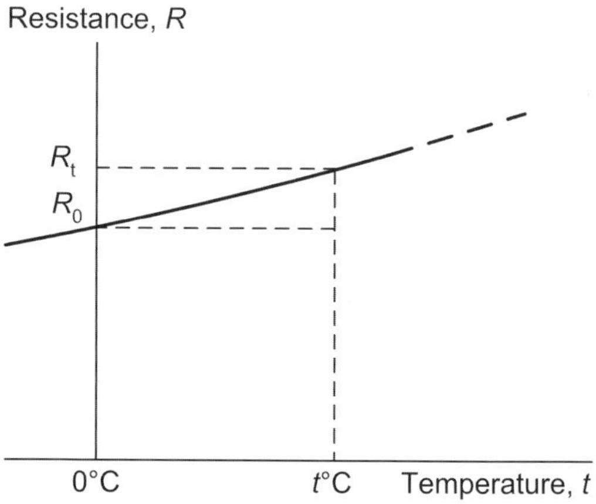

Figure 2.15 Straight line approximation of Fig. 2.14

Hence

$$R_0 = \frac{680}{1 + 0.01} = 673.3 \ \Omega$$

Now

$$R_t = R_0(1 + \alpha t)$$

thus

$$R_{90} = 673.3 \times (1 + (0.0005 \times 90))$$

Hence

$$R_{90} = 673.3 \times 1.045 = 704 \ \Omega$$

Example 2.17

A resistor has a resistance of 40 Ω at at 0°C and 44 Ω at 100°C. Determine the resistor's temperature coefficient.

Solution

First we need to make α the subject of the formula:

$$R_t = R_0(1 + \alpha\, t)$$

Now

$$\alpha = \frac{1}{t}\left(\frac{R_t}{R_0} - 1\right) = \frac{1}{100}\left(\frac{44}{40} - 1\right)$$

from which

$$\alpha = \frac{1}{100}(1.1 - 1) = \frac{1}{100} \times 0.1 = 0.001 \ /°C$$

Table 2.4 Temperature coefficient of resistance

Material	Temperature coefficient of resistance, α (/°C)
Platinum	+0.0034
Silver	+0.0038
Copper	+0.0043
Iron	+0.0065
Carbon	−0.0005

Thermistors

With conventional resistors we would normally require resistance to remain the same over a wide range of temperatures (i.e. α should be zero). On the other hand, there are applications in which we could use the effect of varying resistance to detect a temperature change. Components that allow us to do this are known as **thermistors.** The resistance of a thermistor changes markedly with temperature and these components are widely used in temperature sensing and temperature compensating applications. Two basic types of thermistor are available, NTC and PTC (see Fig. 2.16).

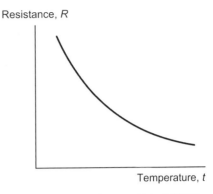

(a) Negative temperature coefficient (NTC)

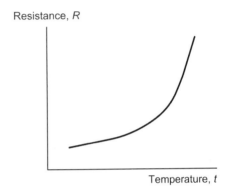

(b) Positive temperature coefficient (PTC)

Figure 2.16 Characteristics of (a) NTC and (b) PTC thermistors

Typical NTC thermistors have resistances that vary from a few hundred (or thousand) ohms at 25°C to a few tens (or hundreds) of ohms at 100°C. PTC thermistors, on the other hand, usually have a resistance-temperature characteristic which remains substantially flat (typically at around 100 Ω) over the range 0°C to around 75°C. Above this, and at a critical temperature (usually in the range 80°C to 120°C) their resistance rises very rapidly to values of up to, and beyond, 10 kΩ (see Fig. 2.16).

A typical application of PTC thermistors is over-current protection. Provided the current passing through the thermistor remains below the threshold current, the effects of self-heating will remain negligible and the resistance of the thermistor will remain low (i.e. approximately the same as the resistance quoted at 25°C). Under fault conditions, the current exceeds the threshold value by a considerable margin and the thermistor starts to self-heat. The resistance then increases rapidly and, as a consequence, the current falls to the rest value. Typical values of threshold and rest currents are 200 mA and 8 mA, respectively, for a device which exhibits a nominal resistance of 25 Ω at 25°C.

Light-dependent resistors

Light-dependent resistors (LDR) use a semiconductor material (i.e. a material that is neither a conductor nor an insulator) whose electrical characteristics vary according to the amount of incident light. The two semiconductor materials used for the manufacture of LDRs are cadmium sulphide (CdS) and cadmium selenide (CdSe). These materials are most sensitive to light in the visible spectrum, peaking at about 0.6 μm for CdS and 0.75 μm for CdSe. A typical CdS LDR exhibits a resistance of around 1 MΩ in complete darkness and less than 1 kΩ when placed under a bright light source (see Fig. 2.17).

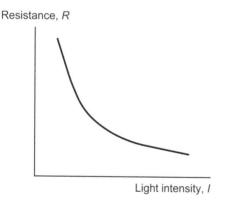

Figure 2.17   Characteristic of a light-dependent resistor (LDR)

Voltage dependent resistors

The resistance of a voltage dependent resistor (VDR) falls very rapidly when the voltage across it exceeds a nominal value in either direction (see Fig. 2.18). In normal operation, the current flowing in a VDR is negligible, however, when the resistance falls, the current will become appreciable and a significant amount of energy will be absorbed. VDRs are used as a means of 'clamping' the voltage in a circuit to a pre-determined level. When connected across the supply rails to a circuit (either a.c or d.c.) they are able to offer a measure of protection against voltage surges.

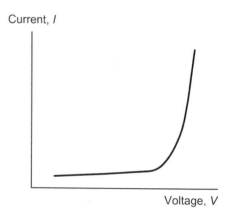

Figure 2.18   Characteristic of a voltage dependent resistor (VDR)

Variable resistors

Variable resistors are available in several forms including those which use carbon tracks and those which use a wirewound resistance element. In either case, a moving slider makes contact with the resistance element. Most variable resistors have three (rather than two) terminals and as such are more correctly known as **potentiometers**. Carbon potentiometers are available with linear or semi-logarithmic law tracks (see Fig. 2.19) and in rotary or slider formats. Ganged controls, in which several potentiometers are linked together by a common control shaft, are also available. Figure 2.20 shows a selection of variable resistors.

You will also encounter various forms of preset resistors that are used to make occasional adjustments (e.g. for calibration). Various forms of preset resistor are commonly used including open carbon track skeleton presets and fully encapsulated carbon and multi-turn cermet types, as shown in Fig. 2.21.

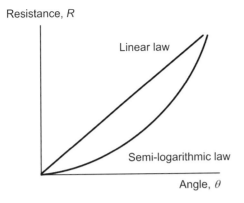

Figure 2.19 Characteristics for linear and semi-logarithmic law variable resistors

Figure 2.20 A selection of common types of carbon and wirewound variable resistors/potentiometers

Figure 2.21 A selection of common types of standard and miniature preset resistors/potentiometers

Capacitors

A capacitor is a device for storing electric charge. In effect, it is a reservoir into which charge can be deposited and then later extracted. Typical applications include reservoir and smoothing capacitors for use in power supplies, coupling a.c. signals between the stages of amplifiers, and decoupling supply rails (i.e. effectively grounding the supply rails as far as a.c. signals are concerned).

A capacitor can consist of nothing more than two parallel metal plates as shown in Fig. 1.10 on page 11. To understand what happens when a capacitor is being charged and discharged take a look at Fig. 2.22. If the switch is left open (position A), no charge will appear on the plates and in this condition there will be no electric field in the space between the plates nor will there be any charge stored in the capacitor.

When the switch is moved to position B, electrons will be attracted from the positive plate to the positive terminal of the battery. At the same time, a similar number of electrons will move from the negative terminal of the battery to the negative plate. This sudden movement of electrons will manifest itself in a momentary surge of current (conventional current will flow from the positive terminal of the battery towards the positive terminal of the capacitor).

Eventually, enough electrons will have moved to make the e.m.f. between the plates the same as that of the battery. In this state, the capacitor is said to be **fully charged** and an electric field will be present in the space between the two plates.

If, at some later time the switch is moved back to position A, the positive plate will be left with a deficiency of electrons whilst the negative plate will be left with a surplus of electrons. Furthermore, since there is no path for current to flow between the two plates the capacitor will remain charged and a potential difference will be maintained between the plates.

Now assume that the switch is moved to position C. The excess electrons on the negative plate will flow through the resistor to the positive plate until a neutral state once again exists (i.e. until there is no excess charge on either plate). In this state the capacitor is said to be **fully discharged** and the electric field between the plates will rapidly collapse. The movement of electrons during the discharging of the capacitor will again result in a momentary surge of current (current will flow from the positive terminal of the capacitor and into the resistor).

Figure 2.23 shows the direction of current flow in the circuit of Fig. 2.22 during charging (switch in position B) and discharging (switch in position C). It should be noted that current flows momentarily

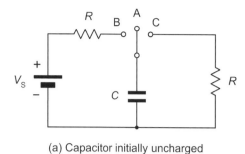

(a) Capacitor initially uncharged

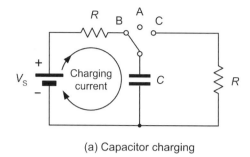

(a) Capacitor charging

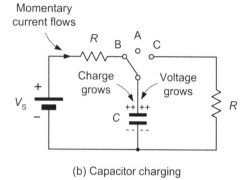

(b) Capacitor charging

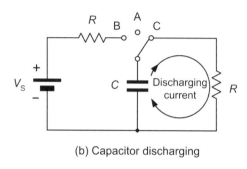

(b) Capacitor discharging

Figure 2.23   Current flow during charging and discharging

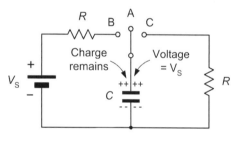

(c) Capacitor remains charged

in both circuits even though you may think that the circuit is broken by the gap between the capacitor plates!

Capacitance

The unit of capacitance is the farad (F). A capacitor is said to have a capacitance of 1 F if a current of 1 A flows in it when a voltage changing at the rate of 1 V/s is applied to it. The current flowing in a capacitor will thus be proportional to the product of the capacitance, C, and the rate of change of applied voltage. Hence:

$i = C \times$ (rate of change of voltage)

Note that we've used a small i to represent the current flowing in the capacitor. We've done this because the current is changing and doesn't remain constant.

The rate of change of voltage is often represented by the expression dv/dt where dv represents a very small change in voltage and dt

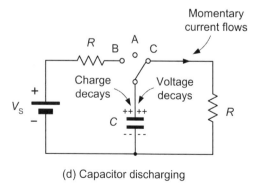

(d) Capacitor discharging

Figure 2.22   Capacitor charging and discharging

represents the corresponding small change in time. Expressing this mathematically gives:

$$i = C \frac{dV}{dt}$$

Example 2.18

A voltage is changing at a uniform rate from 10 V to 50 V in a period of 0.1 s. If this voltage is applied to a capacitor of 22 µF, determine the current that will flow.

Solution

Now the current flowing will be given by:

$$i = C \times (\text{rate of change of voltage})$$

Thus

$$i = C \left(\frac{\text{change in voltage}}{\text{change in time}} \right) = 22 \times 10^{-6} \times \left(\frac{50 - 10}{0.1} \right)$$

From which

$$i = 22 \times 10^{-6} \times \left(\frac{40}{0.1} \right) = 22 \times 10^{-6} \times 400$$

so

$$i = 8.8 \times 10^{-3} = 8.8 \text{ mA}$$

Charge, capacitance and voltage

The charge or **quantity of electricity** that can be stored in the electric field between the capacitor plates is proportional to the applied voltage and the capacitance of the capacitor. Thus:

$$Q = CV$$

where Q is the charge (in coulombs), C is the capacitance (in farads), and V is the potential difference (in volts).

Example 2.19

A 10 uF capacitor is charged to a potential of 250 V. Determine the charge stored.

Solution

The charge stored will be given by:

$$Q = CV = 10 \times 10^{-6} \times 250 = 2.5 \text{ mC}$$

Energy storage

The energy stored in a capacitor is proportional to the product of the capacitance and the square of the potential difference. Thus:

$$W = \frac{1}{2} C V^2$$

where W is the energy (in Joules), C is the capacitance (in Farads), and V is the potential difference (in Volts).

Example 2.20

A capacitor of 47 µF is required to store 4J of energy. Determine the potential difference that must be applied to the capacitor.

Solution

The foregoing formula can be re-arranged to make V the subject as follows:

$$V = \sqrt{\frac{E}{0.5 C}} = \sqrt{\frac{2E}{C}} = \sqrt{\frac{2 \times 4}{47 \times 10^{-6}}}$$

from which

$$V = \sqrt{\frac{8}{47 \times 10^{-6}}} = \sqrt{0.170 \times 10^6} = 0.412 \times 10^3 = 412 \text{ V}$$

Capacitance and physical dimensions

The capacitance of a capacitor depends upon the physical dimensions of the capacitor (i.e. the size of the plates and the separation between them) and the dielectric material between the plates. The capacitance of a conventional parallel plate capacitor is given by:

$$C = \frac{\varepsilon_0 \varepsilon_r A}{d}$$

where C is the capacitance (in farads), ε_0 is the permittivity of free space, ε_r is the **relative permittivity** of the dielectric medium between the plates), and d is the separation between the plates (in metres).

Example 2.21

A capacitor of 1 nF is required. If a dielectric material of thickness 0.1 mm and relative permittivity 5.4 is available, determine the required plate area.

Solution

Re-arranging the formula

$$C = \frac{\varepsilon_0 \varepsilon_r A}{d}$$

to make A the subject gives:

$$A = \frac{Cd}{\varepsilon_0 \varepsilon_r} = \frac{1 \times 10^{-9} \times 0.1 \times 10^{-3}}{8.854 \times 10^{-12} \times 5.4}$$

from which

$$A = \frac{0.1 \times 10^{-12}}{47.8116 \times 10^{-12}}$$

thus

$$A = 0.00209 \text{ m}^2 \text{ or } 20.9 \text{ cm}^2$$

Figure 2.24   A multi-plate capacitor

In order to increase the capacitance of a capacitor, many practical components employ multiple plates (see Fig. 2.24). The capacitance is then given by:

$$C = \frac{\varepsilon_0 \varepsilon_r (n-1) A}{d}$$

where C is the capacitance (in farads), ε_0 is the permittivity of free space, ε_r is the **relative permittivity** of the dielectric medium between the plates), and d is the separation between the plates (in metres) and n is the total number of plates.

Example 2.22

A capacitor consists of six plates each of area 20 cm^2 separated by a dielectric of relative permittivity

4.5 and thickness 0.2 mm. Determine the value of capacitance.

Solution

Using

$$C = \frac{\varepsilon_0 \varepsilon_r (n-1) A}{d}$$

gives:

$$C = \frac{8.854 \times 10^{-12} \times 4.5 \times (6-1) \times 20 \times 10^{-4}}{0.2 \times 10^{-3}}$$

from which

$$C = \frac{3,984.3 \times 10^{-16}}{0.2 \times 10^{-3}} = 19.921 \times 10^{-13} = 190 \times 10^{-12}$$

thus

$$C = 190 \times 10^{-12} \text{ F or 190 pF}$$

Capacitor specifications

The specifications for a capacitor usually include the value of capacitance (expressed in microfarads, nanofarads or picofarads), the voltage rating (i.e. the maximum voltage which can be continuously applied to the capacitor under a given set of conditions), and the accuracy or tolerance (quoted as the maximum permissible percentage deviation from the marked value).

Other practical considerations when selecting capacitors for use in a particular application include temperature coefficient, leakage current, stability and ambient temperature range.

Table 2.5 summarizes the properties of five of the most common types of capacitor. Note that electrolytic capacitors require the application of a polarizing voltage in order to the chemical action on which they depend for their operation. The polarizing voltages used for electrolytic capacitors can range from as little as 1 V to several hundred volts depending upon the working voltage rating for the component in question.

Figure 2.25 shows some typical non-electrolytic capacitors (including polyester, polystyrene, ceramic and mica types) whilst Fig. 2.26 shows a selection of electrolytic (polarized) capacitors. An air-spaced variable capacitor is shown later in Fig. 2.34 on page 38.

Table 2.5   Characteristics of common types of capacitor

Property	Capacitor type				
	Ceramic	Electrolytic	Polyester	Mica	Polystyrene
Capacitance range (F)	2.2 p to 100 n	100 n to 10 m	10 n to 2.2 μ	0.47 to 22 k	10 p to 22 n
Typical tolerance (%)	±10 and ±20	−10 to +50	±10	±1	±5
Typical voltage rating (W)	50 V to 200 V	6.3 V to 400 V	100 V to 400 V	350 V	100 V
Temperature coefficient (ppm/°C)	+100 to −4700	+1000 typical	+100 to +200	+50	+250
Stability	Fair	Poor	Good	Excellent	Good
Ambient temperature range (°C)	−85 to +85	−40 to +80	−40 to +100	−40 to +125	−40 to +100
Typical applications	High-frequency and low-cost	Smoothing and decoupling	General purpose	Tuned circuits and oscillators	General purpose

Figure 2.25   A typical selection of non-electrolytic capacitors (including polyester, polystyrene, ceramic and mica types) with values ranging from 10 pF to 470 nF and working voltages from 50 V to 250 V

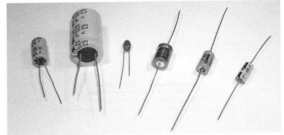

Figure 2.26   A typical selection of electrolytic (polarized) capacitors with values ranging from 1 μF to 470 μF and working voltages from 10 V to 63 V

Capacitor markings

The vast majority of capacitors employ written markings which indicate their values, working voltages, and tolerance. The most usual method of marking resin dipped polyester (and other) types of capacitor involves quoting the value (μF, nF or pF), the tolerance (often either 10% or 20%), and the working voltage (often using _ and ~ to indicate d.c. and a.c. respectively). Several manufacturers use two separate lines for their capacitor markings and these have the following meanings:

First line: capacitance (pF or μF) and tolerance (K = 10%, M = 20%)

Second line: rated d.c. voltage and code for the dielectric material

A three-digit code is commonly used to mark monolithic ceramic capacitors. The first two digits of this code correspond to the first two digits of the value while the third digit is a multipler which gives the number of zeros to be added to give the value in picofarads. Other capacitor may use a colour code similar to that used for marking resistor values (see Fig. 2.28).

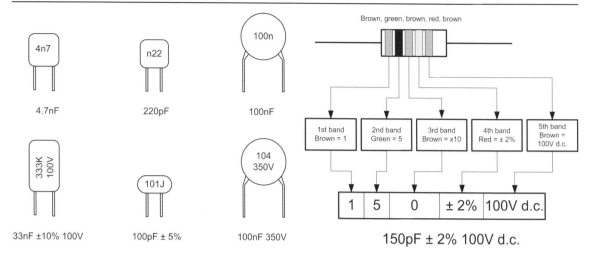

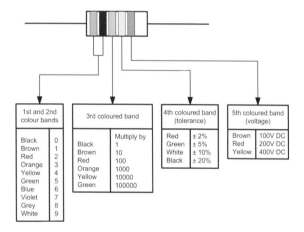

Figure 2.27   Examples of capacitor markings

Figure 2.29   See Example 2.23

Figure 2.28   Capacitor colour code

Example 2.23

A monolithic ceramic capacitor is marked with the legend '103K'. What is its value?

Solution

The value (pF) will be given by the first two digits (10) followed by the number of zeros indicated by the third digit (3). The value of the capacitor is thus 10,000 pF or 10 nF. The final letter (K) indicates that the capacitor has a tolerance of 10%.

Example 2.24

A tubular capacitor is marked with the following coloured stripes: brown, green, brown, red, brown. What is its value, tolerance, and working voltage?

Solution

See Fig. 2.29.

Series and parallel combination of capacitors

In order to obtain a particular value of capacitance, fixed capacitors may be arranged in either series or parallel (Figs 2.30 and 2.31). The reciprocal of the effective capacitance of each of the series circuits shown in Fig. 2.30 is equal to the sum of the reciprocals of the individual capacitances. Hence, for Fig. 2.30(a):

$$\frac{1}{C} = \frac{1}{C_1} + \frac{1}{C_2}$$

while for Fig. 2.30(b):

$$\frac{1}{C} = \frac{1}{C_1} + \frac{1}{C_2} + \frac{1}{C_3}$$

In the former case, the formula can be more conveniently re-arranged as follows:

$$C = \frac{C_1 \times C_2}{C_1 + C_2}$$

You can remember this as the *product* of the two capacitor values *divided by* the *sum* of the two values—just as you did for two resistors in parallel.

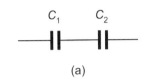

(a)

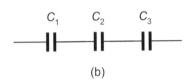

(b)

Figure 2.30   Capacitors in series

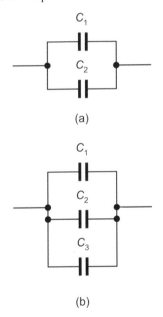

(a)

(b)

Figure 2.31   Capacitors in parallel

For a parallel arrangement of capacitors, the effective capacitance of the circuit is simply equal to the sum of the individual capacitances. Hence, for Fig. 2.31(a):

$$C = C_1 + C_2$$

while for Fig. 2.31(b)

$$C = C_1 + C_2 + C_3$$

Example 2.25

Determine the effective capacitance of the circuit shown in Fig. 2.32.

Solution

The circuit of Fig. 2.32 can be progressively simplified as shown in Fig. 2.33. The stages in this simplification are:

(a) C_1 and C_2 are in parallel and they can be replaced by a single capacitor (C_A) of (2 nF + 4 nF) = 6 nF.

(b) C_A appears in series with C_3. These two resistors can be replaced by a single capacitor (C_B) of (6 nF × 2 nF)/(6 nF + 2 nF) = 1.5 nF.

(c) C_B appears in parallel with C_4. These two capacitors can be replaced by a single capacitance (C) of (1.5 nF + 4 nF) = 5.5 nF.

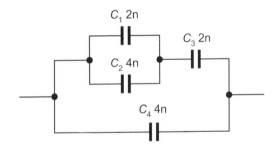

Figure 2.32   See Example 2.25

Example 2.26

A capacitance of 50 μF (rated at 100 V) is required. What series combination of preferred value capacitors will satisfy this requirement? What voltage rating should each capacitor have?

Solution

Two 100 μF capacitors wired in series will provide a capacitance of 50 μF, as follows:

$$C = \frac{C_1 \times C_2}{C_1 + C_2} = \frac{100 \times 100}{100 + 100} = \frac{10,000}{200} = 50 \ \mu F$$

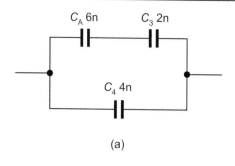

C_A 6n C_3 2n

C_4 4n

(a)

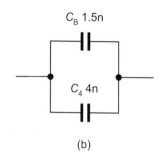

C_B 1.5n

C_4 4n

(b)

$C = 5.5n$

(c)

Figure 2.33   See Example 2.25

Since the capacitors are of equal value, the applied d.c. potential will be shared equally between them. Thus each capacitor should be rated at 50 V. Note that, in a practical circuit, we could take steps to ensure that the d.c. voltage was shared equally between the two capacitors by wiring equal, high-value (e.g. 100 kΩ) resistors across each capacitor.

Variable capacitors

By moving one set of plates relative to the other, a capacitor can be made variable. The dielectric material used in a variable capacitor can be either air (see Fig. 2.34) or plastic (the latter tend to be more compact). Typical values for variable capacitors tend to range from about 25 pF to 500 pF. These components are commonly used for tuning radio receivers.

Figure 2.34   An air-spaced variable capacitor. This component (used for tuning an AM radio) has two separate variable capacitors (each of 500 pF maximum) operated from a common control shaft.

Inductors

Inductors provide us with a means of storing electrical energy in the form of a magnetic field. Typical applications include chokes, filters and (in conjunction with one or more capacitors) frequency selective circuits. The electrical characteristics of an inductor are determined by a number of factors including the material of the core (if any), the number of turns, and the physical dimensions of the coil. Figure 2.35 shows the construction of a typical toroidal inductor wound on a ferrite (high permeability) core.

In practice every coil comprises both inductance (L) and a small resistance (R). The circuit of Fig. 2.36 shows these as two discrete components. In reality the inductance and the resistance (we often refer to this as a **loss resistance** because it's something that we don't actually want) are both

Figure 2.35   A practical coil contains inductance *and* resistance

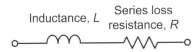

Figure 2.36 A practical coil contains inductance *and* a small amount of series loss resistance

distributed throughout the component but it is convenient to treat the inductance and resistance as separate components in the analysis of the circuit.

To understand what happens when a changing current flows through an inductor, take a look at the circuit shown in Fig. 2.37(a). If the switch is left open, no current will flow and no magnetic flux will be produced by the inductor. If the switch is closed, as shown in Fig. 2.37(b), current will begin to flow as energy is taken from the supply in order to establish the magnetic field. However, the change in magnetic flux resulting from the appearance of current creates a voltage (an **induced e.m.f.**) across the coil which opposes the applied e.m.f. from the battery.

The induced e.m.f. results from the changing flux and it effectively prevents an instantaneous rise in current in the circuit. Instead, the current increases slowly to a maximum at a rate which depends upon the ratio of inductance (L) to resistance (R) present in the circuit.

After a while, a steady state condition will be reached in which the voltage across the inductor will have decayed to zero and the current will have reached a maximum value determined by the ratio of V to R (i.e. Ohm's Law). This is shown in Fig. 2.37(c).

If, after this steady state condition has been achieved, the switch is opened, as shown in Fig. 2.37(d), the magnetic field will suddenly collapse and the energy will be returned to the circuit in the form of an induced **back e.m.f.** which will appear across the coil as the field collapses. For large values of magnetic flux and inductance this back e.m.f. can be extremely large!

Inductance

Inductance is the property of a coil which gives rise to the opposition to a change in the value of current flowing in it. Any change in the current applied to a

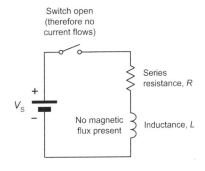

(a) No current in the inductor

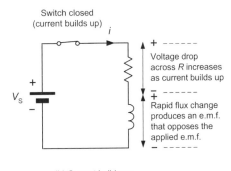

(b) Current builds up

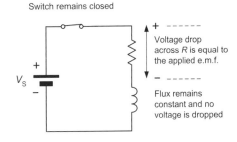

(c) Current remains constant

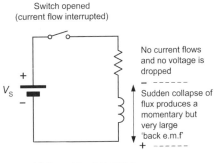

(d) Current flow interrupted

Figure 2.37 Flux and e.m.f. generated when a changing current is applied to an inductor

coil/inductor will result in an induced voltage appearing across it. The unit of inductance is the henry (H) and a coil is said to have an inductance of 1H if a voltage of 1V is induced across it when a current changing at the rate of 1 A/s is flowing in it.

The voltage induced across the terminals of an inductor will thus be proportional to the product of the inductance (L) and the rate of change of applied current. Hence:

$e = -L \times$ (rate of change of current)

Note that the minus sign indicates the polarity of the voltage, i.e. opposition to the change.

The rate of change of current is often represented by the expression di/dt where di represents a very small change in current and dt represents the corresponding small change in time. Using mathematical notation to write this we arrive at:

$$e = -L\frac{di}{dt}$$

You might like to compare this with the similar relationship that we obtained for the current flowing in a capacitor shown on page 33.

Example 2.27

A current increases at a uniform rate from 2 A to 6 A in a period of 250 ms. If this current is applied to an inductor of 600 mH, determine the voltage induced.

Solution

Now the induced voltage will be given by:

$e = -L \times$ (rate of change of current)

Thus

$$e = -L\left(\frac{\text{change in current}}{\text{change in time}}\right) = -60 \times 10^{-3} \times \left(\frac{6-2}{250 \times 10^{-3}}\right)$$

From which

$$e = -600 \times 10^{-3} \times \left(\frac{4}{0.25}\right) = -0.6 \times 10^{-3} \times 16$$

so

$$\varepsilon = -9.6 \text{ V}$$

Energy storage

The energy stored in an inductor is proportional to the product of the inductance and the square of the current flowing in it. Thus:

$W = \frac{1}{2}LI^2$

where W is the energy (in Joules), L is the capacitance (in Henries), and I is the current flowing in the inductor (in Amps).

Example 2.28

An inductor of 20 mH is required to store 2.5 J of energy. Determine the current that must be applied.

Solution

The foregoing formula can be re-arranged to make I the subject as follows:

$$I = \sqrt{\frac{E}{0.5L}} = \sqrt{\frac{2E}{L}} = \sqrt{\frac{2 \times 2.5}{20 \times 10^{-3}}}$$

From which

$$I = \sqrt{\frac{5}{20 \times 10^{-3}}} = \sqrt{0.25 \times 10^{-3}} = \sqrt{250} = 15.81 \text{ A}$$

Inductance and physical dimensions

The inductance of an inductor depends upon the physical dimensions of the inductor (e.g. the length and diameter of the winding), the number of turns, and the permeability of the material of the core. The inductance of an inductor is given by:

$$L = \frac{\mu_0 \mu_r n^2 A}{l}$$

where L is the inductance (in Henries), μ_0 is the permeability of free space, μ_r is the relative permeability of the magnetic core, l is the mean length of the core (in metres), and A is the cross-sectional area of the core (in square metres).

Example 2.29

An inductor of 100 mH is required. If a closed magnetic core of length 20 cm, cross-sectional area 15 cm^2 and relative permeability 500 is available, determine the number of turns required.

Solution

First we must re-arrange the formula

$$L = \frac{\mu_0 \mu_r n^2 A}{l}$$

in order to make n the subject:

$$n = \sqrt{\frac{L \times l}{\mu_0 \mu_r n^2 A}} = \sqrt{\frac{100 \times 10^{-3} \times 20 \times 10^{-2}}{12.57 \times 10^{-7} \times 500 \times 15 \times 10^{-4}}}$$

From which

$$n = \sqrt{\frac{2 \times 10^{-2}}{94,275 \times 10^{-11}}} = \sqrt{21,215} = 146$$

Hence the inductor requires 146 turns of wire.

Inductor specifications

Inductor specifications normally include the value of inductance (expressed in henries, millihenries or microhenries), the current rating (i.e. the maximum current which can be continuously applied to the inductor under a given set of conditions), and the accuracy or tolerance (quoted as the maximum permissible percentage deviation from the marked value). Other considerations may include the temperature coefficient of the inductance (usually expressed in parts per million, p.p.m., per unit temperature change), the stability of the inductor, the d.c. resistance of the coil windings (ideally zero), the Q-factor (quality factor) of the coil, and the recommended working frequency range. Table 2.6 summarizes the properties of four common types of inductor. Some typical small inductors are shown in Fig. 2.38. These have values of inductance ranging from 15 μH to 1 mH.

Inductor markings

As with capacitors, the vast majority of inductors use written markings to indicate values, working current, and tolerance. Some small inductors are marked with coloured stripes to indicate their value and tolerance (in which case the standard colour values are used and inductance is normally expressed in microhenries).

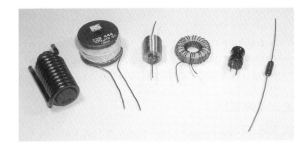

Figure 2.38   A selection of small inductors with values ranging from 15 μH to 1 mH

Table 2.6   Characteristics of common types of inductor

Property	Inductor type			
	Air cored	*Ferrite cored*	*Ferrite pot cored*	*Iron cored*
Core material	Air	Ferrite rod	Ferrite pot	Laminated steel
Inductance range (H)	50 n to 100 μ	10 μ to 1 m	1 m to 100 m	20 m to 20
Typical d.c. resistance (Ω)	0.05 to 5	0.1 to 10	5 to 100	10 to 200
Typical tolerance (%)	±5	±10	±10	±20
Typical Q-factor	60	80	40	20
Typical frequency range (Hz)	1 M to 500 M	100 k to 100 M	1 k to 10 M	50 to 10 k
Typical applications	Tuned circuits and filters	Filters and HF transformers	LF and MF filters and transformers	Smoothing chokes and filters

Series and parallel combinations of inductors

In order to obtain a particular value of inductance, fixed inductors may be arranged in either series or parallel as shown in Figs 2.39 and 2.40.

The effective inductance of each of the series circuits shown in Fig. 2.39 is simply equal to the sum of the individual inductances. So, for the circuit shown in Fig. 2.39(a):

$$L = L_1 + L_2$$

while for Fig. 2.39 (b)

$$L = L_1 + L_2 + L_3$$

Turning to the parallel inductors shown in Fig. 2.40, the reciprocal of the effective inductance of each circuit is equal to the sum of the reciprocals of the individual inductances. Hence, for Fig. 2.40(a):

$$\frac{1}{L} = \frac{1}{L_1} + \frac{1}{L_2}$$

while for Fig. 2.40(b)

$$\frac{1}{L} = \frac{1}{L_1} + \frac{1}{L_2} + \frac{1}{L_3}$$

In the former case, the formula can be more conveniently re-arranged as follows:

$$L = \frac{L_1 \times L_2}{L_1 + L_2}$$

You can remember this as the *product* of the two inductance values *divided by* the *sum* of the two inductance values.

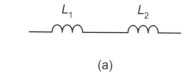

(a)

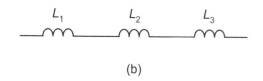

(b)

Figure 2.39 Inductors in series

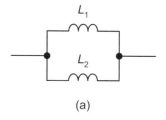

(a)

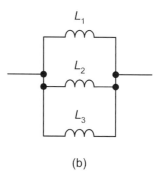

(b)

Figure 2.40 Inductors in parallel

Example 2.30

An inductance of 5 mH (rated at 2 A) is required. What parallel combination of preferred value inductors will satisfy this requirement?

Solution

Two 10 mH inductors may be wired in parallel to provide an inductance of 5 mH as shown below:

$$L = \frac{L_1 \times L_2}{L_1 + L_2} = \frac{10 \times 10}{10 + 10} = \frac{100}{20} = 5 \text{ mH}$$

Since the inductors are identical, the applied current will be shared equally between them. Hence each inductor should have a current rating of 1 A.

Example 2.31

Determine the effective inductance of the circuit shown in Fig. 2.41.

Solution

The circuit can be progressively simplified as shown in Fig. 2.42. The stages in this simplification are as follows:

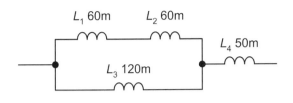

Figure 2.40 See Example 2.31

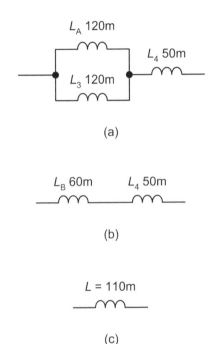

(a)

(b)

$L = 110m$

(c)

Figure 2.42 See Example 2.31

(a) L_1 and L_2 are in series and they can be replaced by a single inductance (L_A) of (60 + 60) = 120 mH.
(b) L_A appears in parallel with L_2. These two inductors can be replaced by a single inductor (L_B) of (120 × 120)/(120 + 120) = 60 mH.
(c) L_B appears in series with L_4. These two inductors can be replaced by a single inductance (L) of (60 + 50) = 110 mH.

Variable inductors

A ferrite cored inductor can be made variable by moving its core in or out of the former onto which the coil is wound. Many small inductors have threaded ferrite cores to make this possible (see Fig. 2.43). Such inductors are often used in radio and high-frequency applications where precise tuning is required.

Figure 2.43 An adjustable ferrite cored inductor

Surface mounted components (SMC)

Surface mounting technology (SMT) is now widely used in the manufacture of printed circuit boards for electronic equipment. SMT allows circuits to be assembled in a much smaller space than would be possible using components with conventional wire leads and pins that are mounted using through-hole techniques. It is also possible to mix the two technologies, i.e. some through-hole mounting of components and some surface mounted components present on the same circuit board. The following combinations are possible:

- Surface mounted components (SMC) on both sides of a printed circuit board
- SMC on one side of the board and conventional **through-hole components** (THC) on the other
- A mixture of SMC and THC on both sides of the printed circuit board.

Surface mounted components are supplied in packages that are designed for mounting directly on the surface of a PCB. To provide electrical contact with the PCB, some SMC have contact pads on their surface. Other devices have contacts which extend beyond the outline of the package itself but which terminate on the surface of the PCB rather than making contact through a hole (as is the case with a conventional THC). In general, passive components (such as resistors, capacitors and inductors) are configured leadless for surface mounting, whilst active devices (such as transistors and integrated circuits) are available in both surface mountable types as well as lead as well as in leadless terminations suitable for making direct contact to the pads on the surface of a PCB.

Most surface mounted components have a flat rectangular shape rather than the cylindrical shape that we associate with conventional wire leaded components. During manufacture of a PCB, the various SMC are attached using re-flow soldering paste (and in some cases adhesives) which consists of particles of solder and flux together with binder, solvents and additives. They need to have good 'tack' in order to hold the components in place and remove oxides without leaving obstinate residues.

The component attachment (i.e. soldering!) process is completed using one of several techniques including convection ovens in which the PCB is passed, using a conveyor belt, through a convection oven which has separate zones for preheating, flowing and cooling, and infra-red re-flow in which infra-red lamps are used to provide the source of heat.

Surface mounted components are generally too small to be marked with colour codes. Instead, values may be marked using three digits. For example, the first two digits marked on a resistor normally specify the first two digits of the value whilst the third digit gives the number of zeros that should be added.

Example 2.32

In Fig. 1.18, R88 is marked '102'. What is its value?

Solution

R88 will have a value of 1,000 Ω (i.e. 10 followed by two zeros).

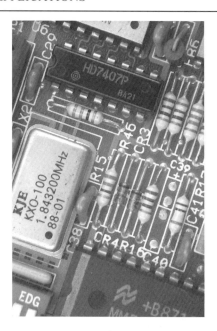

Figure 2.44 Conventional components mounted on a printed circuit board. Note that components such as C38, R46, etc. have leads that pass through holes in the printed circuit board

Figure 2.45 Surface mounted components (note the appearance of capacitors C35, C52, and C53, and resistors, R87, R88, R91, etc.)

Practical investigation

Objective

To investigate the resistance of series and parallel combinations of resistors.

Components and test equipment

Breadboard, digital or analogue meter with d.c. current ranges, 9 V d.c. power source (either a 9 V battery or an a.c. mains adapter with a 9 V 400 mA output), test leads, resistors of 100 Ω, 220 Ω, 330 Ω, 470 Ω, 680 Ω and 1k Ω, connecting wire.

Procedure

Connect the resistor network and power supply (or battery) as shown in Fig. 2.46. Select the 20 V d.c. voltage range on the multimeter then measure and record the supply voltage (this should be approximately 9 V). Now break the positive connection to the circuit, change the range on the multimeter to the 20 mA d.c. current range and measure the current supplied to the circuit.

Next measure and record the voltage dropped across each resistor (don't forget to change ranges on the multimeter when making each measurement).

Finally, break the circuit at one end of each of resistor in turn, then measure and record the current flowing. Repeat the procedure for the other two resistor networks circuits shown in Figs. 2.47 and 2.48.

Measurements and calculations

Record your results in a table for each network. Use the recorded values of current and voltage for each resistor to calculate the value of resistance and compare this with the marked value. Check that the measured value lies within the tolerance band for each resistor.

Calculate the resistance of each network (looking in at the supply terminals) and compare this with the resistance calculated by dividing the supply voltage by the supply current.

Conclusion

Comment on the results. Did your measured values agree with the marked values? Were these within the tolerance range for the resistors used in the investigation? If the readings were not in agreement can you suggest why?

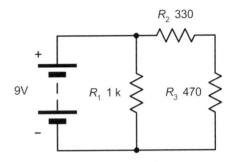

Figure 2.46 Circuit diagram—first network

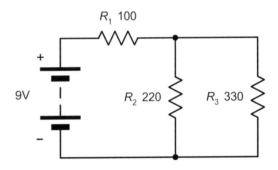

Figure 2.47 Circuit diagram—second network

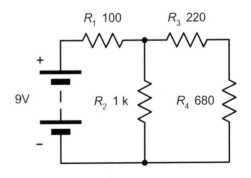

Figure 2.48 Circuit diagram—third network

Important formulae introduced in this chapter

Component tolerance:
(page 18)

$$\text{Tolerance} = \frac{\text{error}}{\text{marked value}} \times 100\%$$

Resistors in series:
(page 7)

$$R = R_1 + R_2 + R_3$$

Resistors in parallel:
(page 5)

$$\frac{1}{R} = \frac{1}{R_1} + \frac{1}{R_2} + \frac{1}{R_3}$$

Two resistors in parallel:
(page 8)

$$R = \frac{R_1 \times R_2}{R_1 + R_2}$$

Resistance and temperature:
(page 9)

$$R_t = R_0(1 + \alpha\, t)$$

Current flowing in a capacitor:
(page 13)

$$i = C\frac{dV}{dt}$$

Charge in a capacitor:
(page 15)

$$Q = C V$$

Energy stored in a capacitor:
(page 13)

$$W = \tfrac{1}{2}\, C\, V^2$$

Capacitance of a capacitor:
(page 15)

$$C = \frac{\varepsilon_0 \varepsilon_r A}{d}$$

Capacitors in series:
(page 14)

$$\frac{1}{C} = \frac{1}{C_1} + \frac{1}{C_2} + \frac{1}{C_3}$$

Two capacitors in series:
(page 14)

$$C = \frac{C_1 \times C_2}{C_1 + C_2}$$

Capacitors in parallel:
(page 36)

$$C = C_1 + C_2 + C_3$$

Induced e.m.f. in an inductor:
(page 39)

$$e = -L\frac{di}{dt}$$

Energy stored in an inductor:
(page 39)

$$W = \tfrac{1}{2}\, L\, I^2$$

Inductance of an inductor:
(page 39)

$$L = \frac{\mu_0 \mu_r n^2 A}{l}$$

Inductors in series:
(page 40)

$$L = L_1 + L_2 + L_3$$

Inductors in parallel:
(page 40)

$$\frac{1}{L} = \frac{1}{L_1} + \frac{1}{L_2} + \frac{1}{L_3}$$

Two inductors in parallel:
(page 40)

$$L = \frac{L_1 \times L_2}{L_1 + L_2}$$

Symbols introduced in this chapter

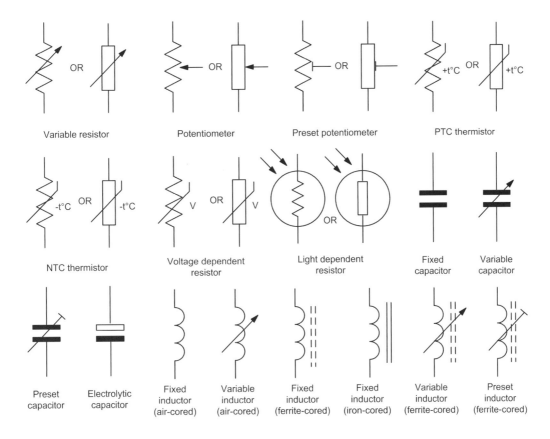

Figure 2.49   Circuit symbols introduced in this chapter

Problems

2.1   A power supply rated at 15 V, 0.25 A is to be tested at full rated output. What value of load resistance is required and what power rating should it have? What type of resistor is most suitable for this application and why?

2.2   Determine the value and tolerance of resistors marked with the following coloured bands:

(a) red, violet, yellow, gold
(b) brown, black, black, silver
(c) blue, grey, green, gold
(d) orange, white, silver, gold
(e) red, red, black, brown, red.

2.3   A batch of resistors are all marked yellow, violet, black, gold. If a resistor is selected from this batch, within what range would you expect its value to be?

2.4   Resistors of 27 Ω, 33 Ω, 56 Ω and 68 Ω are available. How can two or more of these be arranged to realize the following resistance values?

(a) 60 Ω
(b) 14.9 Ω
(b) 124 Ω
(b) 11.7 Ω
(b) 128 Ω.

2.5   Three 100 Ω resistors are connected as shown in Fig. 2.50. Determine the effective resistance of the circuit.

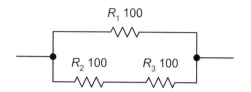

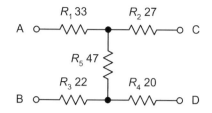

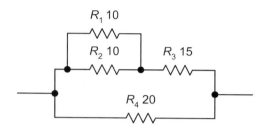

Figure 2.50 See Question 2.5

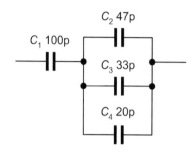

Figure 2.52 See Question 2.7

Figure 2.51 See Question 2.6

Figure 2.53 See Question 2.13

2.6   Determine the effective resistance of the circuit shown in Fig. 2.51.

2.7   Determine the resistance of the resistor network shown in Fig. 2.52 looking into terminal A and B with (a) terminals C and D left open circuit and (b) terminals C and D short circuit.

2.8   A resistor has a temperature coefficient of 0.0008/°C. If the resistor has a resistance of 390 Ω at 0°C, determine its resistance at 55°C.

2.9   A resistor has a temperature coefficient of 0.004/°C. If the resistor has a resistance of 82 kΩ at 20°C, what will its resistance be at 75°C?

2.10   A resistor has a resistance of 218 Ω at 0°C and 225 Ω at 100°C. Determine the resistor's temperature coefficient.

2.11   Capacitors of 1 μF, 3.3 μF, 4.7 μF and 10 μF are available. How can two or more of these capacitors be arranged to realize the following capacitance values?

(a) 8 μF
(b) 11 μF
(c) 19 μF
(d) 0.91 μF
(e) 1.94 μF

2.12   Three 180 pF capacitors are connected (a) in series and (b) in parallel. Determine the effective capacitance of each combination.

2.13   Determine the effective capacitance of the circuit shown in Fig. 2.53.

2.14   A capacitor of 330 μF is charged to a potential of 63 V. Determine the quantity of energy stored.

2.15   A parallel plate capacitor has plates of 0.02 m². Determine the capacitance of the capacitor if the plates are separated by a dielectric of thickness 0.5 mm and relative permittivity 5.6.

2.16   A capacitor is required to store 0.5 J of energy when charged from a 120 V d.c. supply. Determine the value of capacitance required.

2.17   The current in a 2.5 H inductor increases uniformly from zero to 50 mA in 400 ms. Determine the e.m.f. induced.

2.18   An inductor has 200 turns of wire wound on a closed magnetic core of mean length 24 cm, cross-sectional 10 cm² and relative permeability 650. Determine the inductance of the inductor.

2.19   A current of 4 A flows in a 60 mH inductor. Determine the energy stored.

2.20   Inductors of 22 mH and 68 mH are connected (a) in series and (b) in parallel. Determine the effective inductance in each case.

Answers to these problems appear on page 374.

3

D.C. circuits

In many cases, Ohm's Law alone is insufficient to determine the magnitude of the voltages and currents present in a circuit. This chapter introduces several techniques that simplify the task of solving complex circuits. It also introduces the concept of exponential growth and decay of voltage and current in circuits containing capacitance and resistance and inductance and resistance. It concludes by showing how humble C–R circuits can be used for shaping the waveforms found in electronic circuits. We start by introducing two of the most useful laws of electronics.

Kirchhoff's Laws

Kirchhoff's Laws relate to the algebraic sum of currents at a junction (or **node**) or voltages in a network (or **mesh**). The term 'algebraic' simply indicates that the polarity of each current or voltage drop must be taken into account by giving it an appropriate sign, either positive (+) or negative (–).

Kirchhoff's Current Law states that the algebraic sum of the currents present at a junction (node) in a circuit is zero (see Fig. 3.1).

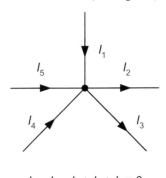

$$I_1 - I_2 - I_3 + I_4 + I_5 = 0$$

Convention:
Current flowing towards the junction is positive (+)
Current flowing away from the junction is negative (–)

Figure 3.1 Kirchhoff's Current Law

Example 3.1

In Fig. 3.2, use Kirchhoff's Current Law to determine:

(a) the value of current flowing between A and B
(b) the value of I_3.

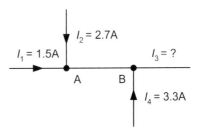

Figure 3.2 See Example 3.1

Solution

(a) I_1 and I_2 both flow towards Node A so, applying our polarity convention, they must both be positive. Now, assuming that a current I_5 flows between A and B and that this current flows away from the junction (obvious because I_1 and I_2 both flow towards the junction) we arrive at the following Kirchhoff's Current Law equation:

$$+I_1 + I_2 - I_5 = 0$$

From which:

$$I_5 = I_1 + I_2 = 1.5 + 2.7 = 4.2 \text{ A}$$

(b) Moving to Node B, let's assume that I_3 flows outwards so we can say that:

$$+I_4 + I_5 - I_3 = 0$$

From which:

$$I_3 = I_4 + I_5 = 3.3 + 4.2 = 7.5 \text{ A}$$

Finally, it's worth checking these results with the Current Law equation (i):

$$+I_1 + I_2 - I_3 = 0$$

Inserting our values for I_1, I_2 and I_3 gives:

$$+0.068 + 0.136 - 204 = 0$$

Since the left and right hand sides of the equation are equal we can be reasonably confident that our results are correct.

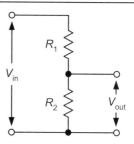

Figure 3.8   Potential divider circuit

The potential divider

The potential divider circuit (see Fig. 3.8) is commonly used to reduce voltages in a circuit. The output voltage produced by the circuit is given by:

$$V_{out} = V_{in} \frac{R_2}{R_1 + R_2}$$

It is, however, important to note that the output voltage (V_{out}) will fall when current is drawn from the arrangement.

Figure 3.9 shows the effect of **loading** the potential divider circuit. In the loaded potential divider (Fig. 3.9) the output voltage is given by:

$$V_{out} = V_{in} \frac{R_p}{R_1 + R_p}$$

where:

$$R_p = \frac{R_2 \times R_L}{R_2 + R_L}$$

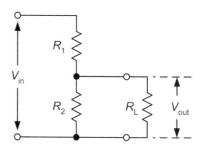

Figure 3.9   Loaded potential divider circuit

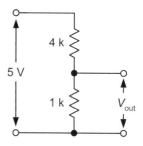

Figure 3.10   See Example 3.4

Example 3.4

The potential divider shown in Fig. 3.10 is used as a simple **voltage calibrator**. Determine the output voltage produced by the circuit:

(a) when the output terminals are left open-circuit (i.e. when no load is connected); and
(b) when the output is loaded by a resistance of 10 kΩ.

Solution

(a) In the first case we can simply apply the formula:

$$V_{out} = V_{in} \frac{R_2}{R_1 + R_2}$$

where V_{in} = 5 V, R_1= 4 kΩ and R_2= 1 kΩ.

Hence:

$$V_{out} = 5 \times \frac{1}{4+1} = 1 \text{ V}$$

(b) In the second case we need to take into account the effect of the 10 kΩ resistor connected to the output terminals of the potential divider.

First we need to find the equivalent resistance of the parallel combination of R_2 and R_L:

$$R_p = \frac{R_2 \times R_L}{R_2 + R_L} = \frac{1 \times 10}{1 + 10} = \frac{10}{11} = 0.909 \text{ k}\Omega$$

Then we can determine the output voltage from:

$$V_{out} = V_{in} \frac{R_p}{R_1 + R_p} = 5 \times \frac{0.909}{4 + 0.909} = 0.925 \text{ V}$$

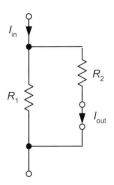

Figure 3.11   Current divider circuit

The current divider

The current divider circuit (see Fig. 3.11) is used to divert a known proportion of the current flowing in a circuit. The output current produced by the circuit is given by:

$$I_{out} = I_{in} \frac{R_1}{R_1 + R_2}$$

It is, however, important to note that the output current (I_{out}) will fall when the load connected to the output terminals has any appreciable resistance.

Example 3.5

A moving coil meter requires a current of 1 mA to provide full-scale deflection. If the meter coil has a resistance of 100 Ω and is to be used as a milliammeter reading 5 mA full-scale, determine the value of parallel shunt resistor required.

Solution

This problem may sound a little complicated so it is worth taking a look at the **equivalent circuit** of the meter (Fig. 3.12) and comparing it with the current divider shown in Fig. 3.11.

We can apply the current divider formula, replacing I_{out} with I_m (the meter full-scale deflection current) and R_2 with R_m (the meter resistance). R_1 is the required value of shunt resistor, R_s, Hence:

$$I_{out} = I_{in} \frac{R_s}{R_s + R_m}$$

Re-arranging the formula gives:

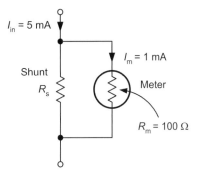

Figure 3.12   See Example 3.5

$$I_m \times (R_s + R_m) = I_{in} \times R_s$$

thus

$$I_m R_s + I_m R_m = I_{in} R_s$$

or

$$I_{in} R_s - I_m R_s = I_m R_m$$

from which

$$R_s(I_{in} - I_m) = I_m R_m$$

so

$$R_s = \frac{I_m R_m}{I_{in} - I_m}$$

Now I_{in} = 1 mA, R_m = 100 Ω and I_{in} = 5 mA, thus:

$$R_s = \frac{1 \times 100}{5 - 1} = \frac{100}{4} = 25 \ \Omega$$

The Wheatstone bridge

The Wheatstone bridge forms the basis of a number of useful electronic circuits including several that are used in instrumentation and measurement.

The basic form of Wheatstone bridge is shown in Fig. 3.13. The voltage developed between A and B will be zero when the voltage between A and Y is the same as that between B and Y. In effect, R_1 and R_2 constitute a potential divider as do R_3 and R_4.

The bridge will be **balanced** (and $V_{AB} = 0$) when the ratio of $R_1:R_2$ is the same as the ratio $R_3:R_4$. Hence, at balance:

$$\frac{R_1}{R_2} = \frac{R_3}{R_4}$$

A practical form of Wheatstone bridge that can be used for measuring unknown resistances is shown in Fig. 3.14.

In this practical form of Wheatstone bridge, R_1 and R_2 are called the **ratio arms** while one arm (that occupied by R_3 in Fig. 3.13) is replaced by a calibrated variable resistor. The unknown resistor, R_x, is connected in the fourth arm. At balance:

$$\frac{R_1}{R_2} = \frac{R_v}{R_x} \quad \text{thus} \quad R_x = \frac{R_2}{R_1} \times R_v$$

Example 3.6

A Wheatstone bridge is based on the circuit shown in Fig. 3.14. If R_1 and R_2 can each be switched so that they have values of either 100 Ω or 1 kΩ and R_V is variable between 10 Ω and 10 kΩ, determine the range of resistance values that can be measured.

Solution

The maximum value of resistance that can be measured will correspond to the largest ratio of $R_2:R_1$ (i.e. when R_2 is 1 kΩ and R_1 is 100 Ω) and the highest value of RV (i.e. 10 kΩ). In this case:

$$R_x = \frac{1,000}{100} \times 10,000 = 100,000 = 100 \text{ k}\Omega$$

The minimum value of resistance that can be measured will correspond to the smallest ratio of $R_2:R_1$ (i.e. when R_1 is 100 Ω and R_1 is 1 kΩ) and the smallest value of RV (i.e. 10 Ω). In this case:

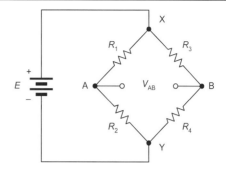

Figure 3.13   Basic Wheatstone bridge circuit

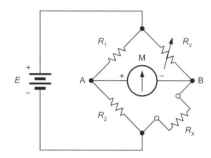

Figure 3.14   See Example 3.6

$$R_x = \frac{100}{1,000} \times 10 = 0.1 \times 10 = 1 \ \Omega$$

Hence the range of values that can be measured extends from 1 Ω to 100 kΩ.

Thévenin's Theorem

Thévenin's Theorem allows us to replace a complicated network of resistances and voltage sources with a simple equivalent circuit comprising a single **voltage source** connected in series with a single resistance (see Fig. 3.15).

The single voltage source in the Thévenin equivalent circuit, V_{oc}, is simply the voltage that appears between the terminals when nothing is connected to it. In other words, it is the *open-circuit* voltage that would appear between A and B.

The single resistance that appears in the Thévenin equivalent circuit, R, is the resistance that would be seen *looking into* the network between A and B when all of the voltage sources (assumed perfect) are replaced by *short-circuit* connections. Note that if the voltage sources are not perfect (i.e. if they have some internal resistance) the equivalent circuit must be constructed on the basis that each voltage source is replaced by its own internal resistance.

Once we have values for V_{oc} and R, we can determine how the network will behave when it is connected to a load (i.e. when a resistor is connected across the terminals A and B).

Example 3.7

Figure 3.16 shows a Wheatstone bridge. Determine the current that will flow in a 100 Ω load connected between terminals A and B.

Solution

First we need to find the Thévenin equivalent of the circuit. To find V_{oc} we can treat the bridge arrangement as two potential dividers.

The voltage across R_2 will be given by:

$$V = 10 \times \frac{R_2}{R_1 + R_2} = 10 \times \frac{600}{500 + 600} = 5.454 \text{ V}$$

Hence the voltage at A relative to Y, V_{AY}, will be 5.454 V.

The voltage across R_4 will be given by:

$$V = 10 \times \frac{R_4}{R_3 + R_4} = 10 \times \frac{400}{500 + 400} = 4.444 \text{ V}$$

Hence the voltage at B relative to Y, V_{BY}, will be 4.444 V.

The voltage V_{AB} will be the difference between V_{AY} and V_{BY}. This, the open-circuit output voltage, V_{AB}, will be given by:

$$V_{AB} = V_{AY} - V_{BY} = 5.454 - 4.444 = 1.01 \text{ V}$$

Next we need to find the Thévenin equivalent resistance looking in at A and B. To do this, we can redraw the circuit, replacing the battery (connected between X and Y) with a short-circuit, as shown in Fig. 3.17.

The Thévenin equivalent resistance is given by the relationship:

$$R = \frac{R_1 \times R_2}{R_1 + R_2} + \frac{R_3 \times R_4}{R_3 + R_4} = \frac{500 \times 600}{500 + 600} + \frac{500 \times 400}{500 + 400}$$

From which:

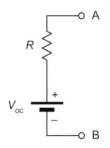

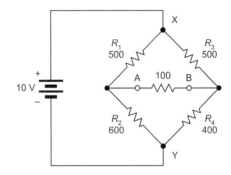

Figure 3.15 Thévenin equivalent circuit

Figure 3.16 See Example 3.7

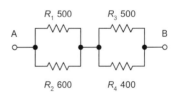

Figure 3.17 See Example 3.7

$$R = \frac{300,000}{1,100} + \frac{200,000}{900} = 272.7 + 222.2 = 494.9 \text{ Ω}$$

The Thévenin equivalent circuit is shown in Fig. 3.18. To determine the current in a 100 Ω load connected between A and B, we can simply add a 100 Ω load to the Thévenin equivalent circuit, as shown in Fig. 3.19. By applying Ohm's Law in Fig. 3.19 we get:

$$I = \frac{V_{oc}}{R + 100} = \frac{1.01}{494.9 + 100} = \frac{1.01}{594.9} = 1.698 \text{ mA}$$

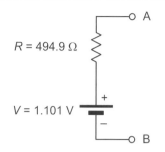

Figure 3.18   Thévenin equivalent of Fig. 3.16

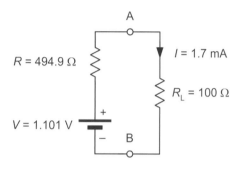

Figure 3.19   Determining the current when the Thévenin equivalent circuit is loaded

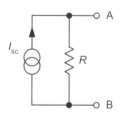

Figure 3.20   Norton equivalent circuit

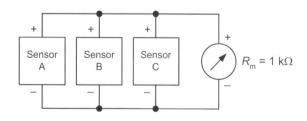

Figure 3.21   See Example 3.8

Norton's Theorem

Norton's Theorem provides an alternative method of reducing a complex network to a simple equivalent circuit. Unlike Thévenin's Theorem, Norton's Theorem makes use of a current source rather than a voltage source. The Norton equivalent circuit allows us to replace a complicated network of resistances and voltage sources with a simple equivalent circuit comprising a single constant current source connected in parallel with a single resistance (see Fig. 3.20).

The constant current source in the Norton equivalent circuit, I_{sc}, is simply the *short-circuit* current that would flow if A and B were to be linked directly together. The resistance that appears in the Norton equivalent circuit, R, is the resistance that would be seen *looking into* the network between A and B when all of the voltage sources are replaced by *short-circuit* connections. Once again, it is worth noting that, if the voltage sources have any appreciable internal resistance, the equivalent circuit must be constructed on the basis that each voltage source is replaced by its own internal resistance.

As with the Thévenin equivalent, we can determine how a network will behave by obtaining values for I_{sc} and R.

Example 3.8

Three temperature sensors having the following characteristics shown in the table below are connected in parallel as shown in Fig. 3.21:

Sensor	A	B	C
Output voltage (open circuit)	20 mV	30 mV	10 mV
Internal resistance	5 kΩ	3 kΩ	2 kΩ

Determine the voltage produced when the arrangement is connected to a moving-coil meter having a resistance of 1 kΩ.

Solution

First we need to find the Norton equivalent of the circuit. To find I_{sc} we can determine the short-circuit current from each sensor and add them together.

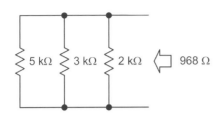

Figure 3.22 Determining the equivalent resistance in Fig. 3.21

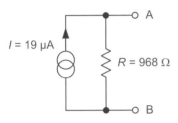

Figure 3.23 Norton equivalent of the circuit in Fig. 3.21

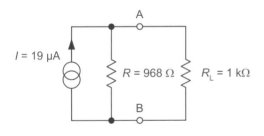

Figure 3.24 Determining the output voltage when the Norton equivalent circuit is loaded with 1 kΩ

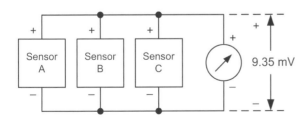

Figure 3.25 The voltage drop across the meter is found to be 9.35 mV

For sensor A:

$$I = \frac{V}{R} = \frac{20\ \text{mV}}{5\ \text{k}\Omega} = 4\ \mu\text{A}$$

For sensor B:

$$I = \frac{V}{R} = \frac{30\ \text{mV}}{3\ \text{k}\Omega} = 10\ \mu\text{A}$$

For sensor C:

$$I = \frac{V}{R} = \frac{10\ \text{mV}}{2\ \text{k}\Omega} = 5\ \mu\text{A}$$

The total current, I_{sc}, will be given by:

$$I_{sc} = 4\ \mu\text{A} + 10\ \mu\text{A} + 5\ \mu\text{A} = 19\ \mu\text{A}$$

Next we need to find the Norton equivalent resistance. To do this, we can redraw the circuit showing each sensor replaced by its internal resistance, as shown in Fig. 3.22.

The equivalent resistance of this arrangement (think of this as the resistance seen *looking into* the circuit in the direction of the arrow shown in Fig. 3.22) is given by:

$$\frac{1}{R} = \frac{1}{R_1} + \frac{1}{R_2} + \frac{1}{R_3} = \frac{1}{5,000} + \frac{1}{3,000} + \frac{1}{2,000}$$

where $R_1 = 5\ \text{k}\Omega$, $R_2 = 3\ \text{k}\Omega$, $R_3 = 2\ \text{k}\Omega$. Hence:

$$\frac{1}{R} = \frac{1}{R_1} + \frac{1}{R_2} + \frac{1}{R_3} = \frac{1}{5,000} + \frac{1}{3,000} + \frac{1}{2,000}$$

or

$$\frac{1}{R} = 0.0002 + 0.00033 + 0.0005 = 0.00103$$

from which:

$$R = 968\ \Omega$$

The Norton equivalent circuit is shown in Fig. 3.23. To determine the voltage in a 1 kΩ moving coil meter connected between A and B, we can make use of the Norton equivalent circuit by simply adding a 1 kΩ resistor to the circuit and applying Ohm's Law, as shown in Fig. 3.24.

The voltage appearing across the moving coil meter in Fig. 3.25 will be given by:

$$V = I_{sc} \times \frac{R \times R_m}{R + R_m} = 19\ \mu\text{A} \times \frac{1,000 \times 968}{1,000 + 968}$$

hence:

$$V = 19\ \mu\text{A} \times 492\ \Omega = 9.35\ \text{mV}$$

C–R circuits

Networks of capacitors and resistors (known as C–R circuits) form the basis of many timing and pulse shaping circuits and are thus often found in practical electronic circuits.

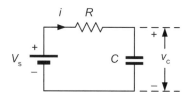

Figure 3.26   A C–R circuit in which C is charged through R

Charging

A simple C–R circuit is shown in Fig. 3.26 . In this circuit C is charged through R from the constant voltage source, V_s. The voltage, v_c, across the (initially uncharged) capacitor voltage will rise exponentially as shown in Fig. 3.27. At the same time, the current in the circuit, i, will fall, as shown in Fig. 3.28.

The rate of growth of voltage with time (and decay of current with time) will be dependent upon the product of capacitance and resistance. This value is known as the **time constant** of the circuit. Hence:

Time constant, $t = C \times R$

where C is the value of capacitance (F), R is the resistance (Ω), and t is the time constant (s).

The voltage developed across the charging capacitor, v_c, varies with time, t, according to the relationship:

$$v_c = V_s \left(1 - e^{-\frac{t}{CR}}\right)$$

where v_c is the capacitor voltage, V_s is the d.c. supply voltage, t is the time, and CR is the time constant of the circuit (equal to the product of capacitance, C, and resistance, R).

The capacitor voltage will rise to approximately 63% of the supply voltage, V_s, in a time interval equal to the time constant.

At the end of the next interval of time equal to the time constant (i.e. after an elapsed time equal to 2CR) the voltage will have risen by 63% of the remainder, and so on. In theory, the capacitor will **never** become fully charged. However, after a period of time equal to 5CR, the capacitor voltage will to all intents and purposes be equal to the supply voltage. At this point the capacitor voltage will have risen to 99.3% of its final value and we can consider it to be fully charged.

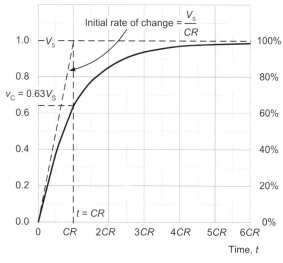

Figure 3.27   Exponential growth of capacitor voltage, v_c, in Fig. 3.26

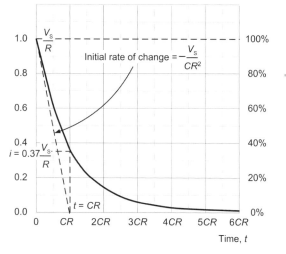

Figure 3.28   Exponential decay of current, i, in Fig. 3.26

During charging, the current in the capacitor, i, varies with time, t, according to the relationship:

$$i = \frac{V_s}{R} e^{-\frac{t}{CR}}$$

where V_s is the d.c. supply voltage, t is the time, R is the series resistance and C is the value of capacitance.

The current will fall to approximately 37% of the initial current in a time equal to the time constant. At the end of the next interval of time equal to the time constant (i.e. after a total time of $2CR$ has elapsed) the current will have fallen by a further 37% of the remainder, and so on.

Example 3.9

An initially uncharged 1 μF capacitor is charged from a 9 V d.c. supply via a 3.3 MΩ resistor. Determine the capacitor voltage 1 s after connecting the supply.

Solution

The formula for exponential growth of voltage in the capacitor is:

$$v_c = V_s \left(1 - e^{-\frac{t}{CR}}\right)$$

Here we need to find the capacitor voltage, v_c, when $V_s = 9$ V, $t = 1$ s, $C = 1$ μF and $R = 3.3$ MΩ. The time constant, CR, will be given by:

$$CR = 1 \times 10^{-6} \times 3.3 \times 10^{6} = 3.3 \text{ s}$$

Thus:

$$v_c = 9 \left(1 - e^{-\frac{t}{3.3}}\right)$$

and

$$v_c = 9 \left(1 - 0.738\right) = 9 \times 0.262 = 2.358 \text{ V}$$

Example 3.10

A 100 μF capacitor is charged from a 350 V d.c. supply through a series resistance of 1 kΩ. Determine the initial charging current and the current that will flow 50 ms and 100 ms after connecting the supply. After what time is the capacitor considered to be fully charged?

Solution

At $t = 0$ the capacitor will be uncharged ($v_c = 0$) and all of the supply voltage will appear across the series resistance. Thus, at $t = 0$:

$$i = \frac{V_s}{R} = \frac{350}{1,000} = 0.35 \text{ A}$$

When $t = 50$ ms, the current will be given by:

$$i = \frac{V_s}{R} e^{-\frac{t}{CR}}$$

Where $V_s = 350$ V, $t = 50$ ms, C = 100 μF, $R = 1$ kΩ. Hence:

$$i = \frac{350}{1,000} e^{-\frac{0.05}{0.1}} = 0.35\, e^{-0.5} = 0.35 \times 0.607 = 0.21 \text{ A}$$

When $t = 100$ ms (using the same equation but with $t = 0.1$ s) the current is given by:

$$i = \frac{350}{1,000} e^{-\frac{0.1}{0.1}} = 0.35\, e^{-1} = 0.35 \times 0.368 = 0.129 \text{ A}$$

The capacitor can be considered to be fully charged when $t = 5CR = 5 \times 100 \times 10^{-6} \times 1 \times 10^{3} = 0.5$ s. Note that, at this point the capacitor voltage will have reached 99% of its final value.

Figure 3.29   C–R circuits are widely used in electronics. In this oscilloscope, for example, a rotary switch is used to select different C–R combinations in order to provide the various timebase ranges (adjustable from 500 ms/cm to 1 μs/cm). Each C–R time constant corresponds to a different timebase range.

Discharge

Having considered the situation when a capacitor is being charged, let's consider what happens when an already charged capacitor is discharged.

When the fully charged capacitor from Fig. 3.24 is connected as shown in Fig. 3.30, the capacitor will discharge through the resistor, and the capacitor voltage, v_C, will fall exponentially with time, as shown in Fig. 3.31.

The current in the circuit, i, will also fall, as shown in Fig. 3.32. The rate of discharge (i.e. the rate of decay of voltage with time) will once again be governed by the time constant of the circuit, $C \times R$.

The voltage developed across the discharging capacitor, v_C, varies with time, t, according to the relationship:

$$v_c = V_s \, e^{-\frac{t}{CR}}$$

where V_s, is the supply voltage, t is the time, C is the capacitance, and R is the resistance.

The capacitor voltage will fall to approximately 37% of the initial voltage in a time equal to the time constant. At the end of the next interval of time equal to the time constant (i.e. after an elapsed time equal to $2CR$) the voltage will have fallen by 37% of the remainder, and so on.

In theory, the capacitor will **never** become fully discharged. However, after a period of time equal to $5CR$, however, the capacitor voltage will to all intents and purposes be zero.

At this point the capacitor voltage will have fallen below 1% of its initial value. At this point we can consider it to be fully discharged.

As with charging, the current in the capacitor, i, varies with time, t, according to the relationship:

$$i = \frac{V_s}{R} \, e^{-\frac{t}{CR}}$$

where V_s, is the supply voltage, t is the time, C is the capacitance, and R is the resistance.

The current will fall to approximately 37% of the initial value of current, V_s/R, in a time equal to the time constant.

At the end of the next interval of time equal to the time constant (i.e. after a total time of $2CR$ has elapsed) the voltage will have fallen by a further 37% of the remainder, and so on.

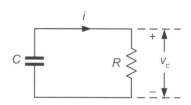

Figure 3.30   A C–R circuit in which C is initially charged and then discharges through R

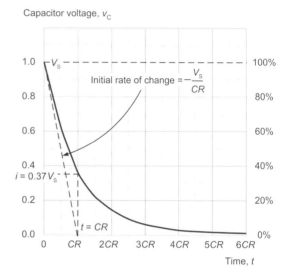

Figure 3.31   Exponential decay of capacitor voltage, v_c, in Fig. 3.30

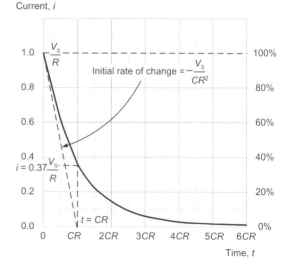

Figure 3.32   Exponential decay of current, i, in Fig. 3.30

Example 3.11

A 10 μF capacitor is charged to a potential of 20 V and then discharged through a 47 kΩ resistor. Determine the time taken for the capacitor voltage to fall below 10 V.

Solution

The formula for exponential decay of voltage in the capacitor is:

$$v_c = V_s \, e^{-\frac{t}{CR}}$$

where V_s = 20 V and CR = 10 μF × 47 kΩ = 0.47 s.
   We need to find t when v_C = 10 V. Rearranging the formula to make t the subject gives:

$$t = -CR \times \ln\left(\frac{v_C}{V_S}\right)$$

thus

$$t = -0.47 \times \ln\left(\frac{10}{20}\right) = -0.47 \times \ln(0.5)$$

or

$$t = -0.47 \times -0.693 = 0.325 \text{ s}$$

In order to simplify the mathematics of exponential growth and decay, Table 3.1 provides an alternative tabular method that may be used to determine the voltage and current in a C–R circuit.

Example 3.12

A 150 μF capacitor is charged to a potential of 150 V. The capacitor is then removed from the charging source and connected to a 2 MΩ resistor. Determine the capacitor voltage 1 minute later.

Solution

We will solve this problem using Table 3.1 rather than the exponential formula.
   First we need to find the time constant:

$C \times R$ = 150 μF × 2 MΩ = 300 s

Next we find the ratio of t to CR:
   After 1 minute, t = 60 s therefore the ratio of t to CR is 60/300 or 0.2. Table 3.1 shows that when t/CR = 0.2, the ratio of instantaneous value to final value (k in Table 3.1) is 0.8187.

Thus

v_c / V_s = 0.8187

or

$v_c = 0.8187 \times V_s = 0.8187 \times 150\text{V} = 122.8\text{V}$

Table 3.1   Exponential growth and decay

t/CR or $t/(L/R)$	k (growth)	k (decay)
0.0	0.0000	1.0000
0.1	0.0951	0.9048
0.2	0.1812	0.8187 (1)
0.3	0.2591	0.7408
0.4	0.3296	0.6703
0.5	0.3935	0.6065
0.6	0.4511	0.5488
0.7	0.5034	0.4965
0.8	0.5506	0.4493
0.9	0.5934	0.4065
1.0	0.6321	0.3679
1.5	0.7769	0.2231
2.0	0.8647 (2)	0.1353
2.5	0.9179	0.0821
3.0	0.9502	0.0498
3.5	0.9698	0.0302
4.0	0.9817	0.0183
4.5	0.9889	0.0111
5.0	0.9933	0.0067

Notes: (1) See Example 3.12
         (2) See Example 3.16
         k is the ratio of the value at time, t, to the final value (e.g. v_c / V_s)

Waveshaping with C–R networks

One of the most common applications of C–R networks is in waveshaping circuits. The circuits shown in Figs 3.33 and 3.35 function as simple square-to-triangle and square-to-pulse converters by, respectively, **integrating** and **differentiating** their inputs.

The effectiveness of the simple integrator circuit shown in Fig. 3.33 depends on the ratio of time constant, $C \times R$, to periodic time, t. The larger this ratio is, the more effective the circuit will be as an integrator. The effectiveness of the circuit of Fig. 3.33 is illustrated by the input and output waveforms shown in Fig. 3.34.

Similarly, the effectiveness of the simple differentiator circuit shown in Fig. 3.35 also depends on the ratio of time constant $C \times R$, to periodic time, t. The smaller this ratio is, the more effective the circuit will be as a differentiator.

The effectiveness of the circuit of Fig. 3.35 is illustrated by the input and output waveforms shown in Fig. 3.36.

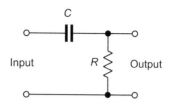

Figure 3.35   A C–R differentiating circuit

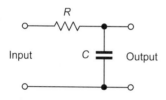

Figure 3.33   A C–R integrating circuit

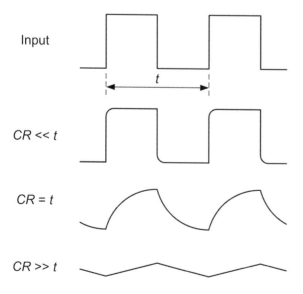

Figure 3.34   Typical input and output waveforms for the integrating circuit shown in Figure 3.33

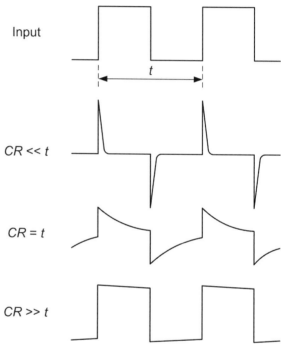

Figure 3.36   Typical input and output waveforms for the integrating circuit shown in Fig. 3.33

Example 3.13

A circuit is required to produce a train of alternating positive and negative pulses of short duration from a square wave of frequency 1 kHz. Devise a suitable C–R circuit and specify suitable values.

Solution

Here we require the services of a differentiating circuit along the lines of that shown in Fig. 3.35. In order that the circuit operates effectively as a differentiator, we need to make the time constant, $C \times R$, very much less than the periodic time of the input waveform (1 ms).

Assuming that we choose a medium value for R of, say, 10 kΩ, the maximum value which we could allow C to have would be that which satisfies the equation:

$C \times R = 0.1\, t$

where R = 10 kΩ and t = 1 ms. Thus

$$C = \frac{0.1t}{R} = \frac{0.1 \times 1 \text{ ms}}{10 \text{ k}\Omega} = 0.1 \times 10^{-3} \times 10^{-4} = 1 \times 10^{-8} \text{ F}$$

or

$C = 10 \times 10^{-9}$ F = 10 nF

In practice, any value equal or less than 10 nF would be adequate. A very small value (say less than 1 nF) will, however, generate pulses of a very narrow width.

Example 3.14

A circuit is required to produce a triangular waveform from a square wave of frequency 1 kHz. Devise a suitable C–R arrangement and specify suitable values.

Solution

This time we require an integrating circuit like that shown in Fig. 3.33. In order that the circuit operates effectively as an integrator, we need to make the time constant, $C \times R$, very much less than the periodic time of the input waveform (1 ms).

Assuming that we choose a medium value for R of, say, 10 kΩ, the minimum value which we could allow C to have would be that which satisfies the equation:

$C \times R = 10\, t$

where R = 10 kΩ and t = 1 ms. Thus

$$C = \frac{10t}{R} = \frac{10 \times 1 \text{ ms}}{10 \text{ k}\Omega} = 10 \times 10^{-3} \times 10^{-4} = 1 \times 10^{-6} \text{ F}$$

or

$C = 1 \times 10^{-6}$ F = 1 μF

In practice, any value equal or greater than 1 μF would be adequate. A very large value (say more

than 10 μF) will, however, generate a triangular wave which has a very small amplitude. To put this in simple terms, although the waveform might be what you want there's not a lot of it!

L–R circuits

Networks of inductors and resistors (known as L–R circuits) can also be used for timing and pulse shaping. In comparison with capacitors, however, inductors are somewhat more difficult to manufacture and are consequently more expensive.

Inductors are also prone to losses and may also require screening to minimize the effects of stray magnetic coupling. Inductors are, therefore, generally unsuited to simple timing and waveshaping applications.

Figure 3.37 shows a simple L–R network in which an inductor is connected to a constant voltage supply. When the supply is first connected, the current, i, will rise exponentially with time, as shown in Fig. 3.38. At the same time, the inductor voltage V_L, will fall, as shown in Fig. 3.39). The rate of change of current with time will depend upon the ratio of inductance to resistance and is known as the **time constant**. Hence:

Time constant, $t = L/R$

where L is the value of inductance (H), R is the resistance (Ω), and t is the time constant (s).

The current flowing in the inductor, i, varies with time, t, according to the relationship:

$$i = \frac{V_s}{R}\left(1 - e^{-\frac{tR}{L}}\right)$$

where V_s is the d.c. supply voltage, R is the resistance of the inductor, and L is the inductance.

The current, i, will initially be zero and will rise to approximately 63% of its maximum value (i.e. V_s/R) in a time interval equal to the time constant. At the end of the next interval of time equal to the time constant (i.e. after a total time of $2L/R$ has elapsed) the current will have risen by a further 63% of the remainder, and so on.

In theory, the current in the inductor will never become equal to V_s/R. However, after a period of time equal to $5L/R$, the current will to all intents and purposes be equal to V_s/R. At this point the

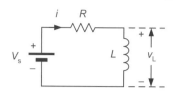

Figure 3.37   A *C–R* circuit in which *C* is initially charged and then discharges through *R*

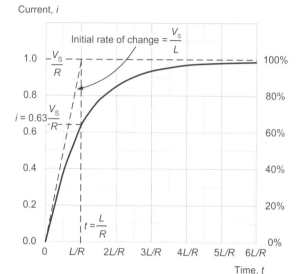

Figure 3.38   Exponential growth of current, *i*, in Fig. 3.37

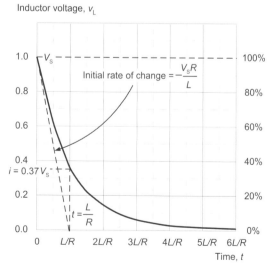

Figure 3.39   Exponential decay of voltage, v_L, in Fig. 3.37

current in the inductor will have risen to 99.3% of its final value.

The voltage developed across the inductor, v_L, varies with time, *t*, according to the relationship:

$$v_L = V_s \, e^{-\frac{tR}{L}}$$

where V_s is the d.c. supply voltage, *R* is the resistance of the inductor, and *L* is the inductance.

The inductor voltage will fall to approximately 37% of the initial voltage in a time equal to the time constant.

At the end of the next interval of time equal to the time constant (i.e. after a total time of $2L/R$ has elapsed) the voltage will have fallen by a further 37% of the remainder, and so on.

Example 3.15

A coil having inductance 6 H and resistance 24 Ω is connected to a 12 V d.c. supply. Determine the current in the inductor 0.1 s after the supply is first connected.

Solution

The formula for exponential growth of current in the coil is:

$$i = \frac{V_s}{R} \left(1 - e^{-\frac{tR}{L}} \right)$$

where $V_s = 12$ V, $L = 6$ H and $R = 24$ Ω.
We need to find *i* when $t = 0.1$ s

$$i = \frac{12}{24} \left(1 - e^{-\frac{0.1 \times 24}{6}} \right) = 0.5 \left(1 - e^{-0.4} \right) = 0.5 \, (1 - 0.67)$$

thus

$$i = 0.5 \times 0.33 = 0.165 \text{ A}$$

In order to simplify the mathematics of exponential growth and decay, Table 3.1 provides an alternative tabular method that may be used to determine the voltage and current in an *L–R* circuit.

Example 3.16

A coil has an inductance of 100 mH and a resistance of 10 Ω. If the inductor is connected to a 5 V d.c.

supply, determine the inductor voltage 20 ms after the supply is first connected.

Solution

We will solve this problem using Table 3.1 rather than the exponential formula.

First we need to find the time constant:

$L/R = 0.1 \text{ H}/10 \ \Omega = 0.01 \text{ s}$

Next we find the ratio of t to L/R.

When $t = 20$ ms the ratio of t to L/R is 0.02/0.01 or 2. Table 3.1 shows that when $t/(L/R) = 2$, the ratio of instantaneous value to final value (k) is 0.8647. Thus

$v_L/V_s = 0.8647$

or

$v_L = 0.8647 \times V_s = 0.8647 \times 5 \text{ V} = 4.32\text{V}$

Practical investigation

Objective

To investigate the charge and discharge of a capacitor.

Components and test equipment

Breadboard, 9V DC power source (either a PP9 9V battery or an AC mains adapter with a 9V 400mA output), digital multimeter with test leads, resistors of 100 kΩ, 220 kΩ and 47 kΩ, capacitor of 1,000 µF, insulated wire links (various lengths), assorted crocodile leads, short lengths of black, red, and green insulated solid wire. A watch or clock with a seconds display will also be required for timing.

Procedure

Connect the charging circuit shown in Fig. 3.40 with $R = 100$ kΩ and $C = 1,000$ µF. Place a temporary shorting link across the capacitor. Set the meter to the 20 V DC range and remove the shorting link. Measure and record the capacitor voltage at 25 s intervals over the range 0 to 250 s after removing the shorting link. Record your result in a table showing capacitor voltage against time. Repeat with $R = 220$ kΩ and $R = 47$ kΩ.

Connect the discharging circuit shown in Fig. 3.41 with $R = 100$ kΩ and $C = 1,000$ µF. Leave the link in place for a few seconds after the supply voltage has been switched on, then remove the link. Measure and record the capacitor voltage at 25 s intervals over the range 0 to 250 s from removing the link. Record your result in a table showing capacitor voltage against time. Repeat with $R = 220$ kΩ and $R = 47$ kΩ.

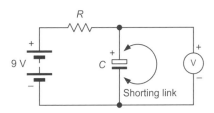

Figure 3.40 Circuit diagram—charging

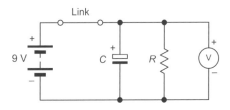

Figure 3.41 Circuit diagram—discharging

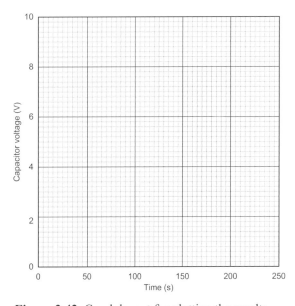

Figure 3.42 Graph layout for plotting the results

Measurements and calculations

Plot graphs of voltage (on the vertical axis) against time (on the horizontal axis) using the graph layout shown in Fig. 3.42.

Calculate the time constant for each combination of resistance and capacitance that you used in the investigation.

Conclusion

Comment on the shape of the graphs. Is this what you would expect? For each combination of resistance and capacitance estimate the time constant from the graph. Compare these values with the calculated values. If they are not the same suggest possible reasons for the difference.

Formulae introduced in this chapter

Kirchhoff's current law:
(page 49)

Algebraic sum of currents = 0

Kirchhoff's voltage law:
(page 50)

Algebraic sum of e.m.f.s = algebraic sum of voltage drops

Potential divider:
(page 52)

$$V_{out} = V_{in} \frac{R_2}{R_1 + R_2}$$

Current divider:
(page 53)

$$I_{out} = I_{in} \frac{R_1}{R_1 + R_2}$$

Wheatstone bridge:
(page 54)

$$\frac{R_1}{R_2} = \frac{R_3}{R_4}$$

and

$$R_x = \frac{R_2}{R_1} \times R_v$$

Time constant of a C–R circuit:
(page 58)

$$t = CR$$

Capacitor voltage (charge):
(page 58)

$$v_c = V_s \left(1 - e^{-\frac{t}{CR}} \right)$$

Capacitor current (charge):
(page 59)

$$i = \frac{V_s}{R} e^{-\frac{t}{CR}}$$

Capacitor voltage (discharge):
(page 60)

$$v_c = V_s e^{-\frac{t}{CR}}$$

Capacitor current (discharge):
(page 60)

$$i = \frac{V_s}{R} e^{-\frac{t}{CR}}$$

Time constant of an L–R circuit:
(page 63)

$$t = L / R$$

Inductor current (growth):
(page 63)

$$i = \frac{V_s}{R} \left(1 - e^{-\frac{tR}{L}} \right)$$

Inductor voltage (decay):
(page 64)

$$v_L = V_s e^{-\frac{tR}{L}}$$

Symbols introduced in this chapter

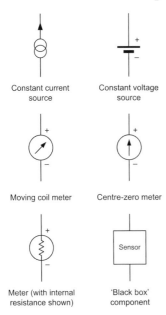

Constant current source

Constant voltage source

Moving coil meter

Centre-zero meter

Meter (with internal resistance shown)

'Black box' component

Figure 3.43   Circuit symbols introduced in this chapter

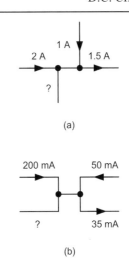

(a)

(b)

Figure 3.44   See Question 3.3

Problems

3.1   A power supply is rated at 500 mA maximum output current. If the supply delivers 150 mA to one circuit and 75 mA to another, how much current would be available for a third circuit?

3.2   A 15 V d.c. supply delivers a total current of 300 mA. If this current is shared equally between four circuits, determine the resistance of each circuit.

3.3   Determine the unknown current in each circuit shown in Fig. 3.44.

3.4   Determine the unknown voltage in each circuit shown in Fig. 3.45.

3.5   Determine all currents and voltages in Fig. 3.46.

3.6   Two resistors, one of 120 Ω and one of 680 Ω are connected as a potential divider across a 12 V supply. Determine the voltage developed across each resistor.

3.7   Two resistors, one of 15 Ω and one of 5 Ω are connected in parallel. If a current of 2 A is applied to the combination, determine the current flowing in each resistor.

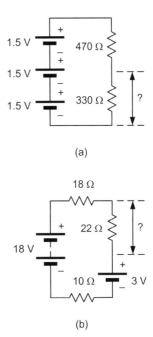

(a)

(b)

Figure 3.45   See Question 3.4

3.8   A switched attenuator comprises five 1 kΩ resistors wired in series across a 5V d.c. supply. If the output voltage is selected by means of a single-pole four-way switch, sketch a circuit and determine the voltage produced for each switch position.

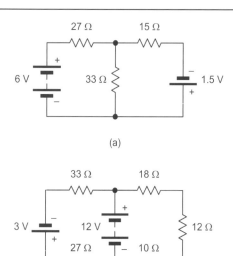

(a)

(b)

Figure 3.46 See Question 3.5

3.9   A capacitor of 1 μF is charged from a 15 V d.c. supply via a 100 kΩ resistor. How long will it take for the capacitor voltage to reach 5 V?

3.10   A capacitor of 22 μF is charged to a voltage of 50 V. If the capacitor is then discharged using a resistor of 100 kΩ, determine the time taken for the capacitor voltage to reach 10 V.

3.11   An initially uncharged capacitor is charged from a 200 V d.c. supply through a 2 MΩ resistor. If it takes 50 s for the capacitor voltage to reach 100 V, determine the value of capacitance.

3.12   A coil has an inductance of 2.5 H and a resistance of 10 Ω. If the coil is connected to a 5 V d.c. supply, determine the time taken for the current to grow to 200 mA.

3.13   Determine the Thévenin equivalent of the circuit shown in Fig. 3.47.

3.14   Determine the Norton equivalent of the circuit shown in Fig. 3.48.

3.15   The Thévenin equivalent of a network is shown in Fig. 3.49. Determine (a) the short-circuit output current and (b) the output voltage developed across a load of 200 Ω.

3.16   The Norton equivalent of a network is

shown in Fig. 3.50. Determine (a) the open-circuit output voltage and (b) the output voltage developed across a load of 5 kΩ.

Answers to these problems appear on page 374.

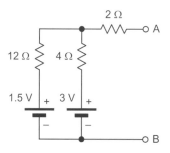

Figure 3.47   See Question 3.13

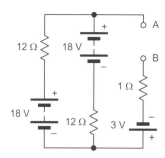

Figure 3.48   See Question 3.14

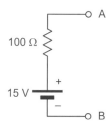

Figure 3.49   See Question 3.15

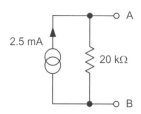

Figure 3.50   See Question 3.16

4

Alternating voltage and current

This chapter introduces basic alternating current theory. We discuss the terminology used to describe alternating waveforms and the behaviour of resistors, capacitors, and inductors when an alternating current is applied to them. The chapter concludes by introducing another useful component, the transformer.

Alternating versus direct current

Direct currents are currents which, even though their magnitude may vary, essentially flow only in one direction. In other words, direct currents are **unidirectional**. Alternating currents, on the other hand, are **bidirectional** and continuously reverse their direction of flow. The polarity of the e.m.f. which produces an alternating current must consequently also be changing from positive to negative, and vice versa.

Alternating currents produce alternating potential differences (voltages) in the circuits in which they flow. Furthermore, in some circuits, alternating voltages may be superimposed on direct voltage levels (see Fig. 4.1). The resulting voltage may be unipolar (i.e. always positive or always negative) or bipolar (i.e. partly positive and partly negative).

Waveforms and signals

A graph showing the variation of voltage or current present in a circuit is known as a **waveform.** There are many common types of waveform encountered in electrical circuits including sine (or sinusoidal), square, triangle, ramp or sawtooth (which may be either positive or negative going), and pulse.

Complex waveforms, like speech and music, usually comprise many components at different frequencies. **Pulse waveforms** are often categorized as either repetitive or non-repetitive (the former comprises a pattern of pulses that repeats regularly while the latter comprises pulses which constitute a unique event). Some common waveforms are shown in Fig. 4.2.

Signals can be conveyed using one or more of the properties of a waveform and sent using wires, cables, optical and radio links. Signals can also be processed in various ways using amplifiers, modulators, filters, etc. Signals are also classified as either **analogue** (continuously variable) or **digital** (based on discrete states).

Frequency

The frequency of a repetitive waveform is the number of cycles of the waveform which occur in unit time. Frequency is expressed in Hertz, Hz, and a frequency of 1 Hz is equivalent to one cycle per second. Hence, if a voltage has a frequency of 400 Hz, 400 cycles of it will occur in every second.

The equation for the voltage shown in Fig. 4.1(a) at a time, t, is:

$$v = V_{max} \sin (2\pi t)$$

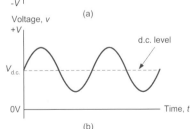

Figure 4.1 (a) Bipolar sine wave; (b) unipolar sine wave (superimposed on a **d.c. level**)

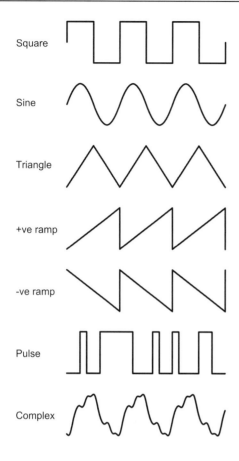

Figure 4.2   Common waveforms

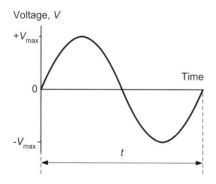

Figure 4.3   One cycle of a sine wave voltage showing its **periodic time**

whilst that in Fig. 4.1(b) is:

$$v = V_{d.c.} + V_{max} \sin (2\pi t)$$

where v is the instantaneous voltage, V_{max} is the maximum (or peak) voltage of the sine wave, $V_{d.c.}$, is the d.c. offset (where present), and f is the frequency of the sine wave.

Example 4.1

A sine wave voltage has a maximum value of 20 V and a frequency of 50 Hz. Determine the instantaneous voltage present (a) 2.5 ms and (b) 15 ms from the start of the cycle.

Solution

We can find the voltage at any instant of time using:

$$v = V_{max} \sin (2\pi t)$$

where $V_{max} = 20$ V and $f = 50$ Hz.

In (a), $t = 2.5$ ms, hence:

$$v = 20 \sin (2\pi \times 50 \times 0.0025) = 20 \sin (0.785)$$

$$= 20 \times 0.707 = 14.14 \text{ V}$$

In (b), $t = 15$ ms, hence:

$$v = 20 \sin (2\pi \times 50 \times 0.0015) = 20 \sin (4.71)$$

$$= 20 \times -1 = -20 \text{ V}$$

Periodic time

The periodic time (or **period**) of a waveform is the time taken for one complete cycle of the wave (see Fig. 4.3). The relationship between periodic time and frequency is thus:

$$t = 1/f \text{ or } f = 1/t$$

where t is the periodic time (in s) and f is the frequency (in Hz).

Example 4.2

A waveform has a frequency of 400 Hz. What is the periodic time of the waveform?

Solution

$$t = 1/f = 1/400 = 0.0025 \text{ s (or 2.5 ms)}$$

Example 4.3

A waveform has a periodic time of 40 ms. What is its frequency?

Solution

$$f = \frac{1}{t} = \frac{1}{40 \times 10^{-3}} = \frac{1}{0.04} = 25 \text{ Hz}$$

Average, peak, peak-peak, and r.m.s. values

The **average value** of an alternating current which swings symmetrically above and below zero will be zero when measured over a long period of time. Hence average values of currents and voltages are invariably taken over one complete half-cycle (either positive or negative) rather than over one complete full-cycle (which would result in an average value of zero).

The **amplitude** (or **peak value**) of a waveform is a measure of the extent of its voltage or current excursion from the resting value (usually zero).

The **peak-to-peak value** for a wave which is symmetrical about its resting value is twice its peak value (see Fig. 4.4).

The **r.m.s.** (or **effective**) **value** of an alternating voltage or current is the value which would produce the same heat energy in a resistor as a direct voltage or current of the same magnitude. Since the r.m.s. value of a waveform is very much

Table 4.1   Multiplying factors for average, peak, peak-peak and r.m.s. values

Given quantity	Wanted quantity			
	Average	*Peak*	*Peak–peak*	*r.m.s.*
Average	1	1.57	3.14	1.11
Peak	0.636	1	2	0.707
Peak– peak	0.318	0.5	1	0.353
r.m.s.	0.9	1.414	2.828	1

dependent upon its shape, values are only meaningful when dealing with a waveform of known shape. Where the shape of a waveform is not specified, r.m.s. values are normally assumed to refer to sinusoidal conditions.

For a given waveform, a set of fixed relationships exist between average, peak, peak-peak, and r.m.s. values. The required multiplying factors are summarized for sinusoidal voltages and currents in Table 4.1.

Example 4.4

A sinusoidal voltage has an r.m.s. value of 240 V. What is the peak value of the voltage?

Solution

The corresponding multiplying factor (found from Table 4.1) is 1.414. Hence:

$$V_{\text{pk}} = 1.414 \times V_{\text{r.m.s.}} = 1.414 \times 240 = 339.4 \text{ V}$$

Example 4.5

An alternating current has a peak-peak value of 50 mA. What is its r.m.s. value?

Solution

The corresponding multiplying factor (found from

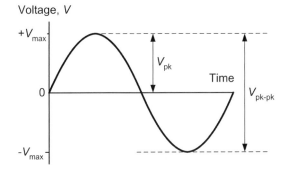

Figure 4.4   One cycle of a sine wave voltage showing its **peak** and **peak-peak** values

Table 4.1) is 0.353. Hence:

$$I_{\text{r.m.s.}} = 0.353 \times V_{\text{pk-pk}} = 0.353 \times 0.05 = 0.0177 \text{ A}$$

(or 17.7mA).

Example 4.6

A sinusoidal voltage 10 V pk-pk is applied to a resistor of 1 kΩ What value of r.m.s. current will flow in the resistor?

Solution

This problem must be solved in two stages. First we will determine the peak-peak current in the resistor and then we shall convert this value into a corresponding r.m.s. quantity.

Since $I = \dfrac{V}{R}$ we can infer that $I_{\text{pk-pk}} = \dfrac{V_{\text{pk-pk}}}{R}$

From which $I_{\text{pk-pk}} = \dfrac{10}{1,000} = 0.01 = 10$ mA pk-pk

The required multiplying factor (peak-peak to r.m.s.) is 0.353. Thus:

$$I_{\text{r.m.s.}} = 0.353 \times I_{\text{pk-pk}} = 0.353 \times 10 = 3.53 \text{ mA}$$

Reactance

When alternating voltages are applied to capacitors or inductors the magnitude of the current flowing will depend upon the value of capacitance or inductance and on the frequency of the voltage. In effect, capacitors and inductors oppose the flow of current in much the same way as a resistor. The important difference being that the effective resistance (or reactance) of the component varies with frequency (unlike the case of a resistor where the magnitude of the current does not change with frequency).

Capacitive reactance

The reactance of a capacitor is defined as the ratio of applied voltage to current and, like resistance, it is measured in Ohms. The reactance of a capacitor is inversely proportional to both the value of capacitance and the frequency of the applied voltage. Capacitive reactance can be found by applying the following formula:

$$X_{\text{c}} = \frac{1}{2\pi fC}$$

where X_{c} is the reactance (in Ohms), f is the frequency (in Hertz), and C is the capacitance (in Farads).

Capacitive reactance falls as frequency increases, as shown in Fig. 4.5. The applied voltage, V_{c}, and current, I_{c}, flowing in a pure capacitive reactance will differ in phase by an angle of 90° or $\pi/2$ radians (the **current leads the voltage**). This relationship is illustrated in the current and voltage waveforms (drawn to a common time scale) shown in Fig. 4.6 and as a **phasor diagram** shown in Fig. 4.7.

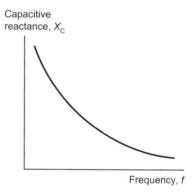

Figure 4.5   Variation of reactance with frequency for a capacitor

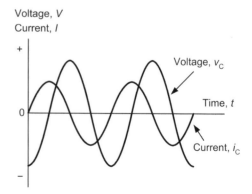

Figure 4.6   Voltage and current waveforms for a pure capacitor (the current leads the voltage by 90°)

Figure 4.7   Phasor diagram for a pure capacitor

Example 4.7

Determine the reactance of a 1 μF capacitor at (a) 100 Hz and (b) 10 kHz.

Solution

This problem is solved using the expression:

$$X_C = \frac{1}{2\pi f C}$$

(a) At 100 Hz

$$X_C = \frac{1}{2\pi \times 100 \times 1 \times 10^{-6}} = \frac{0.159}{10^{-4}} = 1.59 \times 10^3$$

or

$$X_C = 1.59 \text{ k}\Omega$$

(b) At 10 kHz

$$X_C = \frac{1}{2\pi \times 1 \times 10^4 \times 1 \times 10^{-6}} = \frac{0.159}{10^{-2}} = 0.159 \times 10^2$$

or

$$X_C = 15.9 \ \Omega$$

Example 4.8

A 100 nF capacitor is to form part of a filter connected across a 240 V 50 Hz mains supply. What current will flow in the capacitor?

Solution

First we must find the reactance of the capacitor:

$$X_C = \frac{1}{2\pi \times 50 \times 100 \times 10^{-9}} = 31.8 \times 10^3 = 31.8 \text{ k}\Omega$$

The r.m.s. current flowing in the capacitor will thus be:

$$I_C = \frac{V_C}{X_C} = \frac{240}{31.8 \times 10^3} = 7.5 \times 10^{-3} = 7.5 \text{ mA}$$

Inductive reactance

The reactance of an inductor is defined as the ratio of applied voltage to current and, like resistance, it is measured in ohms. The reactance of an inductor is directly proportional to both the value of inductance and the frequency of the applied voltage. Inductive reactance can be found by applying the formula:

$$X_L = 2\pi f L$$

where X_L is the reactance in Ω, f is the frequency in Hz, and L is the inductance in H.

Inductive reactance increases linearly with frequency as shown in Fig. 4.8. The applied voltage, V_L, and current, I_L, developed across a pure inductive reactance will differ in phase by an angle of 90° or π/2 radians (the **current lags the voltage**). This relationship is illustrated in the current and voltage waveforms (drawn to a common time scale) shown in Fig. 4.9 and as a **phasor diagram** shown in Fig. 4.10.

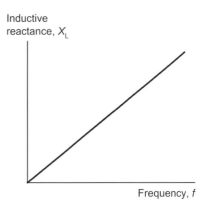

Figure 4.8   Variation of reactance with frequency for an inductor

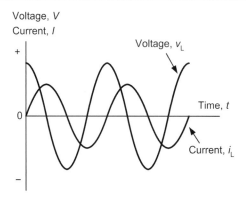

Figure 4.9 Voltage and current waveforms for a pure inductor (the voltage leads the current by 90°)

Figure 4.10 Phasor diagram for a pure inductor

Example 4.9

Determine the reactance of a 10 mH inductor at (a) 100 Hz and (b) at 10 kHz.

Solution

(a) at 100 Hz

$$X_L = 2\pi \times 100 \times 10 \times 10^{-3} = 6.28 \ \Omega$$

(b) At 10 kHz

$$X_L = 2\pi \times 10 \times 10^3 \times 10 \times 10^{-3} = 628 \ \Omega$$

Example 4.10

A 100 mH inductor of negligible resistance is to form part of a filter which carries a current of 20 mA at 400 Hz. What voltage drop will be developed across the inductor?

Solution

The reactance of the inductor will be given by:

$$X_L = 2\pi \times 400 \times 100 \times 10^{-3} = 251 \ \Omega$$

The r.m.s. voltage developed across the inductor will be given by:

$$V_L = I_L \times X_L = 20 \ \text{mA} \times 251 \ \Omega = 5.02 \ \text{V}$$

In this example, it is important to note that we have assumed that the d.c. resistance of the inductor is negligible by comparison with its reactance. Where this is not the case, it will be necessary to determine the **impedance** of the component and use this to determine the voltage drop.

Impedance

Figure 4.11 shows two circuits which contain both resistance and reactance. These circuits are said to exhibit impedance (a combination of resistance and reactance) which, like resistance and reactance, is measured in ohms.

The impedance of the circuits shown in Fig. 4.11 is simply the ratio of supply voltage, V_S, to supply current, I_S. The impedance of the simple C–R and L–R circuits shown in Fig. 4.11 can be found by using the impedance triangle shown in Fig. 4.12. In either case, the impedance of the circuit is given by:

$$Z = \sqrt{R^2 + X^2}$$

and the phase angle (between V_S and I_S) is given by:

$$\phi = \tan^{-1}\left(\frac{X}{R}\right)$$

where Z is the impedance (in Ohms), X is the reactance, either capacitive or inductive (expressed in ohms), R is the resistance (in Ohms), and ϕ is the phase angle in radians.

Example 4.11

A 2 μF capacitor is connected in series with a 100 Ω resistor across a 115 V 400 Hz a.c. supply. Determine the impedance of the circuit and the current taken from the supply.

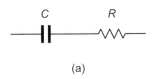

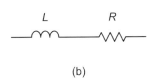

(a)

(b)

Figure 4.11 (a) C and R in series (b) L and R in series (note that both circuits exhibit an **impedance**)

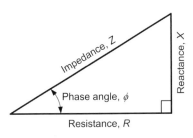

Figure 4.12 The impedance triangle

Solution

First we must find the reactance of the capacitor, X_C:

$$X_C = \frac{1}{2\pi fC} = \frac{1}{6.28 \times 400 \times 2 \times 10^{-6}} = \frac{10^6}{5{,}024} = 199\ \Omega$$

Now we can find the impedance of the C–R series circuit:

$$Z = \sqrt{R^2 + X^2} = \sqrt{199^2 + 100^2} = \sqrt{49{,}601} = 223\ \Omega$$

The current taken from the supply can now be found:

$$I_S = \frac{V_S}{Z} = \frac{115}{223} = 0.52\ \text{A}$$

Power factor

The power factor in an a.c. circuit containing resistance and reactance is simply the ratio of true power to apparent power. Hence:

$$\text{power factor} = \frac{\text{true power}}{\text{apparent power}}$$

The **true power** in an a.c. circuit is the power which is actually dissipated in the resistive component. Thus:

$$\text{true power} = I_S^2 \times R \qquad \text{(Watts)}$$

The **apparent power** in an a.c. circuit is the power which is apparently consumed by the circuit and is the product of the supply current and supply voltage (note that this is not the same as the power which is actually dissipated as heat). Hence:

$$\text{apparent power} = I_S \times V_S \qquad \text{(Volt-Amperes)}$$

Hence

$$\text{power factor} = \frac{I_S^2 \times R}{I_S \times V_S} = \frac{I_S^2 \times R}{I_S \times (I_S \times Z)} = \frac{R}{Z}$$

From Fig. 4.12, $\dfrac{R}{Z} = \cos \phi$

Hence the power factor of a series a.c. circuit can be found from the cosine of the phase angle.

Example 4.12

A **choke** (a form of inductor) having an inductance of 150 mH and resistance of 250 Ω is connected to a 115 V 400 Hz a.c. supply. Determine the power factor of the choke and the current taken from the supply.

Solution

First we must find the reactance of the inductor,

$$X_L = 2\pi \times 400 \times 0.15 = 376.8\ \Omega$$

We can now determine the power factor from:

$$\text{power factor} = \frac{R}{Z} = \frac{250}{376.8} = 0.663$$

The impedance of the choke, Z, will be given by:

of the parallel tuned circuit, the Q-factor will increase as the resistance, R, increases. The response of a tuned circuit can be modified by incorporating a resistance of appropriate value either to 'dampen' (low-Q) or 'sharpen' (high-Q) the response. The relationship between bandwidth and Q-factor is:

$$\text{Bandwidth} = f_2 - f_1 = \frac{f_0}{Q} \quad \text{and} \quad Q = \frac{2\pi f_0 L}{R}$$

where f_2 and f_1 are respectively the upper and lower cut-off (or **half-power**) frequencies (in Hertz), f_0 is the resonant frequency (in Hertz), and Q is the Q-factor.

Example 4.13

A parallel L–C circuit is to be resonant at a frequency of 400 Hz. If a 100 mH inductor is available, determine the value of capacitance required.

Solution

Re-arranging the formula

$$f = \frac{1}{2\pi\sqrt{LC}}$$

to make C the subject gives:

$$C = \frac{1}{f_0^2 (2\pi)^2 L}$$

Thus

$$C = \frac{1}{400^2 \times 39.4 \times 100 \times 10^{-3}} = 1.58 \times 10^{-6} = 1.58 \ \mu\text{F}$$

This value can be made from preferred values using a 2.2 µF capacitor connected in series with a 5.6 µF capacitor.

Example 4.14

A series L–C–R circuit comprises an inductor of 20 mH, a capacitor of 10 nF, and a resistor of 100 Ω. If the circuit is supplied with a sinusoidal signal of 1.5 V at a frequency of 2 kHz, determine the current supplied and the voltage developed across the resistor.

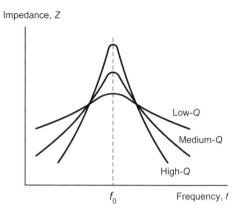

Figure 4.16   Effect of Q-factor on the response of a parallel resonant circuit (the response is similar, but inverted, for a series resonant circuit)

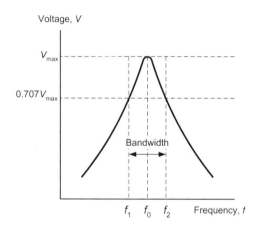

Figure 4.17   Bandwidth of a tuned circuit

Solution

First we need to determine the values of inductive reactance, X_L, and capacitive reactance X_C:

$$X_L = 2\pi f L = 6.28 \times 2 \times 10^3 \times 20 \times 10^{-3} = 251 \ \Omega$$

$$X_C = \frac{1}{2\pi f C} = \frac{1}{6.28 \times 2 \times 10^3 \times 100 \times 10^{-9}} = 796.2 \ \Omega$$

The impedance of the series circuit can now be calculated:

$$Z = \sqrt{R^2 + \left(X_L - X_C\right)^2} = \sqrt{100^2 + \left(251.2 - 796.2\right)^2}$$

From which:

$$Z = \sqrt{10,000 + 297,025} = \sqrt{307,025} = 554 \ \Omega$$

The current flowing in the series circuit will be given by:

$$I = \frac{V}{Z} = \frac{1.5}{554} = 0.0027 = 2.7 \ \text{mA}$$

The voltage developed across the resistor can now be calculated using:

$$V = IR = 2.7 \ \text{mA} \times 100 \ \Omega = 270 \ \text{mV}$$

Transformers

Transformers provide us with a means of coupling a.c. power or signals from one circuit to another. Voltage may be **stepped-up** (secondary voltage greater than primary voltage) or **stepped-down** (secondary voltage less than primary voltage). Since no increase in power is possible (transformers are passive components like resistors, capacitors and inductors) an increase in secondary voltage can only be achieved at the expense of a corresponding reduction in secondary current, and vice versa (in fact, the secondary power will be very slightly less than the primary power due to losses within the transformer). Typical applications for transformers include stepping-up or stepping-down mains voltages in power supplies, coupling signals in AF amplifiers to achieve impedance matching and to isolate d.c. potentials associated with active components.

The electrical characteristics of a transformer are determined by a number of factors including the core material and physical dimensions. The specifications for a transformer usually include the rated primary and secondary voltages and current the required power rating (i.e. the maximum power, usually expressed in volt-amperes, VA) which can be continuously delivered by the transformer under a given set of conditions, the frequency range for the component (usually stated as upper and lower

Figure 4.18   A selection of transformers with power ratings from 0.1 VA to 100 VA

Table 4.2   Characteristics of common types of transformer

Property	Transformer core type			
	Air cored	Ferrite cored	Iron cored (audio)	Iron cored (power)
Core material/construction	Air	Ferrite ring or pot	Laminated steel	Laminated steel
Typical frequency range (Hz)	30 M to 1 G	10 k to 10 M	20 to 20 k	50 to 400
Typical power rating (VA)	(see note)	1 to 200	0.1 to 50	3 to 500
Typical regulation	(see note)	(see note)	(see note)	5% to 15%
Typical applications	RF tuned circuits and filters	Filters and HF transformers, switched mode power supplies	Smoothing chokes and filters, audio matching	Power supplies

Note: Not usually important for this type of transformer

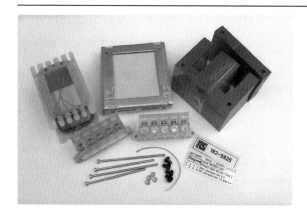

Figure 4.19   Parts of a typical iron-cored power transformer prior to assembly

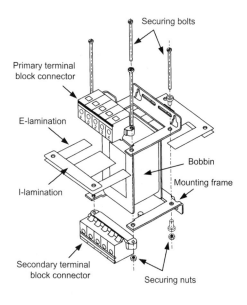

Figure 4.20   Construction of a typical iron-cored transformer

working frequency limits), and the **regulation** of a transformer (usually expressed as a percentage of full-load). This last specification is a measure of the ability of a transformer to maintain its rated output voltage under load.

Table 4.2 summarizes the properties of three common types of transformer. Figure 4.20 shows the construction of a typical iron-cored power transformer.

Voltage and turns ratio

The principle of the transformer is illustrated in Fig. 4.21. The primary and secondary windings are wound on a common low-reluctance magnetic core. The alternating flux generated by the primary winding is therefore coupled into the secondary winding (very little flux escapes due to leakage). A sinusoidal current flowing in the primary winding produces a sinusoidal flux. At any instant the flux in the transformer is given by the equation:

$$\phi = \phi_{max} \sin\left(2\pi ft\right)$$

where ϕ_{max} is the maximum value of flux (in Webers), f is the frequency of the applied current (in Hertz), and t is the time in seconds.

The r.m.s. value of the primary voltage, V_P, is given by:

$$V_P = 4.44 fN_P\phi_{max}$$

Similarly, the r.m.s. value of the secondary voltage, V_S, is given by:

$$V_S = 4.44 fN_P\phi_{max}$$

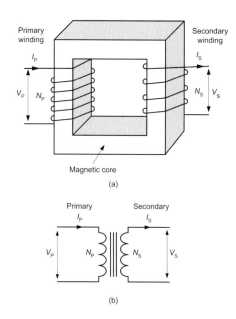

Figure 4.21   The transformer principle

Now

$$\frac{V_P}{V_S} = \frac{N_P}{N_S}$$

where N_P/N_s is the **turns ratio** of the transformer.

Assuming that the transformer is loss-free, primary and secondary powers (P_P and P_S respectively) will be identical. Hence:

$$P_P = P_S \text{ thus } V_P \times I_P = V_S \times I_P$$

Hence

$$\frac{V_P}{V_S} = \frac{I_S}{I_P} \text{ and } \frac{I_S}{I_P} = \frac{N_P}{N_S}$$

Finally, it is sometimes convenient to refer to a **turns-per-volt** rating for a transformer. This rating is given by:

$$\text{turns-per-volt} = \frac{N_P}{V_P} = \frac{N_P}{V_S}$$

Figure 4.22   Resonant air-cored transformer arrangement. The two inductors are tuned to resonance at the operating frequency (145 MHz) by means of the two small preset capacitors

Example 4.15

A transformer has 2,000 primary turns and 120 secondary turns. If the primary is connected to a 220 V r.m.s. a.c. mains supply, determine the secondary voltage.

Solution

Rearranging $\dfrac{V_P}{V_S} = \dfrac{N_P}{N_S}$ gives:

$$V_S = \frac{N_S \times V_P}{N_P} = \frac{120 \times 220}{2,000} = 13.2 \text{ V}$$

Example 4.16

A transformer has 1,200 primary turns and is designed to operate with a 200 V a.c. supply. If the transformer is required to produce an output of 10 V, determine the number of secondary turns required. Assuming that the transformer is loss-free, determine the input (primary) current for a load current of 2.5 A.

Figure 4.23   This small 1:1 ratio toroidal transformer forms part of a noise filter connected in the input circuit of a switched mode power supply. The transformer is wound on a ferrite core and acts as a choke, reducing the high-frequency noise that would otherwise be radiated from the mains supply wiring

Solution

Rearranging $\dfrac{V_P}{V_S} = \dfrac{N_P}{N_S}$ gives:

$$N_S = \frac{N_P \times V_S}{V_P} = \frac{1,200 \times 10}{200} = 60 \text{ turns}$$

Rearranging $\dfrac{I_S}{I_P} = \dfrac{N_P}{N_S}$ gives:

$$N_S = \frac{N_S \times I_S}{N_P} = \frac{200 \times 2.5}{1,200} = 0.42 \text{ A}$$

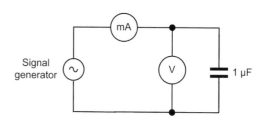

Figure 4.24 Circuit diagram—capacitive reactance

Practical investigation

Objective

To investigate reactance in an a.c. circuit.

Components and test equipment

Breadboard, digital or analogue meters with a.c. voltage and current ranges, sine wave signal generator (with an output impedance of 50 Ω, or less), 1 µF capacitor, 60 mH inductor (with low series loss resistance), test leads, connecting wire.

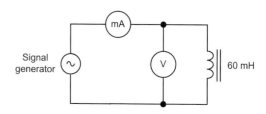

Figure 4.25 Circuit diagram—inductive reactance

Procedure

Connect the circuit shown in Fig. 4.24 (capacitive reactance). Set the voltmeter and ammeter respectively to the 2 V and 20 mA ranges. Set the signal generator to produce a sine wave output at 100 Hz.

Adjust the signal generator output voltage so that the voltmeter reads exactly 1 V before measuring and recording the current (this should be less than 1 mA). Repeat this measurement at frequencies from 200 Hz to 1 kHz in steps of 100 Hz. At each step, check that the voltage is exactly 1 V (adjust, if necessary).

Connect the circuit shown in Fig. 4.25 (inductive reactance). As before, set the voltmeter and ammeter to the 2 V and 20 mA ranges and set the signal generator to produce a sine wave output at 100 Hz.

Adjust the signal generator output for a voltage of exactly 1 V before measuring and recording the current. Repeat this measurement at frequencies over from 200 Hz to 1 kHz in steps of 100 Hz. At

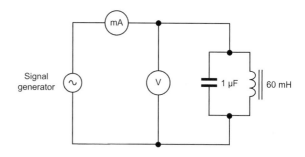

Figure 4.26 Circuit diagram—resonant circuit

each step, check that the voltage is exactly 1 V (adjust the signal generator output if necessary).

Connect the circuit shown in Fig. 4.26 (resonant circuit). As before, set the voltmeter and ammeter respectively to the 2 V and 20 mA ranges. Set the signal generator to produce a sine wave output at 100 Hz.

Adjust the signal generator output voltage so that the voltmeter reads exactly 1 V before measuring and recording the current. Repeat this measurement at frequencies over from 200 Hz to 1 kHz in steps of 100 Hz. At each step, check that the voltage is exactly 1 V (adjust, if necessary). Note the frequency at which the current takes a minimum value.

Measurements and calculations

Record your results in a table showing values of I_C, I_L and I_S, at each frequency from 100 Hz to 1 kHz. Plot graphs showing how the current in each circuit varies over the frequency range 100 Hz to 1 kHz. Calculate the resonant frequency of the L–C circuit shown in Fig. 4.26.

Conclusion

Comment on the shape of each graph. Is this what you would expect (recall that the current flowing in the circuit will be proportional to the reciprocal of the reactance)? Compare the measured resonant frequency with the calculated value. If they are not the same suggest reasons for any difference.

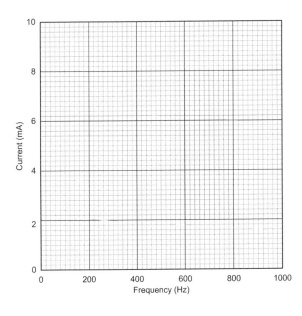

Figure 4.27 Graph layout for plotting the results

Important formulae introduced in this chapter

Sine wave voltage:
(page 69)

$$v = V_{max} \sin (2\pi t)$$

Sine wave voltage superimposed on a d.c. level:
(page 70)

$$v = V_{d.c.} + V_{max} \sin (2\pi t)$$

Frequency and periodic time:
(page 70)

$$t = 1 / f \text{ or } f = 1 / t$$

Peak and r.m.s. values for a sine wave:
(page 71)

$$V_{pk} = 1.414 \times V_{r.m.s}$$

$$V_{r.m.s.} = 0.707 \times V_{pk}$$

Capacitive reactance:
(page 72)

$$X_C = \frac{1}{2\pi f C}$$

Inductive reactance:
(page 73)

$$X_L = 2\pi f L$$

Impedance of C–R or L–R in series:
(page 74)

$$Z = \sqrt{R^2 + X^2}$$

Phase angle for C–R or L–R in series:
(page 74)

$$\phi = \tan^{-1} \left(\frac{X}{R} \right)$$

Power factor:
(page 75)

$$\text{power factor} = \frac{\text{true power}}{\text{apparent power}}$$

$$\text{power factor} = \frac{R}{Z} = \cos\phi$$

Resonant frequency of a tuned circuit:
(page 77)

$$f = \frac{1}{2\pi\sqrt{LC}}$$

Bandwidth of a tuned circuit:
(page 78)

$$\text{Bandwidth} = f_2 - f_1 = \frac{f_0}{Q}$$

Q-factor for a series tuned circuit:
(page 78)

$$Q = \frac{2\pi f_0 L}{R}$$

Flux in a transformer:
(page 80)

$$\phi = \phi_{max} \sin\left(2\pi ft\right)$$

Transformer voltages:
(page 80)

$$V_P = 4.44 f N_P \phi_{max}$$

$$V_S = 4.44 f N_P \phi_{max}$$

Voltage and turns ratio:
(page 81)

$$\frac{V_P}{V_S} = \frac{N_P}{N_S}$$

Current and turns ratio:
(page 81)

$$\frac{I_S}{I_P} = \frac{N_P}{N_S}$$

Turns-per-volt:
(page 81)

$$\text{turns-per-volt} = \frac{N_P}{V_P} = \frac{N_P}{V_S}$$

Symbols introduced in this chapter

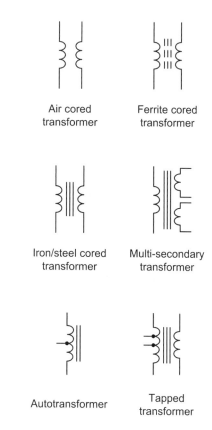

Figure 4.28   Circuit symbols introduced in this chapter

Problems

4.1   A sine wave has a frequency of 250 Hz and an amplitude of 50 V. Determine its periodic time and r.m.s. value.

4.2   A sinusoidal voltage has an r.m.s. value of 240 V and a period of 16.7 ms. What is the frequency and peak value of the voltage?

4.3   Determine the frequency and peak-peak values of each of the waveforms shown in Fig. 4.29.

4.4   A sine wave has a frequency of 100 Hz and in amplitude of 20 V. Determine the instantaneous value of voltage (a) 2 ms and (b) 9 ms from the start of a cycle.

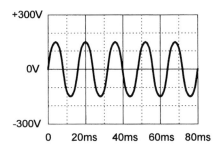

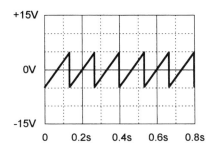

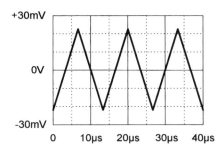

Figure 4.29 See Question 4.3

voltage drop will appear across the inductor when a current of 1.5 A is flowing?

4.10 A 10 uF capacitor is connected in series with a 500 Ω resistor across a 110 V 50 Hz a.c. supply. Determine the impedance of the circuit and the current taken from the supply.

4.11 A choke having an inductance of 1 H and resistance of 250 Ω is connected to a 220 V 60 Hz a.c. supply. Determine the power factor of the choke and the current taken from the supply.

4.12 A series-tuned *L–C* network is to be resonant at a frequency of 1.8 kHz. If a 60 mH inductor is available, determine the value of capacitance required.

4.13 Determine the impedance at 1 kHz of each of the circuits shown in Fig. 4.30.

4.14 A parallel resonant circuit employs a fixed inductor of 22 μH and a variable tuning capacitor. If the maximum and minimum values of capacitance are respectively 20 pF and 365 pF, determine the effective tuning range for the circuit.

4.15 A series *L-C-R* circuit comprises an inductor of 15 mH (with negligible resistance), a capacitor of 220 nF and a resistor of 100 Ω. If the circuit is supplied with a sinusoidal signal of 15 V at a frequency of 2 kHz, determine the current supplied and the voltage developed across the capacitor.

4.16 A 470 μH inductor has a resistance of 20 Ω. If the inductor is connected in series with a capacitor of 680 pF, determine the resonant

4.5 A sinusoidal current of 20 mA pk-pk flows in a resistor of 1.5 kΩ. Determine the r.m.s. voltage applied.

4.6 Determine the reactance of a 220 nF capacitor at (a) 20 Hz and (b) 5 kHz.

4.7 A 47 nF capacitor is connected across the 240 V 50 Hz mains supply. Determine the r.m.s. current flowing in the capacitor.

4.8 Determine the reactance of a 33 mH inductor at (a) 50 Hz and (b) 7 kHz.

4.9 A 10 mH inductor of negligible resistance is used to form part of a filter connected in series with a 50 Hz mains supply. What

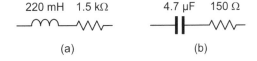

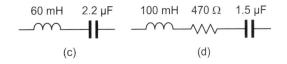

Figure 4.30 See Question 4.13

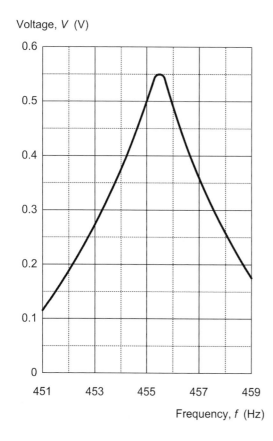

Figure 4.31   See Question 4.17

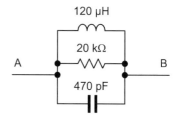

Figure 4.32   See Question 4.18

Figure 4.33   See Question 4.19

frequency, Q-factor, and bandwidth of the circuit.

4.17   The graph shown in Fig. 4.31 was obtained during measurements on a high-Q filter. Determine the following parameters for the filter:
(a) resonant frequency
(b) bandwidth
(c) Q-factor.

4.18   The circuit shown in Fig. 4.32 is fed from a 2 V constant voltage source. At what frequency will the supply current be a minimum and what will the current be at this frequency? What current will flow in the inductor and capacitor at this frequency?

4.19   Identify the component shown in Fig. 4.33. Explain how this component operates and state a typical application for it.

4.20   A transformer has 1,600 primary turns and 120 secondary turns. If the primary is connected to a 240 V r.m.s. a.c. mains supply, determine the secondary voltage.

4.21   A transformer has 800 primary turns and 60 secondary turns. If the secondary is connected to a load resistance of 15 Ω, determine the value of primary voltage required to produce a power of 22.5 W in the load   (assume that the transformer is loss-free).

4.22   A transformer has 440 primary turns and operates from a 110 V a.c. supply. The transformer has two secondary windings each rated at 12 V 20 VA. Determine:
(a) the turns-per-volt rating
(b) the secondary turns (each winding)
(c) the secondary current (each winding)
(c) the full-load primary current.

Answers to these problems appear on page 374.

5

Semiconductors

This chapter introduces devices that are made from materials that are neither conductors nor insulators. These **semiconductor** materials form the basis of diodes, thyristors, triacs, transistors and integrated circuits. We start this chapter with a brief introduction to the principles of semiconductors before going on to examine the characteristics of each of the most common types of semiconductor.

In Chapter 1 we described the simplified structure of an atom and showed that it contains both negative charge carriers (electrons) and positive charge carriers (protons). Electrons each carry a single unit of negative electric charge while protons each exhibit a single unit of positive charge. Since atoms normally contain an equal number of electrons and protons, the net charge present will be zero. For example, if an atom has eleven electrons, it will also contain eleven protons. The end result is that the negative charge of the electrons will be exactly balanced by the positive charge of the protons.

Electrons are in constant motion as they orbit around the nucleus of the atom. Electron orbits are organized into shells. The maximum number of electrons present in the first shell is 2, in the second shell 8, and in the third, fourth and fifth shells it is 18, 32 and 50, respectively. In electronics, only the electron shell furthermost from the nucleus of an atom is important. It is important to note that the movement of electrons only involves those present in the outer **valence shell**.

If the valence shell contains the maximum number of electrons possible the electrons are rigidly bonded together and the material has the properties of an insulator. If, however, the valence shell does not have its full complement of electrons, the electrons can be easily loosened from their orbital bonds, and the material has the properties associated with an electrical conductor.

An isolated silicon atom contains four electrons in its valence shell. When silicon atoms combine to form a solid crystal, each atom positions itself between four other silicon atoms in such a way that

the valence shells overlap from one atom to another. This causes each individual valence electron to be shared by two atoms, as shown in Fig. 5.1. By sharing the electrons between four adjacent atoms each individual silicon atom *appears* to have eight electrons in its valence shell. This sharing of valence electrons is called **covalent bonding**.

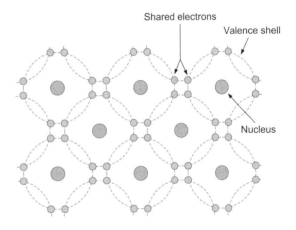

Figure 5.1 Lattice showing covalent bonding

In its pure state, silicon is an insulator because the covalent bonding rigidly holds all of the electrons leaving no free (easily loosened) electrons to conduct current. If, however, an atom of a different element (i.e. an **impurity**) is introduced that has five electrons in its valence shell, a surplus electron will be present, as shown in Fig. 5.2. These **free electrons** become available for use as **charge carriers** and they can be made to move through the lattice by applying an external potential difference to the material.

Similarly, if the impurity element introduced into the pure silicon lattice has three electrons in its valence shell, the absence of the fourth electron needed for proper covalent bonding will produce a

In the case of a reverse biased diode, the P-type material is negatively biased relative to the N-type material. In this case, the negative potential applied to the P-type material attracts the positive charge carriers, drawing them away from the junction. Likewise, the positive potential applied to the N-type material attracts the negative charge carriers away from the junction. This leaves the junction area depleted; virtually no charge carriers exist. Therefore, the junction area becomes an insulator, and current flow is inhibited. The reverse bias potential may be increased to the reverse breakdown voltage for which the particular diode is rated. As in the case of the maximum forward current rating, the reverse breakdown voltage is specified by the manufacturer. The reverse breakdown voltage is usually very much higher than the forward threshold voltage. A typical general-purpose diode may be specified as having a forward threshold voltage of 0.6 V and a reverse breakdown voltage of 200 V. If the latter is exceeded, the diode may suffer irreversible damage. It is also worth noting that, where diodes are designed for use as rectifiers, manufacturers often quote **peak inverse voltage (PIV)** or **maximum reverse repetitive voltage (V_{RRM})** rather than maximum reverse breakdown voltage.

Figure 5.7 shows a test circuit for obtaining diode characteristics (note that the diode must be reverse connected in order to obtain the reverse characteristic).

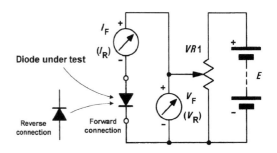

Figure 5.7   Diode test circuit

Example 5.1

The characteristic shown in Fig. 5.8 refers to a germanium diode. Determine the resistance of the diode when (a) the forward current is 2.5 mA and (b) when the forward voltage is 0.65 V.

Solution

(a)   When $I_F = 2.5$ mA the corresponding value of V_F can be read from the graph. This shows that $V_F = 0.43$ V. The resistance of the diode at this point on the characteristic will be given by:

$$R = \frac{V_F}{I_F} = \frac{0.43 \text{ V}}{2.5 \text{ mA}} = 172 \ \Omega$$

(b)   When $V_F = 0.65$ V the corresponding value of I_F can be read from the graph. This shows that $I_F = 7.4$ mA. The resistance of the diode at this point on the characteristic will be given by:

$$R = \frac{V_F}{I_F} = \frac{0.65 \text{ V}}{7.4 \text{ mA}} = 88 \ \Omega$$

This example shows how the resistance of a diode does not remain constant but instead changes according to the point on the characteristic at which it is operating.

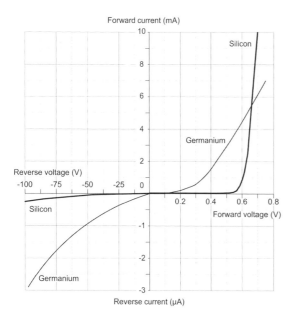

Figure 5.6   Typical diode characteristics

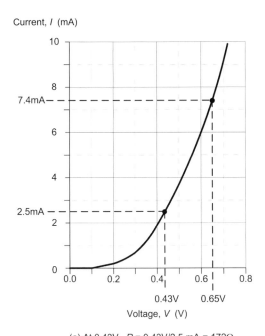

Current, *I* (mA)

(a) At 0.43V, *R* = 0.43V/2.5 mA = 172Ω
(b) At 0.65V, *R* = 0.65V/7.4 mA = 88Ω

Figure 5.8   See Example 5.1

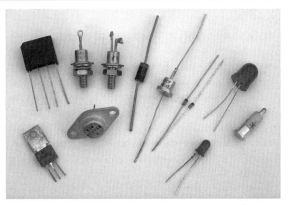

Figure 5.9   A selection of diodes including power and bridge rectifiers, thyristors, signal and zener diodes

forward characteristics with low forward voltage drop. Rectifier diodes need to be able to cope with high values of reverse voltage and large values of forward current, consistency of characteristics is of secondary importance in such applications. Table 5.1 summarizes the characteristics of some common semiconductor diodes whilst a selection of diodes are shown in Fig. 5.9.

Diode types

Diodes are often divided into **signal** or **rectifier** types according to their principal field of application. Signal diodes require consistent

Zener diodes

Zener diodes are heavily doped silicon diodes which, unlike normal diodes, exhibit an abrupt reverse breakdown at relatively low voltages

Table 5.1   Characteristics of some common semiconductor diodes

Device	Material	PIV	I_F max.	I_R max.	Application
1N4148	Silicon	100 V	76 mA	25 nA	General purpose
1N914	Silicon	100 V	75 mA	25 nA	General purpose
AA113	Germanium	60 V	10 mA	200 μA	RF detector
OA47	Germanium	25 V	110 mA	100 μA	Signal detector
OA91	Germanium	115 V	50 mA	275 μA	General purpose
1N4001	Silicon	50 V	1 A	10 μA	Low-voltage rectifier
1N5404	Silicon	400 V	3 A	10 μA	High-voltage rectifier
BY127	Silicon	1,250 V	1 A	10 μA	High-voltage rectifier

(typically less than 6 V). A similar effect occurs in less heavily doped diodes. These **avalanche diodes** also exhibit a rapid breakdown with negligible current flowing below the avalanche voltage and an increasingly large current flowing once the avalanche voltage has been reached. For avalanche diodes, this breakdown voltage usually occurs at voltages above 6 V. In practice, however, both types of diode are referred to as zener diodes. A typical characteristic for a 12 V zener diode is shown in Fig. 5.10.

Whereas reverse breakdown is a highly undesirable effect in circuits that use conventional diodes, it can be extremely useful in the case of zener diodes where the breakdown voltage is precisely known. When a diode is undergoing reverse breakdown *and provided its maximum ratings are not exceeded* the voltage appearing across it will remain substantially constant (equal to the nominal zener voltage) regardless of the current flowing. This property makes the zener diode ideal for use as a **voltage regulator** (see Chapter 6).

Zener diodes are available in various families (according to their general characteristics, encapsulation and power ratings) with reverse breakdown (zener) voltages in the E12 and E24 series (ranging from 2.4 V to 91 V). Table 5.2 summarizes the characteristics of common zener diodes.

Figure 5.11 shows a test circuit for obtaining zener diode characteristics. The circuit is shown with the diode connected in the forward direction and it must be reverse connected in order to obtain the reverse characteristic. Finally, it is important to note that, when used as a voltage regulator, the cathode connection is the more positive terminal.

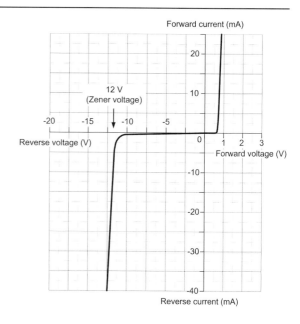

Figure 5.10   Typical characteristic for a 12 V zener diode

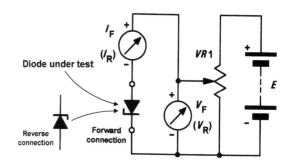

Figure 5.11   Zener diode test circuit

Table 5.2   Characteristics of some common zener diodes

BZY88 series	Miniature glass encapsulated diodes rated at 500 mW (at 25°C). Zener voltages range from 2.7 V to 15 V (voltages are quoted for 5 mA reverse current at 25°C)
BZX61 series	Encapsulated alloy junction rated at 1.3 W (25°C ambient). Zener voltages range from 7.5 V to 72 V
BZX85 series	Medium-power glass-encapsulated diodes rated at 1.3 W and offering zener voltages in the range 5.1 V to 62 V
BZY93 series	High-power diodes in stud mounting encapsulation. Rated at 20 W for ambient temperatures up to 75°C. Zener voltages range from 9.1 V to 75 V
1N5333 series	Plastic encapsulated diodes rated at 5 W. Zener voltages range from 3.3 V to 24 V

Variable capacitance diodes

The capacitance of a reverse-biased diode junction will depend on the width of the depletion layer which, in turn, varies with the reverse voltage applied to the diode. This allows a diode to be used as a voltage controlled capacitor. Diodes that are specially manufactured to make use of this effect (and which produce comparatively large changes in capacitance for a small change in reverse voltage) are known as variable capacitance diodes (or **varicaps**). Such diodes are used (often in pairs) to provide tuning in radio and TV receivers. A typical characteristic for a variable capacitance diode is shown in Fig. 5.12. Table 5.3 summarizes the characteristics of several common variable capacitance diodes.

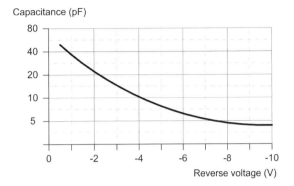

Figure 5.12 Typical characteristic for a variable capacitance diode

Table 5.3 Characteristics of several common types of variable capacitance diode

Type	Capacitance (at 4V)	Capacitance ratio	Q-factor
1N5450	33 pF	2.6 (4 V to 60 V)	350
MV1404	50 pF	>10 (2 V to 10 V)	200
MV2103	10 pF	2 (4 V to 60 V)	400
MV2115	100 pF	2.6 (4 V to 60 V)	100

Thyristors

Thyristors (or **silicon controlled rectifiers**) are three-terminal devices which can be used for switching and a.c. power control. Thyristors can switch very rapidly from a conducting to a non-conducting state. In the off state, the thyristor exhibits negligible leakage current, while in the on state the device exhibits very low resistance. This results in very little power loss within the thyristor even when appreciable power levels are being controlled. Once switched into the conducting state, the thyristor will remain conducting (i.e. it is latched in the on state) until the forward current is removed from the device. In d.c. applications this necessitates the interruption (or disconnection) of the supply before the device can be reset into its non-conducting state. Where the device is used with an alternating supply, the device will automatically become reset whenever the main supply reverses. The device can then be triggered on the next half-cycle having correct polarity to permit conduction. Like their conventional silicon diode counterparts, thyristors have anode and cathode connections; control is applied by means of a gate terminal (see Fig. 5.13). The device is triggered into the conducting (on state) by means of the application of a current pulse to this terminal. The effective triggering of a thyristor requires a **gate trigger** pulse having a fast rise time derived from a low-resistance source. Triggering can

Table 5.4 Characteristics of several common types of thyristor

Type	I_F ave. (A)	V_{RRM} (V)	V_{GT} (V)	I_{GT} (mA)
2N4444	5.1	600	1.5	30
BT106	1	700	3.5	50
BT152	13	600	1	32
BTY79-400R	6.4	400	3	30
TIC106D	3.2	400	1.2	0.2
TIC126D	7.5	400	2.5	20

become erratic when insufficient gate current is available or when the gate current changes slowly. Table 5.4 summarizes the characteristics of several common thyristors.

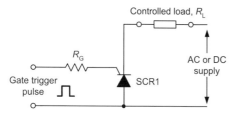

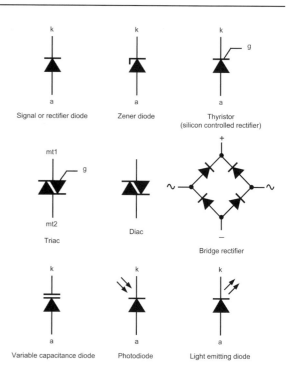

Signal or rectifier diode   Zener diode   Thyristor
(silicon controlled rectifier)

Triac   Diac

Bridge rectifier

Variable capacitance diode   Photodiode   Light emitting diode

Figure 5..13   Method of triggering a thyristor

Triacs

Triacs are a refinement of the thyristor which, when triggered, conduct on both positive and negative half-cycles of the applied voltage. Triacs have three terminals known as main terminal one (MT1), main terminal two (MT2) and gate (G), as shown in Fig. 5.14. Triacs can be triggered by both positive and negative voltages applied between G and MT1 with positive and negative voltages present at MT2 respectively. Triacs thus provide **full-wave control** and offer superior performance in a.c. power control applications when compared with thyristors which only provide **half-wave control**.

Table 5.5 summarizes the characteristics of several common triacs. In order to simplify the design of triggering circuits, triacs are often used in conjunction with diacs (equivalent to a bi-directional zener diode). A typical **diac** conducts heavily when the applied voltage exceeds approximately 30 V in either direction. Once in the conducting state, the resistance of the diac falls to a very low value and thus a relatively large value of current will flow. The characteristic of a typical diac is shown in Fig. 5.15.

Figure 5.14   Symbols used for various types of diode

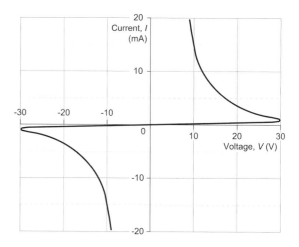

Figure 5.15   Typical diac characteristic

Light emitting diodes

Light emitting diodes (LED) can be used as general-purpose indicators and, compared with conventional filament lamps, operate from significantly smaller voltages and currents. LEDs are also very much more reliable than filament lamps. Most LEDs will provide a reasonable level of light output when a forward current of between 5 mA and 20 mA is applied.

Table 5.5 Characteristics of some common triacs

Type	I_T (A)	V_{RRM} (V)	V_{GT} (V)	I_{GT} (mA)
2N6075	4	600	2.5	5
BT139	15	600	1.5	5
TIC206M	4	600	2	5
TIC226M	8	600	2	50

Light emitting diodes are available in various formats with the round types being most popular. Round LEDs are commonly available in the 3 mm and 5 mm (0.2 inch) diameter plastic packages and also in a 5 mm x 2 mm rectangular format. The viewing angle for round LEDs tends to be in the region of 20° to 40°, whereas for rectangular types this is increased to around 100°. Table 5.6 summarizes the characteristics of several common types of LED.

In order to limit the forward current of an LED to an appropriate value, it is usually necessary to include a fixed resistor in series with an LED indicator, as shown in Fig. 5.16. The value of the resistor may be calculated from:

$$R = \frac{V - V_F}{I}$$

where V_F is the forward voltage drop produced by the LED and V is the applied voltage. Note that it is usually safe to assume that V_F will be 2 V and choose the nearest preferred value for R.

Example 5.2

An LED is to be used to indicate the presence of a 21 V d.c. supply rail. If the LED has a nominal forward voltage of 2.2 V, and is rated at a current of 15 mA, determine the value of series resistor required.

Solution

Here we can use the formula:

$$R = \frac{21 \text{ V} - 2.2 \text{ V}}{15 \text{ mA}} = \frac{18.8 \text{ V}}{15 \text{ mA}} = 1.25 \text{ k}\Omega$$

The nearest preferred value is 1.2 kΩ. The power dissipated in the resistor will be given by:

$$P = I \times V = 15 \text{ mA} \times 18.8 \text{ V} = 280 \text{ mW}$$

Hence the resistor should be rated at 0.33 W, or greater.

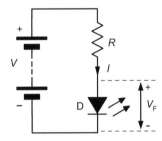

Figure 5.16 Use of a current limiting resistor with an LED

Table 5.6 Characteristics of some common types of LED

Resistance (Ω)	LED type			
	Miniature	Standard	High efficiency	High intensity
Diameter (mm)	3	5	5	5
Maximum forward current (mA)	40	30	30	30
Typical forward current (mA)	12	10	7	10
Typical forward voltage drop (V)	2.1	2.0	1.8	2.2
Maximum reverse voltage (V)	5	3	5	5
Maximum power dissipation (mW)	150	100	27	135
Peak wavelength (nm)	690	635	635	635

Diode coding

The European system for classifying semiconductor diodes involves an alphanumeric code which employs either two letters and three figures (general-purpose diodes) or three letters and two figures (special-purpose diodes). Table 5.7 shows how diodes are coded. Note that the cathode connection of most wire-ended diodes is marked with a stripe.

Example 5.3

Identify each of the following diodes:
(a) AA113
(b) BB105
(c) BZY88C4V7.

Table 5.7 The European system of diode coding

First letter – semiconductor material:
    A Germanium
    B Silicon
    C Gallium arsenide, etc.
    D Photodiodes, etc.
Second letter – application:
    A General-purpose diode
    B Variable-capacitance diode
    E Tunnel diode
    P Photodiode
    Q Light emitting diode
    T Controlled rectifier
    X Varactor diode
    Y Power rectifier
    Z Zener diode
Third letter – in the case of diodes for specialized applications, the third letter does not generally have any particular significance
Zener diodes – zener diodes have an additional letter (which appears after the numbers) which denotes the tolerance of the zener voltage. The following letters are used:
    A ±1%
    B ±2%
    C ±5%
    D ±10%
Zener diodes also have additional characters which indicate the zener voltage (e.g. 9V1 denotes 9.1V).

Solution

Diode (a) is a general-purpose germanium diode.
Diode (b) is a silicon variable capacitance diode.
Diode (c) is a silicon zener diode having ±5% tolerance and 4.7 V zener voltage.

Bipolar junction transistors

Transistor is short for transfer resistor, a term which provides something of a clue as to how the device operates; the current flowing in the output circuit is determined by the current flowing in the input circuit. Since transistors are three-terminal devices, one electrode must remain common to both the input and the output.

Transistors fall into two main categories: **bipolar junction transistors (BJT)** and **field effect transistors (FET)** and are also classified according to the semiconductor material employed (silicon or germanium) and to their field of application (e.g. general-purpose, switching, high-frequency, etc.). Various classes of transistor are available according to the application concerned (see Table 5.8).

BJT operation

Bipolar junction transistors generally comprise NPN or PNP junctions of either **silicon (Si)** or **germanium (Ge)** material (see Figs 5.17 and 5.18). The junctions are, in fact, produced in a single slice of silicon by diffusing impurities through a photographically reduced mask. Silicon transistors are superior when compared with germanium transistors in the vast majority of applications (particularly at high temperatures) and thus germanium devices are very rarely encountered.

Figures 5.19(a) and 5.19(b), respectively, show a simplified representation of NPN and PNP transistors together with their circuit symbols. In either case the electrodes are labelled **collector, base** and **emitter**. Note that each junction within the transistor, whether it be collector-base or base-emitter, constitutes a P-N junction. Figures 5.20(a) and 5.20(b), respectively, show the normal bias voltages applied to NPN and PNP transistors. Note that the base-emitter junction is forward biased and the collector-base junction is reverse biased. The base region is, however, made very narrow so that carriers are swept across it from emitter to collector and only a relatively small current flows in the base. To put this into context, the current flowing in the emitter circuit is typically 100 times greater than that flowing in the base. The direction of conventional current flow is from emitter to collector in the case of a PNP transistor, and collector to emitter in the case of an NPN device. The equation that relates current flow in the collector, base, and emitter currents is:

$$I_E = I_B + I_C$$

where I_E is the emitter current, I_B is the base current, and I_C is the collector current (all expressed in the same units).

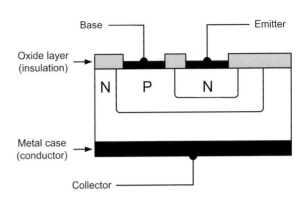

Figure 5.17 NPN transistor construction

Bipolar transistor characteristics

The characteristics of a transistor are often presented in the form of a set of graphs relating voltage and current present at the transistor's terminals.

A typical **input characteristic** (I_B plotted against V_{BE}) for a small-signal general-purpose NPN transistor operating in common-emitter mode (see Chapter 7) is shown in Fig. 5.21. This characteristic shows that very little base current flows until the

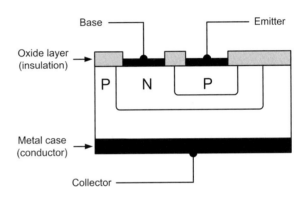

Figure 5.18 PNP transistor construction

Table 5.8 Classes of transistor

Classification	Typical applications
Low-frequency	Transistors designed specifically for audio and low-frequency linear applications (below 100 kHz)
High-frequency	Transistors designed specifically for radio and wideband linear applications (100 kHz and above)
Power	Transistors that operate at significant power levels (such devices are often sub-divided into audio and radio frequency types)
Switching	Transistors designed for switching applications (including power switching)
Low-noise	Transistors that have low-noise characteristics and which are intended primarily for the amplification of low-amplitude signals
High-voltage	Transistors designed specifically to handle high voltages
Driver	Transistors that operate at medium power and voltage levels and which are often used to precede a final (power) stage which operates at an appreciable power level

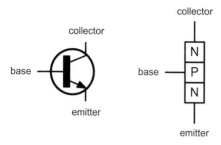

(a) NPN bipolar junction transistor (BJT)

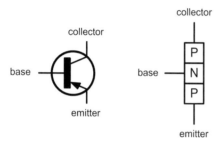

(b) PNP bipolar junction transistor (BJT)

Figure 5.19   Symbols and simplified models for (a) NPN and (b) PNP bipolar junction transistors

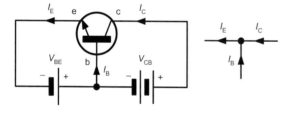

(a) NPN bipolar junction transistor (BJT)

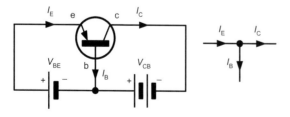

(b) PNP bipolar junction transistor (BJT)

Figure 5.20   Bias voltages and current flow for (a) NPN and (b) PNP bipolar junction transistors

base-emitter voltage (V_{BE}) exceeds 0.6 V. Thereafter, the base current increases rapidly (this characteristic bears a close resemblance to the forward part of the characteristic for a silicon diode, see Fig. 5.5).

Figure 5.22 shows a typical **output characteristic** (I_C plotted against V_{CE}) for a small-signal general-purpose NPN transistor operating in common-emitter mode (see Chapter 7). This characteristic comprises a family of curves, each relating to a different value of base current (I_B). It is worth taking a little time to get familiar with this characteristic as we shall be putting it to good use in Chapter 7. In particular it is important to note the 'knee' that occurs at values of V_{CE} of about 2 V. Also, note how the curves become flattened above this value with the collector current (I_C) not changing very greatly for a comparatively large change in collector-emitter voltage (V_{CE}).

Finally, a typical **transfer characteristic** (I_C plotted against I_B) for a small-signal general-purpose NPN transistor operating in common-emitter mode (see Chapter 7) is shown in Fig. 5.23. This characteristic shows an almost linear relationship between collector current and base current (i.e. doubling the value of base current produces double the value of collector current, and so on). This characteristic is reasonably independent of the value of collector-emitter voltage (V_{CE}) and thus only a single curve is used.

Current gain

The current gain offered by a transistor is a measure of its effectiveness as an amplifying device. The most commonly quoted parameter is that which relates to **common-emitter mode**. In this mode, the input current is applied to the base and the output current appears in the collector (the emitter is effectively common to both the input and output circuits).

The common-emitter current gain is given by:

$$h_{FE} = \frac{I_C}{I_B}$$

where h_{FE} is the **hybrid parameter** which represents **large signal (d.c.) forward current gain**, I_C is the collector current, and I_B is the base current. When small (rather than large) signal

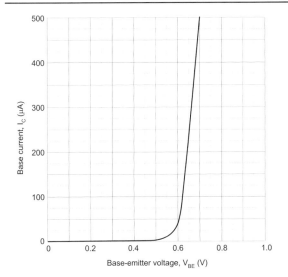

Figure 5.21 Typical input characteristic for a small-signal NPN BJT operating in common-emitter mode

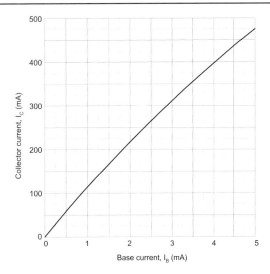

Figure 5.23 Typical transfer characteristic for a small-signal NPN BJT operating in common-emitter mode

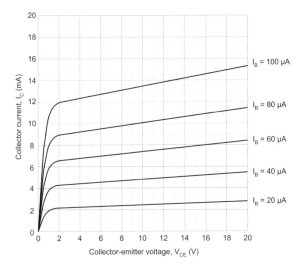

Figure 5.22 Typical family of output (collector) characteristics for a small-signal NPN BJT operating in common-emitter mode

operation is considered, the values of I_C and I_B are incremental (i.e. small changes rather than static values). The current gain is then given by:

$$h_{fe} = \frac{\Delta I_C}{\Delta I_B}$$

where h_{fe} is the **hybrid parameter** which

represents **small signal (a.c.) forward current gain**, ΔI_C is the change in collector current which results from a corresponding change in base current, ΔI_B.

Values of h_{FE} and h_{fe} can be obtained from the transfer characteristic (I_C plotted against I_B) as shown in Figs 5.23 and 5.24. Note that h_{FE} is found from corresponding **static values** while h_{fe} is found by measuring the slope of the graph. Also note that, if the transfer characteristic is linear, there is little (if any) difference between h_{FE} and h_{fe}.

It is worth noting that current gain (h_{fe}) varies with collector current. For most small-signal transistors, h_{fe} is a maximum at a collector current in the range 1 mA and 10 mA. Furthermore, current gain falls to very low values for power transistors when operating at very high values of collector current. Another point worth remembering is that most transistor parameters (particularly common-emitter current gain, h_{fe} are liable to wide variation from one device to the next. It is, therefore, important to design circuits on the basis of the minimum value for h_{fe} in order to ensure successful operation with a variety of different devices.

Transistor parameters are listed in Table 5.9, while Table 5.10 shows the characteristics of several common types of bipolar transistor. Finally, Fig. 5.25 shows a test circuit for obtaining

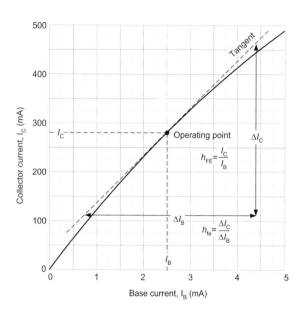

Figure 5.24 Determining the static and small-signal values of current gain (h_{FE} and h_{fe} respectively) from the transfer characteristic. In this particular example $h_{FE} = 280 / 2.5 = 112$ whilst $h_{fe} = 350 / 3.65 = 96$

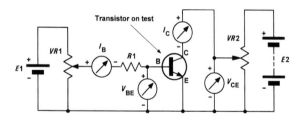

Figure 5.25 NPN transistor test circuit (the arrangement for a PNP transistor is similar but all meters and supplies must be reversed)

NPN transistor characteristics (the arrangement for a PNP transistor is similar but all meters and supplies are reversed). A typical BJT data sheet is shown in Fig. 5.26.

Example 5.4

A transistor operates with $I_C = 30$ mA and $I_B = 600$ µA. Determine the value of I_E and h_{FE}.

Solution

The value of I_E can be calculated from $I_C + I_B$, thus:

$I_E = 30 + 0.6 = 30.6$ mA.

The value of h_{FE} can be found from $h_{FE} = I_C/I_B$, thus:

$h_{FE} = I_C/I_B = 30/0.6 = 50$.

Example 5.5

A transistor operates with a collector current of 97 mA and an emitter current of 98 mA. Determine the value of base current and common emitter current gain.

Solution

Since $I_C = I_C + I_B$, the base current will be given by:

$I_B = I_E - I_C = 98 - 97 = 1$ mA

The common-emitter current gain (h_{FE}) will be given by:

$h_{FE} = I_C/I_B = 97/1 = 97$.

Example 5.6

An NPN transistor is to be used in a regulator circuit in which a collector current of 1.5 A is to be controlled by a base current of 50 mA. What value of h_{FE} will be required?

If the device is to be operated with $V_{CE} = 6$ V, which transistor selected from Table 5.10 would be appropriate for this application and why?

Solution

The required current gain can be found from:

$h_{FE} = I_C/I_B = 1.5$ A / 50 mA $= 1.5 / 0.05 = 30$

The most appropriate device would be the BD131. The only other device capable of operating at a collector current of 1.5 A would be a 2N3055.

The collector power dissipation will be given by:

$P_C = I_C \times V_{CE} = 1.5$ A $\times 6$ V $= 9$ W

However, the 2N3055 is rated at 115 W maximum total power dissipation and this is more than ten times the power required.

2N3702

PNP General Purpose Amplifier
- This device designed for use as general purpose amplifier and switches requiring collector currents to 300mA.
- Sourced from Process 68.
- See PN200 for Characteristics.

TO-92

1. Emitter 2. Collector 3. Base

PNP Epitaxial Silicon Transistor

Absolute Maximum Ratings* T_a=25°C unless otherwise noted

Symbol	Parameter	Value	Units
V_{CEO}	Collector-Emitter Voltage	-25	V
V_{CBO}	Collector-Base Voltage	-40	V
V_{EBO}	Emitter-Base Voltage	-5.0	V
I_C	Collector Current - Continuous	-500	mA
T_J, T_{ST}	Operating and Storage Junction Temperature Range	-55 ~ +150	°C

* These ratings are limiting values above which the serviceability of any semiconductor device may be impaired.

NOTES:
1) These ratings are based on a maximum junction temperature of 150 degrees C.
2) These are steady state limits. The factory should be consulted on applications involving pulsed or low duty cycle operations.

Electrical Characteristics T_a=25°C unless otherwise noted

Symbol	Parameter	Test Condition	Min.	Typ.	Max.	Units
Off Characteristics						
$BV_{(BR)CEO}$	Collector-Emitter Breakdown Voltage	I_C = -10mA, I_B = 0	-25			V
$BV_{(BR)CBO}$	Collector-Base Breakdown Voltage	I_C = -100µA, I_E = 0	-40			V
$BV_{(BR)EBO}$	Emitter-Base Breakdown Voltage	I_E = -100µA, I_C = 0	-5.0			V
I_{CBO}	Collector Cut-off Current	V_{CB} = -20V, I_E = 0			-100	nA
I_{EBO}	Emitter Cut-off Current	V_{EB} = -3.0V, I_C = 0			-100	nA
On Characteristics *						
h_{FE}	DC Current Gain	V_{CE} = -5.0V, I_C = -50mA	60		300	
V_{CE}(sat)	Collector-Emitter Saturation Voltage	I_C = -50mA, I_B = -5.0mA			-0.25	V
V_{BE}(sat)	Base-Emitter Saturation Voltage	V_{CE} = -5.0V, I_C = -50mA	-0.6		-1.0	V
Small Signal Characteristics						
C_{ob}	Current Gain Bandwidth Product	V_{CB} = -10V, f = 1.0MHz			12	pF
f_T	Output Capacitance	I_E = -50mA, V_{CE} = -5.0V	100			MHz

* Pulse Test: Pulse ≤ 300µs, Duty Cycle ≤ 2.0%

Thermal Characteristics T_A=25°C unless otherwise noted

Symbol	Parameter	Max.	Units
P_D	Total Device Dissipation	625	mW
	Derate above 25°C	5.0	mW/°C
$R_{\theta JC}$	Thermal Resistance, Junction to Case	83.3	°C/W
$R_{\theta JA}$	Thermal Resistance, Junction to Ambient	200	°C/W

Rev. B, July 2002

Figure 5.26   Extract from the data sheet for a 2N3702 BJT (courtesy of Fairchild Semiconductor)

Figure 5.30 shows a typical **output characteristic** (I_D plotted against V_{DS}) for a small-signal general-purpose N-channel JFET operating in common-source mode (see Chapter 7). This characteristic comprises a family of curves, each relating to a different value of gate-source voltage (V_{GS}). It is worth taking a little time to get familiar with this characteristic as we shall be using it again in Chapter 7 (you might also like to compare this characteristic with the output characteristic for a transistor operating in common-emitter mode (see Fig. 5.22).

Once again, the characteristic curves have a 'knee' that occurs at low values of V_{DS}. Also, note how the curves become flattened above this value with the drain current (I_D) not changing very greatly for a comparatively large change in drain-source voltage (V_{DS}). These characteristics are, in fact, even flatter than those for a bipolar transistor. Because of their flatness, they are often said to represent a **constant current** characteristic.

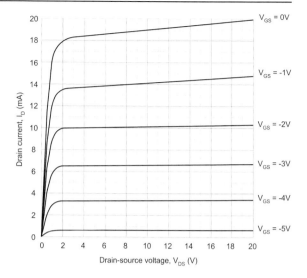

Figure 5.30   A typical family of output characteristics for an N-channel JFET operating in common-source mode

JFET parameters

The gain offered by a field effect transistor is normally expressed in terms of its **forward transfer conductance** (g_{fs} or Y_{fs}) in **common-source** mode. In this mode, the input voltage is applied to the gate and the output current appears in the drain (the source is effectively common to both the input and output circuits).

The common-source forward transfer conductance is given by:

$$g_{fs} = \frac{\Delta I_D}{\Delta V_{GS}}$$

where ΔI_D is the change in drain current resulting from a corresponding change in gate-source voltage (ΔV_{GS}). The units of forward transfer conductance are Siemens (S).

Forward transfer conductance (g_{fs}) varies with drain current collector current. For most small signal devices, g_{fs} is quoted for values of drain current between 1mA and 10 mA. It's also worth noting that most FET parameters (particularly forward transfer conductance) are liable to wide variation from one device to the next. It is, therefore, important to design circuits on the basis of the minimum value for g_{fs} in order to ensure

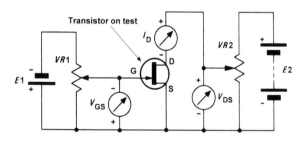

Figure 5.31   N-channel JFET test circuit (the arrangement for a P-channel JFET is similar but all meters and supplies must be reversed)

successful operation with a variety of different devices. JFET parameters are shown in Table 5.11.

The characteristics of several common N-channel field effect transistors are shown in Table 5.12. Figure 5.31 shows a test circuit for obtaining the characteristics of an N-channel FET (the arrangement for a P-channel FET is similar but all meters and supplies must be reversed). A typical N-channel JFET data sheet is shown in Fig. 5.32.

Example 5.8

A FET operates with a drain current of 50 mA and a gate-source bias of −2 V. If the device has a g_{fs} of

2N3819

N-Channel RF Amplifier

- This device is designed for RF amplifier and mixer applications operating up to 450MHz, and for analog switching requiring low capacitance.
- Sourced from process 50.

TO-92

1. Drain 2. Gate 3. Source

Epitaxial Silicon Transistor

Absolute Maximum Ratings* T_C=25°C unless otherwise noted

Symbol	Parameter	Ratings	Units
V_{DG}	Drain-Gate Voltage	25	V
V_{GS}	Gate-Source Voltage	-25	V
I_D	Drain Current	50	mA
I_{GF}	Forward Gate Current	10	mA
T_{STG}	Storage Temperature Range	-55 ~ 150	°C

* This ratings are limiting values above which the serviceability of any semiconductor device may be impaired.

NOTES:
1) These rating are based on a maximum junction temperature of 150 degrees C.
2) These are steady limits. The factory should be consulted on applications involving pulsed or low duty cycle operations.

Electrical Characteristics T_C=25°C unless otherwise noted

Symbol	Parameter	Test Condition	Min.	Typ.	Max.	Units
Off Characteristics						
$V_{(BR)GSS}$	Gate-Source Breakdwon Voltage	I_G = 1.0μA, V_{DS} = 0	25			V
I_{GSS}	Gate Reverse Current	V_{GS} = -15V, V_{DS} = 0			2.0	nA
V_{GS}(off)	Gate-Source Cutoff Voltage	V_{DS} = 15V, I_D = 2.0nA			8.0	V
V_{GS}	Gate-Source Voltage	V_{DS} = 15V, I_D = 200μA	-0.5		-7.5	V
On Characteristics						
I_{DSS}	Zero-Gate Voltage Drain Current	V_{DS} = 15V, V_{GS} = 0	2.0		20	mA
Small Signal Characteristics						
gfs	Forward Transfer Conductance	V_{DS} = 15V, V_{GS} = 0, f = 1.0KHz	2000		6500	μmhos
goss	Output Conductance	V_{DS}= 15V, V_{GS} = 0, f = 1.0KHz			50	μmhos
y_{fs}	Reverse Transfer Admittance	V_{DS}= 15V, V_{GS} = 0, f = 1.0KHz	1600			μmhos
C_{iss}	Input Capacitance	V_{DS} = 15V, V_{GS} = 0, f = 1.0KHz			8.0	pF
C_{rss}	Reverse Transfer Capacitance	V_{DS} = 15V, V_{GS} = 0, f = 1.0KHz			4.0	pF

Thermal Characteristics T_A=25°C unless otherwise noted

Symbol	Parameter	Max.	Units
P_D	Total Device Dissipation	350	mW
	Derate above 25°C	2.8	mW/°C
$R_{\theta JC}$	Thermal Resistance, Junction to Case	125	°C/W
$R_{\theta JA}$	Thermal Resistance, Junction to Ambient	357	°C/W

* Device mounted on FR-4 PCB 1.5" × 1.6" × 0.06"

Rev. A1, December 2002

Figure 5.32 Extract from the data sheet for a 2N3819 JFET (courtesy of Fairchild Semiconductor)

0.025 S, determine the change in drain current if the bias voltage increases to −2.5 V.

Solution

The change in gate-source voltage (ΔV_{gs}) is −0.5 V and the resulting change in drain current can be determined from:

$$\Delta I_D = \Delta V_{GS} \times g_{fs} = 0.025 \text{ S} \times -0.5 \text{ V} = -0.0125 \text{ A}$$

This $\Delta I_D = -12.5$ mA

The new value of drain current will thus be (50 mA − 12.5 mA) or 37.5 mA.

The mutual characteristic shown in Fig. 5.33 illustrates this change in drain current. Note how the drain current falls in response to the change in gate-source voltage.

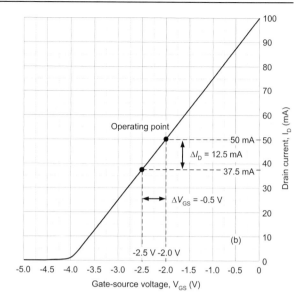

Figure 5.33 See Example 5.8

Table 5.11   FET parameters

Parameter	Meaning
I_D max.	The maximum value of drain current
V_{DS} max.	The maximum value of drain-source voltage
V_{GS} max.	The maximum value of gate-source voltage
P_D max.	The maximum drain power dissipation
g_{fs}	The common-source forward transfer conductance
t_r typ.	The typical output rise time in response to a perfect rectangular pulse input
t_f typ.	The typical output fall time in response to a perfect rectangular pulse input
$R_{DS(on)}$ max.	The maximum value of drain-source resistance when the device is in the conducting (on) state

Table 5.12 Characteristic of some common types of JFET

Device	Type	I_D max.	V_{DS} max.	P_D max.	g_{fs} min.	Application
2N3819	N-channel	10 mA	25 V	200 mW	4 mS	General purpose
2N5457	N-channel	10 mA	25 V	310 mW	1 mS	General purpose
BF244A	N-channel	100 mA	30 V	360 mW	3 mS	RF amplifier
2N3820	P-channel	−15 mA	20 V	200 mW	0.8 mS	General purpose
2N5461	P-channel	−9 mA	40 V	310 mW	1.5 mS	Audio amplifiers
2N5459	N-channel	16 mA	15 V	310 mW	2 mS	General purpose
J310	N-channel	30 mA	25 V	350 mW	8 mS	VHF/UHF amplifier

Transistor packages

A wide variety of packaging styles are used for transistors. Small-signal transistors tend to have either plastic packages (e.g. TO92) or miniature metal cases (e.g. TO5 or TO18). Medium and high-power devices may also be supplied in plastic cases but these are normally fitted with integral metal heat-sinking tabs (e.g. TO126, TO218 or TO220) in order to conduct heat away from the junction. Some older power transistors are supplied in metal cases (either TO3 or TO3). Several popular transistor case styles are shown in Fig. 5.34.

Figure 5.34 Some common transistor packages including TO3, TO220, TO5, TO18 and TO92

Transistor coding

The European system for classifying transistors involves an alphanumeric code which employs either two letters and three figures (general-purpose transistors) or three letters and two figures (special-purpose transistors). Table 5.13 shows how transistors are coded.

Example 5.9

Identify each of the following transistors:
(a) AF115
(b) BC109
(c) BD135
(d) BFY51.

Solution

(a) Transistor (a) is a general-purpose, low-power, high-frequency germanium transistor.
(b) Transistor (b) is a general-purpose, low-power, low-frequency silicon transistor.
(c) Transistor (c) is a general-purpose, high-power, low-frequency silicon transistor.
(d) Transistor (d) is a special-purpose, low-power, high-frequency silicon transistor.

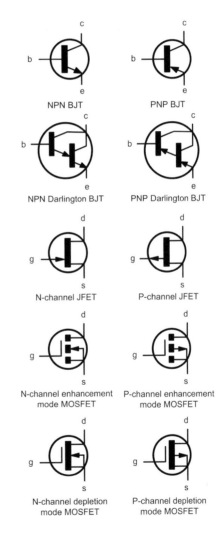

Figure 5.35 Symbols used for different types of transistor including BJT, JFET and MOSFET (metal oxide semiconductor field effect transistor) types

Integrated circuits

Integrated circuits are complex circuits fabricated on a small slice of silicon. Integrated circuits may contain as few as 10 or more than 100,000 active devices (transistors and diodes). With the exception of a few specialized applications (such as amplification at high-power levels) integrated circuits have largely rendered a great deal of conventional discrete circuitry obsolete.

Integrated circuits can be divided into two general classes, linear (analogue) and digital. Typical examples of linear integrated circuits are operational amplifiers (see Chapter 8) whereas typical examples of digital integrated are logic gates (see Chapter 9).

A number of devices bridge the gap between the analogue and digital world. Such devices include analogue to digital converters (ADC), digital to analogue converters (DAC), and timers. For example, the ubiquitous 555 timer contains two operational amplifier stages (configured as voltage comparators) together with a digital bistable stage, a buffer amplifier and an open-collector transistor.

IC packages

As with transistors, a variety of different packages are used for integrated circuits. The most popular form of encapsulation used for integrated circuits is the **dual-in-line** (**DIL**) package which may be fabricated from either plastic or ceramic material (with the latter using a glass hermetic sealant) Common DIL packages have 8, 14, 16, 28 and 40 pins on a 0.1 inch matrix.

Flat package (**flatpack**) construction (featuring both glass-metal and glass-ceramic seals and welded construction) are popular for planar mounting on flat circuit boards. No holes are required to accommodate the leads of such devices which are arranged on a 0.05 inch pitch (i.e. half the pitch used with DIL devices). **Single-in-line** (**SIL**) and **quad-in-line** (**QIL**) packages are also becoming increasingly popular while TO5, TO72, TO3 and TO220 encapsulations are also found (the latter being commonly used for three-terminal voltage regulators). Figure 5.36 shows a variety of common integrated circuit packages whilst Fig. 5.37 shows a modern **LSI** device.

Table 5.13 The European system of transistor coding

First letter – semiconductor material:
  A Germanium
  B Silicon
Second letter – application:
  C Low-power, low-frequency
  D High-power, low-frequency
  F Low-power, high-frequency
  L High-power, high-frequency
Third letter – in the case of transistors for specialized applications, the third letter does not generally have any particular significance

Figure 5.36 Various common integrated circuit packages including DIL (dual-in-line) and QIL (quad-in-line) types

Figure 5.37 A large-scale integrated (LSI) circuit mounted in a socket fitted to a printed circuit board

Practical investigation

Objective

To obtain the common emitter transfer characteristic and small-signal current gain for three different NPN bipolar junction transistors.

Components and test equipment

Breadboard, digital or analogue meters with d.c. current ranges, 9 V d.c. power supply (or battery), three different NPN transistors (e.g. 2N3904, BC548, BFY50, 2N2222), test leads, connecting wire.

Procedure

Connect the circuit shown in Fig. 5.38. Set the base current meter to the 2 mA range and the collector current meter to the 20 mA range.

Set the variable resistor to minimum position and slowly increase base current until it reaches 0.01 mA. Measure and record the collector current. Repeat for base currents up to 0.1 mA in steps of 0.02 m, at each stage measuring and recording the corresponding value of collector current.

Repeat the investigation with at least two further types of transistor.

Measurements and calculations

For each transistor, record your results in a table showing corresponding values of I_C and I_B (see Table 5.14). Plot graphs showing I_C plotted against I_B (see Fig. 5.39). Calculate the value of h_{FE} for each transistor at $I_C = 2$ mA. Compare your calculated results and characteristic graphs with manufacturer's data.

Conclusion

Comment on the shape of each graph. Is this what you would expect? Is the graph linear? If not, what will this imply about the static and small-signal values of current gain? Which of the transistors had the highest value of current gain and which had the least value of current gain?

Table 5.14   Results

Base current (mA)	Collector current (mA)		
	2N3904	BC548	BFY50
0.01			
0.02			
0.03			
0.04			
0.05			
0.06			
0.07			
0.08			
0.09			
0.1			

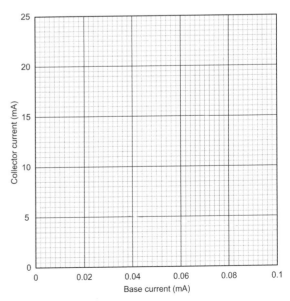

Figure 5.39   Graph layout for plotting the results

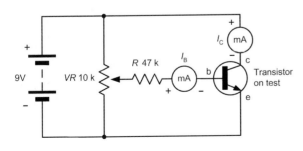

Figure 5.38   Transistor test circuit

Important formulae introduced in this chapter

LED series resistor:
(page 95)

$$R = \frac{V - V_F}{I}$$

Bipolar transistor currents:
(page 97)

$$I_E = I_B + I_C$$

$$I_B = I_E - I_C$$

$$I_C = I_E - I_B$$

Static current gain for a bipolar transistor:
(page 98)

$$h_{FE} = \frac{I_C}{I_B}$$

Small-signal current gain for a bipolar transistor:
(page 99)

$$h_{fe} = \frac{\Delta I_C}{\Delta I_B}$$

Forward transfer conductance for a FET:
(page 104)

$$g_{fs} = \frac{\Delta I_D}{\Delta V_{GS}}$$

Symbols introduced in this chapter

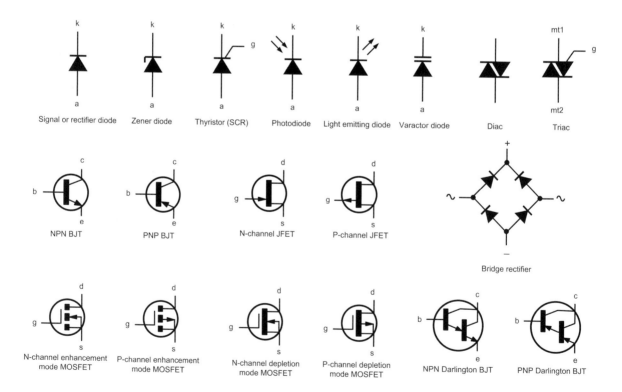

Figure 5.40   Circuit symbols introduced in this chapter

Problems

5.1 Figure 5.41 shows the characteristics of a diode. What type of material is used in this diode? Give a reason for your answer.

5.2 Use the characteristic shown in Fig. 5.41 to determine the resistance of the diode when (a) $V_F = 0.65$ V and (b) $I_F = 4$mA.

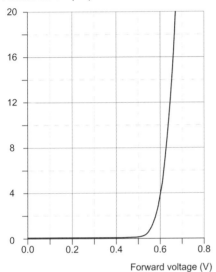

Figure 5.41 See Questions 5.1 and 5.2

5.3 The following data refers to a signal diode:

V_F (V)	I_F (mA)
0.0	0.0
0.1	0.05
0.2	0.02
0.3	1.2
0.4	3.6
0.5	6.5
0.6	10.1
0.7	13.8

Plot the characteristic and use it to determine:
(a) the forward current when $V_F = 350$ mV;
(b) the forward voltage when $I_F = 15$ mA.

5.4 A diode is marked 'BZY88C9V1'. What type of diode is it? What is its rated voltage? State one application for the diode.

5.5 An LED is to be used to indicate the presence of a 5 V d.c. supply. If the LED has a nominal forward voltage of 2 V, and is rated at a current of 12 mA, determine the value of series resistor required.

5.6 Identify each of the following transistors:
(a) AF117   (b) BC184
(c) BD131   (d) BF180.

5.7 A transistor operates with a collector current of 2.5 A and a base current 125 mA. Determine the value of emitter current and static common-emitter current gain.

5.8 A transistor operates with a collector current of 98 mA and an emitter current of 103 mA. Determine the value of base current and the static value of common-emitter current gain.

5.9 A bipolar transistor is to be used in a driver circuit in which a base current of 12 mA is available. If the load requires a current of 200 mA, determine the minimum value of common-emitter current gain required.

5.10 An NPN transistor is to operate with $V_{CE} = 10$ V, $I_C = 50$ mA, and $I_B = 400$ µA. Which of the devices listed in Table 5.10 is most suitable for use in this application?

5.11 A transistor is used in a linear amplifier arrangement. The transistor has small and large signal current gains of 250 and 220, respectively, and bias is arranged so that the static value of collector current is 2 mA. Determine the value of base bias current and the change of output (collector) current that would result from a 5 µA change in input (base) current.

5.12 The transfer characteristic for an NPN transistor is shown in Fig. 5.42. Use this characteristic to determine:
(a) I_C when $I_B = 50$ µA;
(b) h_{FE} when $I_B = 50$ µA;
(c) h_{fe} when $I_C = 75$ mA.

5.13 The output characteristic of an NPN transistor is shown in Fig. 5.43. Use this characteristic to determine:
(a) I_C when $I_B = 100$ µA and $V_{CE} = 4$ V;
(b) V_{CE} when $I_B = 40$ µA and $I_C = 5$ mA;
(c) I_B when $I_C = 7$ mA and $V_{CE} = 6$ V.

5.14 An N-channel FET operates with a drain current of 20 mA and a gate-source bias of

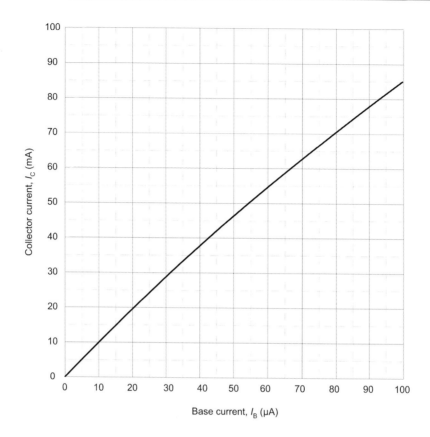

Figure 5.42   See Question 5.12

−1 V. If the device has a g_{fs} of 8 mS, determine the new drain current if the bias voltage increases to −1.5 V.

5.15   The following results were obtained during an experiment on an N-channel FET:

V_{GS} (V)	I_D (mA)
0.0	11.4
−2.0	7.6
−4.0	3.8
−6.0	0.2
−8.0	0.0
−10.0	0.0

Plot the mutual characteristic for the FET and use it to determine g_{fs} when $I_D = 5$ A.

5.16   In relation to integrated circuit packages, what do each of the following abbreviations stand for?
(a) DIL
(b) SIL
(c) QIL
(d) LSI.

5.17   Use the data sheet for a 1N4148 diode in Appendix 8 to determine:
(a) the absolute maximum value of reverse repetitive voltage
(b) the maximum power dissipation for the device
(c) the typical reverse current when a reverse voltage of 70 V is applied
(d) the typical forward voltage for a forward current of 5 mA.

5.18   Identify each of the semiconductor devices shown in Fig. 5.44.

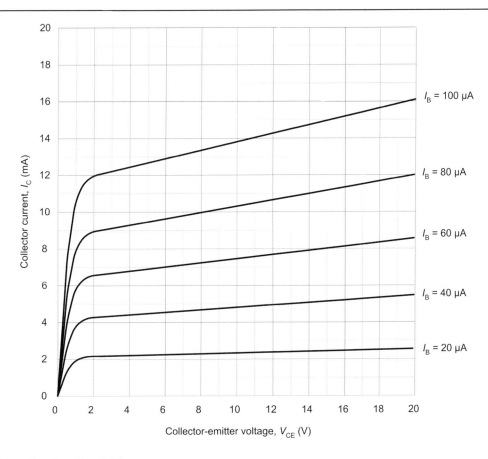

Figure 5.00 See Question 5.13

5.19 Identify each of the semiconductor devices shown in Fig. 5.45.

5.20 Use the data sheet for a 2N3702 transistor shown in Fig. 5.26 to determine:
   (a) The type and class of device.
   (b) The absolute maximum value of collector-base voltage.
   (c) The absolute maximum value of collector current.
   (d) The maximum value of collector-emitter saturation voltage.
   (e) The maximum value of total power dissipation.
   (f) The package style.

5.21 A 2N3702 transistor operates under the following conditions: $V_{BE} = 0.7$ V, $I_B = 0.5$ mA, $V_{CE} = 9$ V and $I_C = 50$ mA. Determine whether or not this exceeds the maximum total device dissipation.

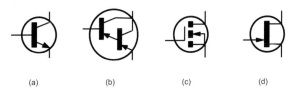

(a)   (b)   (c)   (d)

Figure 5.44 See Question 5.18

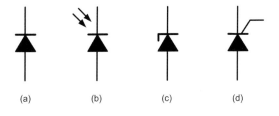

(a)   (b)   (c)   (d)

Figure 5.45 See Question 5.19

5.22 Use the data sheet for a 2N3819 transistor shown in Fig. 5.32 to determine:
 (a) The type and class of device
 (b) The absolute maximum value of drain-gate voltage.
 (c) The absolute maximum value of drain current.
 (d) The maximum value of reverse gate current.
 (e) The minimum value of forward transfer conductance.
 (f) The maximum value of input capacitance.
 (e) The maximum value of total device dissipation.
 (f) The package style.

5.23 The output characteristic of an N-channel JFET is shown in Fig. 5.46. Use this characteristic to determine:

 (a) I_D when $V_{GS} = -2$ V and $V_{DS} = 10$ V;
 (b) V_{DS} when $V_{GS} = -1$ V and $I_D = 13$ mA;
 (c) V_{GS} when $I_D = 18$ mA and $V_{DS} = 11$ V.

5.24 Sketch a circuit diagram showing a test circuit for obtaining the characteristics of a PNP BJT. Label your diagram clearly and indicate the polarity of all supplies and test meters.

5.25 Sketch a circuit diagram showing a test circuit for obtaining the characteristics of a P-channel JFET. Label your diagram clearly and indicate the polarity of all supplies and test meters.

5.26 Sketch a circuit showing how a thyristor can be used to control the current through a resistive load. Label your diagram clearly and explain briefly how the circuit is triggered.

Answers to these problems appear on page 375.

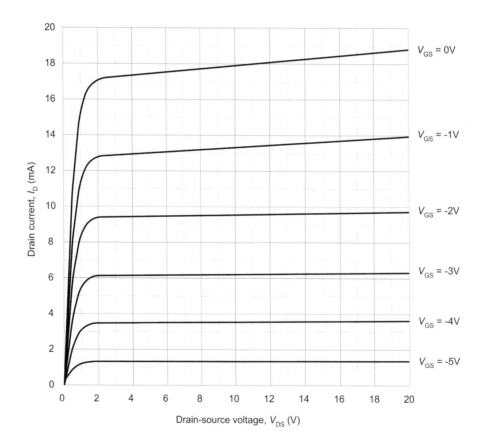

Figure 5.46 See Question 5.23

6

Power supplies

This chapter deals with the unsung hero of most electronic systems, the power supply. Nearly all electronic circuits require a source of well regulated d.c. at voltages of typically between 5 V and 30 V. In some cases, this supply can be derived directly from batteries (e.g. 6 V, 9 V, 12 V) but in many others it is desirable to make use of a standard a.c. mains outlet. This chapter explains how rectifier and smoothing circuits operate and how power supply output voltages can be closely regulated. The chapter concludes with a brief description of some practical power supply circuits.

The block diagram of a d.c. power supply is shown in Fig. 6.1. Since the mains input is at a relatively high voltage, a step-down transformer of appropriate turns ratio is used to convert this to a low voltage. The a.c. output from the transformer secondary is then rectified using conventional silicon rectifier diodes (see Chapter 5) to produce an unsmoothed (sometimes referred to as **pulsating d.c.**) output. This is then smoothed and filtered before being applied to a circuit which will **regulate** (or **stabilize**) the output voltage so that it

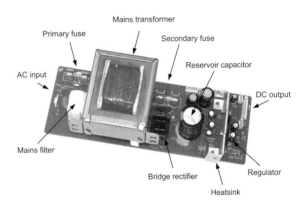

Figure 6.3   A simple d.c. power supply

remains relatively constant in spite of variations in both load current and incoming mains voltage. Figure 6.2 shows how some of the electronic components that we have already met can be used in the realization of the block diagram in Fig. 6.1. The iron-cored step-down transformer feeds a rectifier arrangement (often based on a bridge

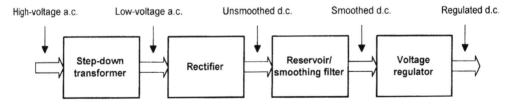

Figure 6.1   Block diagram of a d.c. power supply

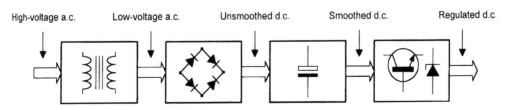

Figure 6.2   Block diagram of a d.c. power supply showing principal components

circuit). The output of the rectifier is then applied to a high-value **reservoir** capacitor. This capacitor stores a considerable amount of charge and is being constantly topped-up by the rectifier arrangement. The capacitor also helps to smooth out the voltage pulses produced by the rectifier. Finally, a stabilizing circuit (often based on a **series transistor regulator** and a zener diode **voltage reference**) provides a constant output voltage. We shall now examine each stage of this arrangement in turn, building up to some complete power supply circuits at the end of the chapter.

Rectifiers

Semiconductor diodes (see Chapter 5) are commonly used to convert alternating current (a.c.) to direct current (d.c), in which case they are referred to as **rectifiers**. The simplest form of rectifier circuit makes use of a single diode and, since it operates on only either positive or negative half-cycles of the supply, it is known as a **half-wave** rectifier.

Figure 6.4 shows a simple half-wave rectifier circuit. Mains voltage (220 to 240 V) is applied to the primary of a step-down transformer (T1). The secondary of T1 steps down the 240 V r.m.s. to 12 V r.m.s. (the turns ratio of T1 will thus be 240/12 or 20:1). Diode D1 will only allow the current to flow in the direction shown (i.e. from cathode to anode). D1 will be forward biased during each positive half-cycle (relative to common) and will effectively behave like a closed switch. When the circuit current tries to flow in the opposite direction, the voltage bias across the diode will be reversed, causing the diode to act like an open switch (see Figs 6.5(a) and 6.5(b), respectively).

The switching action of D1 results in a pulsating output voltage which is developed across the load resistor (R_L). Since the mains supply is at 50 Hz, the pulses of voltage developed across R_L will also be at 50 Hz even if only half the a.c. cycle is present. During the positive half-cycle, the diode will drop the 0.6 V to 0.7 V forward threshold voltage normally associated with silicon diodes. However, during the negative half-cycle the peak a.c. voltage will be dropped across D1 when it is reverse biased. This is an important consideration

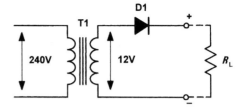

Figure 6.4   A simple half-wave rectifier circuit

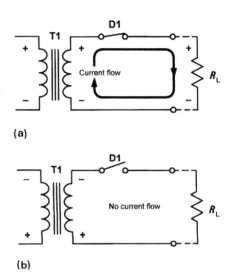

(a)

(b)

Figure 6.5   (a) Half-wave rectifier circuit with D1 conducting (positive-going half-cycles of secondary voltage) (b) half-wave rectifier with D1 not conducting (negative-going half-cycles of secondary voltage)

when selecting a diode for a particular application. Assuming that the secondary of T1 provides 12 V r.m.s., the peak voltage output from the transformer's secondary winding will be given by:

$$V_{pk} = 1.414 \times V_{r.m.s.} = 1.414 \times 12 \text{ V} = 16.97 \text{ V}$$

The peak voltage applied to D1 will thus be approximately 17 V. The negative half-cycles are blocked by D1 and thus only the positive half-cycles appear across R_L. Note, however, that the actual peak voltage across R_L will be the 17 V positive peak being supplied from the secondary on T1, *minus* the 0.7 V forward threshold voltage dropped by D1. In other words, positive half-cycle pulses having a peak amplitude of 16.3 V will appear across R_L.

Example 6.1

A mains transformer having a turns ratio of 44:1 is connected to a 220 V r.m.s. mains supply. If the secondary output is applied to a half-wave rectifier, determine the peak voltage that will appear across a load.

Solution

The r.m.s. secondary voltage will be given by:

$$V_S = V_P / 11 = 220 / 44 = 5 \text{ V}$$

The peak voltage developed after rectification will be given by:

$$V_{PK} = 1.414 \times 5 \text{ V} = 7.07 \text{ V}$$

Assuming that the diode is a silicon device with a forward voltage drop of 0.6 V, the actual peak voltage dropped across the load will be:

$$V_L = 7.07 \text{ V} - 0.6 \text{ V} = 6.47 \text{ V}$$

Reservoir and smoothing circuits

Figure 6.6 shows a considerable improvement to the circuit of Fig. 6.3. The capacitor, $C1$, has been added to ensure that the output voltage remains at, or near, the peak voltage even when the diode is not conducting. When the primary voltage is first applied to T1, the first positive half-cycle output from the secondary will charge $C1$ to the peak value seen across R_L. Hence $C1$ charges to 16.3 V at the peak of the positive half-cycle. Because $C1$ and R_L are in parallel, the voltage across R_L will be the same as that across $C1$.

The time required for $C1$ to charge to the maximum (peak) level is determined by the charging circuit time constant (the series resistance multiplied by the capacitance value). In this circuit, the series resistance comprises the secondary winding resistance together with the forward resistance of the diode and the (minimal) resistance of the wiring and connections. Hence $C1$ charges very rapidly as soon as D1 starts to conduct.

The time required for $C1$ to discharge is, in contrast, very much greater. The discharge time constant is determined by the capacitance value and the load resistance, R_L. In practice, R_L is very much larger than the resistance of the secondary circuit

and hence $C1$ takes an appreciable time to discharge. During this time, D1 will be reverse biased and will thus be held in its non-conducting state. As a consequence, the only discharge path for $C1$ is through R_L.

$C1$ is referred to as a **reservoir** capacitor. It stores charge during the positive half-cycles of secondary voltage and releases it during the negative half-cycles. The circuit of Fig. 6.5 is thus able to maintain a reasonably constant output voltage across R_L. Even so, $C1$ will discharge by a small amount during the negative half-cycle periods from the transformer secondary.

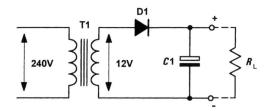

Figure 6.6   A simple half-wave rectifier circuit with reservoir capacitor

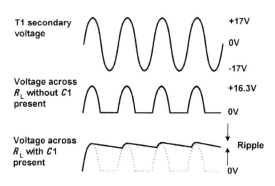

Figure 6.7   A simple half-wave rectifier circuit with reservoir capacitor

Figure 6.7 shows the secondary voltage waveform together with the voltage developed across R_L with and without $C1$ present. This gives rise to a small variation in the d.c. output voltage (known as **ripple**). Since ripple is undesirable we must take additional precautions to reduce it. One obvious method of reducing the amplitude of the ripple is that of simply increasing the discharge time constant. This can be achieved either by increasing the value of C1 or by increasing the resistance value of R_L. In practice, however, the latter is not really an option because R_L is the effective resistance of the circuit being supplied and we don't usually have the ability to change it! Increasing the value of C1 is a more practical alternative and very large capacitor values (often in excess of 4,700 μF) are typical.

Figure 6.8 shows a further refinement of the simple power supply circuit. This circuit employs two additional components, R1 and C1, which act as a filter to remove the ripple. The value of C1 is chosen so that the component exhibits a negligible reactance at the ripple frequency (50 Hz for a half-wave rectifier or 100 Hz for a full-wave rectifier—see later). In effect, R1 and C1 act like a potential divider. The amount of ripple is reduced by an approximate factor equal to:

$$\frac{X_C}{\sqrt{R^2 + X_C^2}}$$

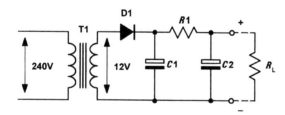

Figure 6.8 Half-wave rectifier circuit with R–C smoothing filter

Example 6.2

The R–C smoothing filter in a 50 Hz mains operated half-wave rectifier circuit consists of R1 = 100 Ω and C2 = 1,000 μF. If 1 V of ripple appears at the input of the circuit, determine the amount of ripple appearing at the output.

Solution

First we must determine the reactance of the capacitor, C1, at the ripple frequency (50 Hz):

$$X_C = \frac{1}{2\pi fC} = \frac{1}{6.28 \times 50 \times 1,000 \times 10^{-6}} = \frac{1,000}{314} = 3.18\ \Omega$$

The amount of ripple at the output of the circuit (i.e. appearing across C1) will be given by:

$$V_{ripple} = 1 \times \frac{X_C}{\sqrt{R^2 + X_C^2}} = 1 \times \frac{3.18}{\sqrt{100^2 + 3.18^2}}$$

From which:

$$V = 0.032\ \text{V} = 32\ \text{mV}$$

Improved ripple filters

A further improvement can be achieved by using an inductor, L1, instead of a resistor in the smoothing circuit. This circuit also offers the advantage that the minimum d.c. voltage is dropped across the inductor (in the circuit of Fig. 6.7, the d.c. output voltage is *reduced* by an amount equal to the voltage drop across R1).

Figure 6.9 shows the circuit of a half-wave power supply with an L–C smoothing circuit. At the ripple frequency, L1 exhibits a high value of inductive reactance while C1 exhibits a low value of capacitive reactance. The combined effect is that of an attenuator which greatly reduces the amplitude of the ripple while having a negligible effect on the direct voltage.

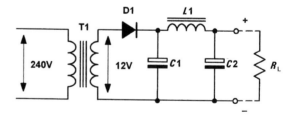

Figure 6.9 Half-wave rectifier circuit with L–C smoothing filter

Example 6.3

The L–C smoothing filter in a 50 Hz mains operated half-wave rectifier circuit consists of L1 = 10 H and C2 = 1,000 μF. If 1 V of ripple appears at the input of the circuit, determine the amount of ripple appearing at the output.

Solution

Once again, the reactance of the capacitor, $C1$, is 3.18 Ω (see Example 6.2). The reactance of $L1$ at 50 Hz can be calculated from:

$$X_L = 2\pi f L = 2 \times 3.14 \times 50 \times 10 = 3,140 \ \Omega$$

The amount of ripple at the output of the circuit (i.e. appearing across $C1$) will be approximately given by:

$$V = 1 \times \frac{X_C}{X_C + X_L} = 1 \times \frac{3.18}{3140 + 3.18} \approx 0.001 \ V$$

Hence the ripple produced by this arrangement (with 1 V of 50 Hz a.c. superimposed on the rectified input) will be a mere 1 mV. It is worth comparing this value with that obtained from the previous example!

Finally, it is important to note that the amount of ripple present at the output of a power supply will increase when the supply is loaded.

Full-wave rectifiers

Unfortunately, the half-wave rectifier circuit is relatively inefficient as conduction takes place only on alternate half-cycles. A better rectifier arrangement would make use of both positive *and* negative half-cycles. These **full-wave rectifier** circuits offer a considerable improvement over their half-wave counterparts. They are not only more efficient but are significantly less demanding in terms of the reservoir and smoothing components. There are two basic forms of full-wave rectifier; the bi-phase type and the bridge rectifier type.

Bi-phase rectifier circuits

Figure 6.10 shows a simple bi-phase rectifier circuit. Mains voltage (240 V) is applied to the primary of step-down transformer (T1) which has two identical secondary windings, each providing 12 V r.m.s. (the turns ratio of T1 will thus be 240/12 or 20:1 for *each* secondary winding).

On positive half-cycles, point A will be positive with respect to point B. Similarly, point B will be

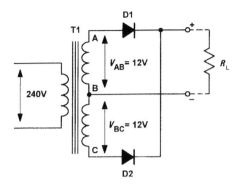

Figure 6.10 Bi-phase rectifier circuit

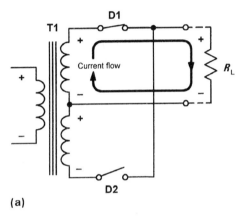

(a)

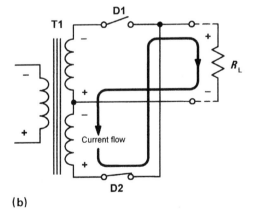

(b)

Figure 6.11 (a) Bi-phase rectifier with D1 conducting and D2 non-conducting (b) bi-phase rectifier with D2 conducting and D1 non-conducting

positive with respect to point C. In this condition D1 will allow conduction (its anode will be positive with respect to its cathode) while D2 will not allow conduction (its anode will be negative with respect to its cathode). Thus D1 alone conducts on positive half-cycles.

On negative half-cycles, point C will be positive with respect to point B. Similarly, point B will be positive with respect to point A. In this condition D2 will allow conduction (its anode will be positive with respect to its cathode) while D1 will not allow conduction (its anode will be negative with respect to its cathode). Thus D2 alone conducts on negative half-cycles.

Figure 6.11 shows the bi-phase rectifier circuit with the diodes replaced by switches. In Fig. 6.11 (a) D1 is shown conducting on a positive half-cycle while in Fig. 6.11(b) D2 is shown conducting. The result is that current is routed through the load *in the same direction* on successive half-cycles. Furthermore, this current is derived alternately from the two secondary windings.

As with the half-wave rectifier, the switching action of the two diodes results in a pulsating output voltage being developed across the load resistor (R_L). However, unlike the half-wave circuit the pulses of voltage developed across R_L will occur at a frequency of 100 Hz (*not* 50 Hz). This doubling of the ripple frequency allows us to use smaller values of reservoir and smoothing capacitor to obtain the same degree of ripple reduction (recall that the reactance of a capacitor is reduced as frequency increases).

As before, the peak voltage produced by each of the secondary windings will be approximately 17 V and the peak voltage across R_L will be 16.3 V (i.e. 17 V less the 0.7 V forward threshold voltage dropped by the diodes).

Figure 6.12 shows how a reservoir capacitor ($C1$) can be added to ensure that the output voltage remains at, or near, the peak voltage even when the diodes are not conducting. This component operates in exactly the same way as for the half-wave circuit, i.e. it charges to approximately 16.3 V at the peak of the positive half-cycle and holds the voltage at this level when the diodes are in their non-conducting states. The time required for C1 to charge to the maximum (peak) level is determined by the charging circuit time constant (the series resistance multiplied by the capacitance value). In this circuit, the series resistance comprises the

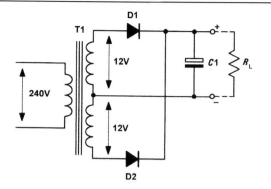

Figure 6.12   Bi-phase rectifier with reservoir capacitor

secondary winding resistance together with the forward resistance of the diode and the (minimal) resistance of the wiring and connections. Hence C1 charges very rapidly as soon as either D1 or D2 starts to conduct. The time required for C1 to discharge is, in contrast, very much greater. The discharge time contrast is determined by the capacitance value and the load resistance, R_L. In practice, R_L is very much larger than the resistance of the secondary circuit and hence C1 takes an appreciable time to discharge. During this time, D1 and D2 will be reverse biased and held in a non-conducting state. As a consequence, the only discharge path for C1 is through R_L. Figure 6.13 shows voltage waveforms for the bi-phase rectifier, with and without C1 present. Note that the ripple frequency (100 Hz) is twice that of the half-wave circuit shown previously in Fig. 6.7.

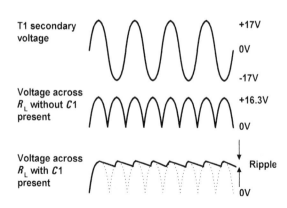

Figure 6.13   Waveforms for the bi-phase rectifier

Bridge rectifier circuits

An alternative to the use of the bi-phase circuit is that of using a four-diode bridge rectifier (see Fig. 6.14) in which opposite pairs of diode conduct on alternate half-cycles. This arrangement avoids the need to have two separate secondary windings.

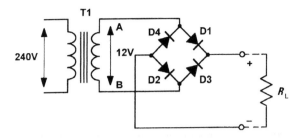

Figure 6.15 Full-wave bridge rectifier circuit

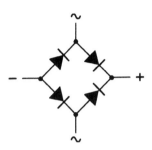

Figure 6.14   Four diodes connected as a bridge

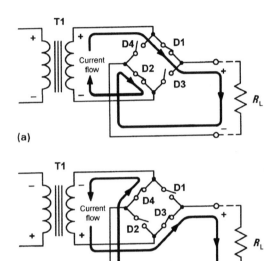

(a)

(b)

Figure 6.16   (a) Bridge rectifier with D1 and D2 conducting, D3 and D4 non-conducting (b) bridge rectifier with D1 and D2 non-conducting, D3 and D4 conducting

A full-wave bridge rectifier arrangement is shown in Fig. 6.15. Mains voltage (240 V) is applied to the primary of a step-down transformer (Tl). The secondary winding provides 12 V r.m.s. (approximately 17 V peak) and has a turns ratio of 20:1, as before. On positive half-cycles, point A will be positive with respect to point B. In this condition Dl and D2 will allow conduction while D3 and D4 will not allow conduction. Conversely, on negative half-cycles, point B will be positive with respect to point A. In this condition D3 and D4 will allow conduction while Dl and D2 will not allow conduction.

Figure 6.16 shows the bridge rectifier circuit with the diodes replaced by four switches. In Fig. 6.16(a) Dl and D2 are conducting on a positive half-cycle while in Fig. 6.16(b) D3 and D4 are conducting. Once again, the result is that current is routed through the load *in the same direction* on successive half-cycles. As with the bi-phase rectifier, the switching action of the two diodes results in a pulsating output voltage being developed across the load resistor (R_L). Once again, the peak output voltage is approximately 16.3 V (i.e. 17 V less the 0.7 V forward threshold voltage).

Figure 6.16 shows how a reservoir capacitor (C1) can be added to maintain the output voltage when the diodes are not conducting. This component operates in exactly the same way as for the bi-phase circuit, i.e. it charges to approximately 16.3 V at the peak of the positive half-cycle and holds the voltage at this level when the diodes are in their non-conducting states. This component operates in exactly the same way as for the bi-phase circuit and the secondary and rectified output waveforms are shown in Fig. 6.18. Once again note that the ripple frequency is twice that of the incoming a.c. supply.

Finally, *R–C* and *L–C* ripple filters can be added to bi-phase and bridge rectifier circuits in exactly the same way as those shown for the half-wave rectifier arrangement (see Figs 6.8 and 6.9).

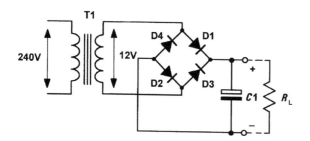

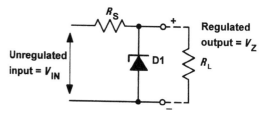

Figure 6.17   Bridge rectifier with reservoir capacitor

Figure 6.19   Bridge rectifier with reservoir capacitor

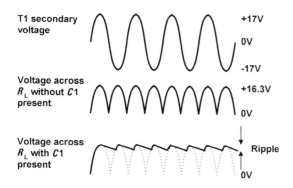

Figure 6.18   Waveforms for the bridge rectifier

Voltage regulators

A simple voltage regulator is shown in Fig. 6.19. R_S is included to limit the zener current to a safe value when the load is disconnected. When a load (R_L) is connected, the zener current (I_Z) will fall as current is diverted into the load resistance (it is usual to allow a minimum current of 2 mA to 5 mA in order to ensure that the diode regulates). The output voltage (V_Z) will remain at the zener voltage until regulation fails at the point at which the potential divider formed by R_S and R_L produces a lower output voltage that is less than V_Z. The ratio of R_S to R_L is thus important. At the point at which the circuit just begins to fail to regulate:

$$V_Z = V_{IN} \times \frac{R_L}{R_L + R_S}$$

where V_{IN} is the unregulated input voltage. Thus the *maximum* value for R_S can be calculated from:

$$R_S \text{ max.} = R_L \times \left(\frac{V_{IN}}{V_{IN}} - 1 \right)$$

The power dissipated in the zener diode, will be given by $P_Z = I_Z \times V_Z$ hence the minimum value for R_S can be determined from the off-load condition when:

$$R_S \text{ min.} = \frac{V_{IN} - V_Z}{I_Z} = \frac{V_{IN} - V_Z}{\left(\dfrac{P_Z \text{ max.}}{V_Z} \right)} = \frac{(V_{IN} - V_Z) \times V_Z}{P_Z \text{ max.}}$$

Thus:

$$R_S \text{ min.} = \frac{V_{IN} V_Z - V_Z^{\,2}}{P_Z \text{ max.}}$$

where P_Z max. is the maximum rated power dissipation for the zener diode.

Example 6.4

A 5V zener diode has a maximum rated power dissipation of 500 mW. If the diode is to be used in a simple regulator circuit to supply a regulated 5 V to a load having a resistance of 400 Ω, determine a suitable value of series resistor for operation in conjunction with a supply of 9 V.

Solution

We shall use an arrangement similar to that shown in Fig. 6.19. First we should determine the maximum value for the series resistor, R_S:

$$R_S \text{ max.} = R_L \times \left(\frac{V_{IN}}{V_{IN}} - 1 \right)$$

thus:

$$R_S \text{ max.} = 400 \times \left(\frac{9}{5} - 1 \right) = 400 \times (1.8 - 1) = 320 \ \Omega$$

Now we need to determine the minimum value for the series resistor, R_S:

$$R_S \text{ min.} = \frac{V_{IN}V_Z - V_Z^2}{P_Z \text{ max.}}$$

thus:

$$R_S \text{ min.} = \frac{(9 \times 5) - 5^2}{0.5} = \frac{45 - 25}{0.5} = 40 \ \Omega$$

Hence a suitable value for R_S would be 150 Ω (roughly mid-way between the two extremes).

Output resistance and voltage regulation

In a perfect power supply, the output voltage would remain constant regardless of the current taken by the load. In practice, however, the output voltage falls as the load current increases. To account for this fact, we say that the power supply has **internal resistance** (ideally this should be zero). This internal resistance appears at the output of the supply and is defined as the change in output voltage divided by the corresponding change in output current. Hence:

$$R_{out} = \frac{\text{change in output voltage}}{\text{change in output current}} = \frac{\Delta V_{out}}{\Delta I_{out}}$$

where ΔI_{out} represents a small change in output (load) current and ΔV_{out} represents a corresponding small change in output voltage.

The **regulation** of a power supply is given by the relationship:

$$\text{Regulation} = \frac{\text{change in output voltage}}{\text{change in line (input) voltage}} \times 100\%$$

Ideally, the value of regulation should be very small. Simple shunt zener diode regulators of the type shown in Fig. 6.19 are capable of producing values of regulation of 5% to 10%. More sophisticated circuits based on discrete components produce values of between 1% and 5% and integrated circuit regulators often provide values of 1% or less.

Example 6.5

The following data was obtained during a test carried out on a d.c. power supply:

(i) *Load test*

Output voltage (no-load) = 12 V
Output voltage (2 A load current) = 11.5 V

(ii) *Regulation test*

Output voltage (mains input, 220 V) = 12 V
Output voltage (mains input, 200 V) = 11.9 V

Determine (a) the equivalent output resistance of the power supply and (b) the regulation of the power supply.

Solution

The output resistance can be determined from the load test data:

$$R_{out} = \frac{\text{change in output voltage}}{\text{change in output current}} = \frac{12 - 11.5}{2 - 0} = 0.25 \ \Omega$$

The regulation can be determined from the regulation test data:

$$\text{Regulation} = \frac{\text{change in output voltage}}{\text{change in line (input) voltage}} \times 100\%$$

thus

$$\text{Regulation} = \frac{12 - 1.9}{220 - 200} \times 100\% = \frac{0.1}{20} \times 100\% = 0.5\%$$

Practical power supply circuits

Figure 6.20 shows a simple power supply circuit capable of delivering an output current of up to 250 mA. The circuit uses a full-wave bridge rectifier arrangement (Dl to D4) and a simple C–R filter. The output voltage is regulated by the shunt-connected 12 V zener diode.

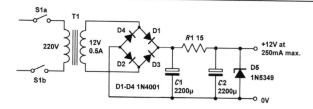

Figure 6.20   Simple d.c. power supply with shunt zener regulated output

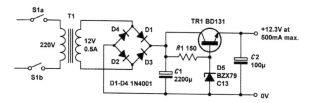

Figure 6.21   Improved regulated d.c. power supply with series-pass transistor

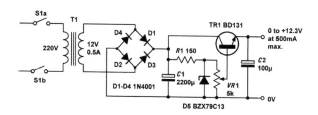

Figure 6.22   Variable d.c. power supply

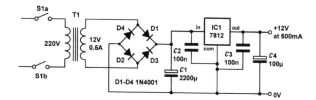

Figure 6.23   Power supply with three-terminal IC voltage regulator

Figure 6.24   This four-diode bridge rectifier arrangement is part of a high-voltage d.c. supply. Each BY253 diode is rated for a reverse repetitive maximum voltage (V_{RRM}) of 600 V, and a maximum forward current, (I_F max.) of 3 A

Figure 6.21 shows an improved power supply in which a transistor is used to provide current gain and minimize the power dissipated in the zener diode (TR1 is sometimes referred to as a **series-pass transistor**). The zener diode, D5, is rated at 13 V and the output voltage will be approximately 0.7 V less than this (i.e. 13 V minus the base-emitter voltage drop associated with TR1). Hence the output voltage is about 12.3 V. The circuit is capable of delivering an output current of up to 500 mA (note that TR1 should be fitted with a small heatsink to conduct away any heat produced). Figure 6.22 shows a variable power supply. The base voltage to the series-pass transistor is derived from a potentiometer connected across the zener diode, D5. Hence the base voltage is variable from 0 V to 13 V. The transistor requires a substantial heatsink (note that TR1's dissipation *increases* as the output voltage is reduced).

Finally, Fig. 6.23 shows a d.c. power supply based on a fixed-voltage **three-terminal integrated circuit voltage regulator**. These devices are available in standard voltage and current ratings (e.g. 5 V, 12 V, 15 V at 1 A, 2 A and 5 A) and they provide excellent performance in terms of output resistance, ripple rejection and voltage regulation.

In addition, such devices usually incorporate overcurrent protection and can withstand a direct short-circuit placed across their output terminals. This is an essential feature in many practical applications!

Voltage multipliers

By adding a second diode and capacitor, we can increase the output of the simple half-wave rectifier that we met earlier. A voltage doubler using this technique is shown in Fig. 6.25. In this arrangement C1 will charge to the positive peak secondary voltage whilst C2 will charge to the negative peak secondary voltage. Since the output is taken from C1 and C2 connected in series the resulting output voltage is twice that produced by one diode alone.

The voltage doubler can be extended to produce higher voltages using the cascade arrangement shown in Fig. 6.26. Here C1 charges to the positive peak secondary voltage, whilst C2 and C3 charge to twice the positive peak secondary voltage. The result is that the output voltage is the sum of the voltages across C1 and C3 which is three times the voltage that would be produced by a single diode. The ladder arrangement shown in Fig. 6.25 can be easily extended to provide even higher voltages but the efficiency of the circuit becomes increasingly impaired and high order voltage multipliers of this type are only suitable for providing relatively small currents.

Switched mode power supplies

Power supplies can be divided into two principal categories, linear and non-linear types. **Linear power supplies** make use of conventional analogue control techniques—the regulating device operates through a continuous range of current and voltage according to the input and load conditions prevailing at the time. **Non-linear power supplies**, on the other hand, use digital techniques where the regulating device is switched rapidly 'on' and 'off' in order to control the mean current and voltage delivered to the load. These non-linear power supplies are commonly referred to as **switched mode power supplies** or just **SMPS**. Note that SMPS can be used to step-up (**boost**) or step-down (**buck**) the input voltage.

Compared with their conventional linear counterparts, the advantages of SMPS are:
- ability to cope with a very wide input voltage range
- very high efficiency (typically 80%, or more)
- compact size and light weight

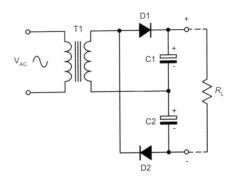

Figure 6.25 A voltage doubler

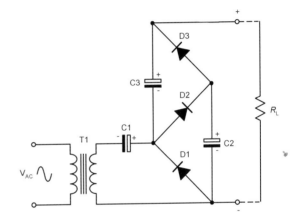

Figure 6.26 A voltage tripler

The disadvantages of SMPS are:
- relatively complex circuitry
- appreciable noise generated (resulting from the high switching frequency switching action).

Fortunately, the principle of the basic step-down type switched mode regulator is quite straightforward. Take a look at the circuit diagram shown in Fig. 6.27. With S1 closed, the switching diode, D, will be reverse biased and will thus be in a non-conducting state. Current will flow through the inductor, L, charging the capacitor, C, and delivering current to the load, R_L.

Current flowing in the inductor, I_L, produces a magnetic flux in its core. When S1 is subsequently opened, the magnetic flux within the inductor rapidly collapses and an e.m.f. is generated across the terminals of the inductor. The polarity of the

Low-noise amplifiers

Low-noise amplifiers are designed so that they contribute negligible noise (signal disturbance) to the signal being amplified. These amplifiers are usually designed for use with very small signal levels (usually less than 10 mV or so).

Gain

One of the most important parameters of an amplifier is the amount of amplification or gain that it provides. Gain is simply the ratio of output voltage to input voltage, output current to input current, or output power to input power (see Fig. 7.2). These three ratios give, respectively, the voltage gain, current gain and power gain. Thus:

Voltage gain, $A_v = \dfrac{V_{out}}{V_{in}}$

Current gain, $A_i = \dfrac{I_{out}}{I_{in}}$

Power gain, $A_p = \dfrac{P_{out}}{P_{in}}$

Note that, since power is the product of current and voltage ($P = I\,V$), we can infer that:

$$A_p = \frac{P_{out}}{P_{in}} = \frac{I_{out} \times V_{out}}{I_{in} \times V_{in}} = \frac{I_{out}}{I_{in}} \times \frac{V_{out}}{V_{in}} = A_i \times A_v$$

Example 7.1

An amplifier produces an output voltage of 2 V for an input of 50 mV. If the input and output currents in this condition are, respectively, 4 mA and 200 mA, determine:

(a) the voltage gain;
(b) the current gain;
(c) the power gain.

Solution

(a) The voltage gain is calculated from:

$$A_v = \frac{V_{out}}{V_{in}} = \frac{2 \text{ V}}{50 \text{ mV}} = 40$$

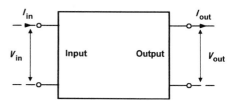

Figure 7.2   Block diagram for an amplifier showing input and output voltages and currents

(b) The current gain is calculated from:

$$A_i = \frac{I_{out}}{I_{in}} = \frac{200 \text{ mA}}{4 \text{ mA}} = 50$$

(c) The power gain is calculated from:

$$A_p = \frac{I_{out} \times V_{out}}{I_{in} \times V_{in}} = \frac{200 \text{ mA} \times 2 \text{ V}}{4 \text{ mA} \times 50 \text{ mV}} = \frac{0.4 \text{ W}}{200 \ \mu\text{W}} = 2{,}000$$

Note that the same result is obtained from:

$$A_p = A_i \times A_v = 50 \times 40 = 2{,}000$$

Class of operation

An important requirement of most amplifiers is that the output signal should be a faithful copy of the input signal, albeit somewhat larger in amplitude. Other types of amplifier are non-linear, in which case their input and output waveforms will not necessarily be similar. In practice, the degree of **linearity** provided by an amplifier can be affected by a number of factors including the amount of bias applied (see later) and the amplitude of the input signal.

It is also worth noting that a linear amplifier will become non-linear when the applied input signal exceeds a threshold value. Beyond this value the amplifier is said to be **overdriven** and the output will become increasingly distorted if the input signal is further increased.

Amplifiers are usually designed to be operated with a particular value of bias supplied to the active devices (i.e. transistors). For linear operation, the

Voltage multipliers

By adding a second diode and capacitor, we can increase the output of the simple half-wave rectifier that we met earlier. A voltage doubler using this technique is shown in Fig. 6.25. In this arrangement C1 will charge to the positive peak secondary voltage whilst C2 will charge to the negative peak secondary voltage. Since the output is taken from C1 and C2 connected in series the resulting output voltage is twice that produced by one diode alone.

The voltage doubler can be extended to produce higher voltages using the cascade arrangement shown in Fig. 6.26. Here C1 charges to the positive peak secondary voltage, whilst C2 and C3 charge to twice the positive peak secondary voltage. The result is that the output voltage is the sum of the voltages across C1 and C3 which is three times the voltage that would be produced by a single diode. The ladder arrangement shown in Fig. 6.25 can be easily extended to provide even higher voltages but the efficiency of the circuit becomes increasingly impaired and high order voltage multipliers of this type are only suitable for providing relatively small currents.

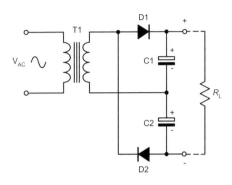

Figure 6.25   A voltage doubler

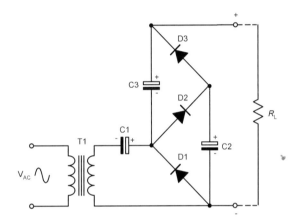

Figure 6.26   A voltage tripler

Switched mode power supplies

Power supplies can be divided into two principal categories, linear and non-linear types. **Linear power supplies** make use of conventional analogue control techniques—the regulating device operates through a continuous range of current and voltage according to the input and load conditions prevailing at the time. **Non-linear power supplies**, on the other hand, use digital techniques where the regulating device is switched rapidly 'on' and 'off' in order to control the mean current and voltage delivered to the load. These non-linear power supplies are commonly referred to as **switched mode power supplies** or just **SMPS**. Note that SMPS can be used to step-up (**boost**) or step-down (**buck**) the input voltage.

Compared with their conventional linear counterparts, the advantages of SMPS are:

- ability to cope with a very wide input voltage range
- very high efficiency (typically 80%, or more)
- compact size and light weight

The disadvantages of SMPS are:

- relatively complex circuitry
- appreciable noise generated (resulting from the high switching frequency switching action).

Fortunately, the principle of the basic step-down type switched mode regulator is quite straightforward. Take a look at the circuit diagram shown in Fig. 6.27. With S1 closed, the switching diode, D, will be reverse biased and will thus be in a non-conducting state. Current will flow through the inductor, L, charging the capacitor, C, and delivering current to the load, R_L.

Current flowing in the inductor, I_L, produces a magnetic flux in its core. When S1 is subsequently opened, the magnetic flux within the inductor rapidly collapses and an e.m.f. is generated across the terminals of the inductor. The polarity of the

Low-noise amplifiers

Low-noise amplifiers are designed so that they contribute negligible noise (signal disturbance) to the signal being amplified. These amplifiers are usually designed for use with very small signal levels (usually less than 10 mV or so).

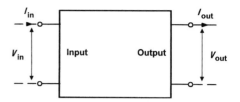

Figure 7.2   Block diagram for an amplifier showing input and output voltages and currents

Gain

One of the most important parameters of an amplifier is the amount of amplification or gain that it provides. Gain is simply the ratio of output voltage to input voltage, output current to input current, or output power to input power (see Fig. 7.2). These three ratios give, respectively, the voltage gain, current gain and power gain. Thus:

Voltage gain, $A_v = \dfrac{V_{out}}{V_{in}}$

Current gain, $A_i = \dfrac{I_{out}}{I_{in}}$

Power gain, $A_p = \dfrac{P_{out}}{P_{in}}$

Note that, since power is the product of current and voltage ($P = I\,V$), we can infer that:

$$A_p = \frac{P_{out}}{P_{in}} = \frac{I_{out} \times V_{out}}{I_{in} \times V_{in}} = \frac{I_{out}}{I_{in}} \times \frac{V_{out}}{V_{in}} = A_i \times A_v$$

Example 7.1

An amplifier produces an output voltage of 2 V for an input of 50 mV. If the input and output currents in this condition are, respectively, 4 mA and 200 mA, determine:

(a) the voltage gain;
(b) the current gain;
(c) the power gain.

Solution

(a) The voltage gain is calculated from:

$$A_v = \frac{V_{out}}{V_{in}} = \frac{2\ \text{V}}{50\ \text{mV}} = 40$$

(b) The current gain is calculated from:

$$A_i = \frac{I_{out}}{I_{in}} = \frac{200\ \text{mA}}{4\ \text{mA}} = 50$$

(c) The power gain is calculated from:

$$A_p = \frac{I_{out} \times V_{out}}{I_{in} \times V_{in}} = \frac{200\ \text{mA} \times 2\ \text{V}}{4\ \text{mA} \times 50\ \text{mV}} = \frac{0.4\ \text{W}}{200\ \mu\text{W}} = 2,000$$

Note that the same result is obtained from:

$$A_p = A_i \times A_v = 50 \times 40 = 2,000$$

Class of operation

An important requirement of most amplifiers is that the output signal should be a faithful copy of the input signal, albeit somewhat larger in amplitude. Other types of amplifier are non-linear, in which case their input and output waveforms will not necessarily be similar. In practice, the degree of **linearity** provided by an amplifier can be affected by a number of factors including the amount of bias applied (see later) and the amplitude of the input signal.

It is also worth noting that a linear amplifier will become non-linear when the applied input signal exceeds a threshold value. Beyond this value the amplifier is said to be **overdriven** and the output will become increasingly distorted if the input signal is further increased.

Amplifiers are usually designed to be operated with a particular value of bias supplied to the active devices (i.e. transistors). For linear operation, the

Voltage multipliers

By adding a second diode and capacitor, we can increase the output of the simple half-wave rectifier that we met earlier. A voltage doubler using this technique is shown in Fig. 6.25. In this arrangement C1 will charge to the positive peak secondary voltage whilst C2 will charge to the negative peak secondary voltage. Since the output is taken from C1 and C2 connected in series the resulting output voltage is twice that produced by one diode alone.

The voltage doubler can be extended to produce higher voltages using the cascade arrangement shown in Fig. 6.26. Here C1 charges to the positive peak secondary voltage, whilst C2 and C3 charge to twice the positive peak secondary voltage. The result is that the output voltage is the sum of the voltages across C1 and C3 which is three times the voltage that would be produced by a single diode. The ladder arrangement shown in Fig. 6.25 can be easily extended to provide even higher voltages but the efficiency of the circuit becomes increasingly impaired and high order voltage multipliers of this type are only suitable for providing relatively small currents.

Switched mode power supplies

Power supplies can be divided into two principal categories, linear and non-linear types. **Linear power supplies** make use of conventional analogue control techniques—the regulating device operates through a continuous range of current and voltage according to the input and load conditions prevailing at the time. **Non-linear power supplies**, on the other hand, use digital techniques where the regulating device is switched rapidly 'on' and 'off' in order to control the mean current and voltage delivered to the load. These non-linear power supplies are commonly referred to as **switched mode power supplies** or just **SMPS**. Note that SMPS can be used to step-up (**boost**) or step-down (**buck**) the input voltage.

Compared with their conventional linear counterparts, the advantages of SMPS are:
- ability to cope with a very wide input voltage range
- very high efficiency (typically 80%, or more)
- compact size and light weight

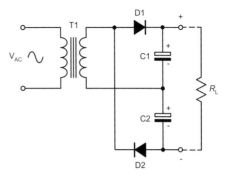

Figure 6.25 A voltage doubler

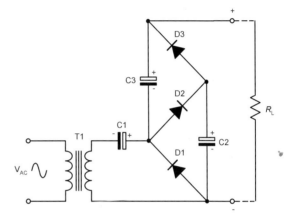

Figure 6.26 A voltage tripler

The disadvantages of SMPS are:
- relatively complex circuitry
- appreciable noise generated (resulting from the high switching frequency switching action).

Fortunately, the principle of the basic step-down type switched mode regulator is quite straightforward. Take a look at the circuit diagram shown in Fig. 6.27. With S1 closed, the switching diode, D, will be reverse biased and will thus be in a non-conducting state. Current will flow through the inductor, L, charging the capacitor, C, and delivering current to the load, R_L.

Current flowing in the inductor, I_L, produces a magnetic flux in its core. When S1 is subsequently opened, the magnetic flux within the inductor rapidly collapses and an e.m.f. is generated across the terminals of the inductor. The polarity of the

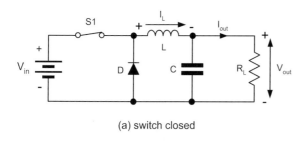

(a) switch closed

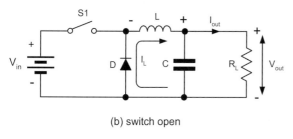

(b) switch open

Figure 6.27 Principle of the switched mode power supply

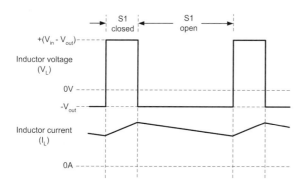

Figure 6.28   Waveforms for the switched mode power supply shown in Fig. 6.27

induced e.m.f. is such that it opposes the original potential and will cause current to continue to flow in the load (i.e. clockwise around the circuit). In this condition, the switching diode, D, will become forward biased, completing the circuit in order to provide a return path for the current.

Waveforms for the circuit of Fig. 6.27 are shown in Fig. 6.28. The inductor voltage, V_L, is alternately positive and negative. When S1 is closed, the voltage dropped across the inductor (from left to right in the diagram) will be equal to $+(V_{in} - V_{out})$. When S1 is open, the voltage across the inductor

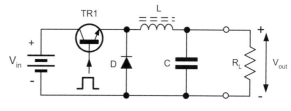

Figure 6.29   Step-down switching regulator using a series switching transistor

will be equal to $-V_{out}$ (less the small forward voltage drop of the switching diode, D). The average current through the inductor, I_L, is equal to the load current, I_{out}. Note that a small amount of ripple voltage appears superimposed on the steady output voltage, V_{out}. This ripple voltage can be reduced by using a relatively large value for the capacitor, C.

In order to control the voltage delivered to the load, V_{out}, it is simply necessary to adjust the ratio of on to off time of the switch. A larger ratio of 'on' to 'off' time (i.e. a larger **duty cycle** or **mark to space ratio**) produces a greater output voltage, and vice versa.

In a practical switched mode power supply, the witch, S1, is replaced by a semiconductor switching device (i.e. a bipolar switching transistor or a MOSFET device). The switching device must have a low 'on' resistance and a high 'off' resistance and must be capable of switching from the 'on' state to the 'off' state in a very short time.

Regardless of whether it is a bipolar transistor or a MOSFET, the switching device is controlled by a train of rectangular pulses applied to its base or gate terminal. The output voltage can be controlled by varying the width of the pulses in this train. In a practical switched mode power supply, a closed-loop feedback path is employed in which the output voltage is sensed and fed back to the control input of a pulse generator. The result is **pulse width modulation (PWM)** of the pulse train to the switching device. This pulse width modulation can be achieved using a handful of discrete components or, more usually, is based on a dedicated switched mode controller chip.

The job of controlling a switched mode power supply is an ideal task for an integrated circuit and Fig. 6.29 shows the internal arrangement of a typical example, the LM78S40. This device contains two operational amplifiers (IC1 and IC2) designed to work as **comparators**, and a two-stage Darlington transistor switch comprising an emitter-follower driver, Q1, and output switch, Q2. The LM78S40 is supplied in a 16-pin dual-in-line (DIL) package.

The LM78S40 can be configured to provide step-up (boost), step-down (buck), and inverting operation. The frequency of the internal current controlled oscillator is set by the value of capacitor connected to pin-12. Oscillator frequencies of between 100 Hz and 100 kHz are possible but most practical applications operate at frequencies between 20 kHz and 50 kHz. The oscillator duty cycle is internally set to 6:1 but can be varied by means of the current sensing circuit which normally senses the current in an external resistor connected between pins-12 and 13.

An internal **band-gap voltage reference** provides a stable voltage reference of 1.3 V at pin-8. The internal reference voltage source is capable of providing a current of up to 10 mA drawn from pin-8. The output transistor, Q2, is capable of carrying a peak current of up to 1.5 A and has a maximum collector-emitter voltage rating of 40 V. The internal power switching diode, D1, is accessible between pins-1 and 2 and this has similar ratings to Q2. Both D1 and Q2 have switching times of between 300 and 500 ns. IC1 is used to compare the 1.3 V voltage reference (pin-10 connected to pin-8) with a proportion of the output (derived from a simple two-resistor potential divider).

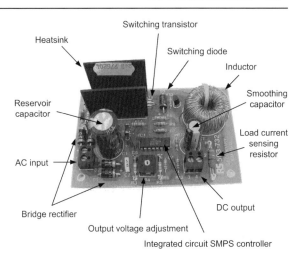

Figure 6.31 A switched mode power supply

Practical investigation

Objective

To investigate the operation of simple voltage regulators.

Components and test equipment

Breadboard, digital or analogue meters with d.c. voltage and current ranges, 9 V d.c. power supply (or battery), 3.9 V zener diode (e.g. BZX85 or BZY88), NPN TO5 transistor (e.g. 2N3053 or BFY50), 48°C/W TO5 clip-on heatsink, 220 Ω 0.3W resistor, 15 Ω 0.3W resistor, 500 Ω and 1 kΩ wirewound variable resistors, connecting wire, test leads, .

Procedure

Connect the simple zener diode shunt regulator shown in Fig. 6.32. Set the variable resistor to produce a load current (I_L) of 10 mA then measure and record the output voltage produced across the load, V_L. Repeat for load currents from 20 mA to 100 mA in 10 mA steps.

Connect the transistor regulator shown in Fig. 6.33. Set the variable resistor to produce a load current (I_L) of 25 mA then measure and record the output voltage produced across the load, V_L. Repeat for load current from 50 mA to 250 mA in 25 mA steps.

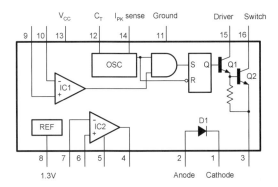

Figure 6.30 The LM78S40 switched mode power supply controller

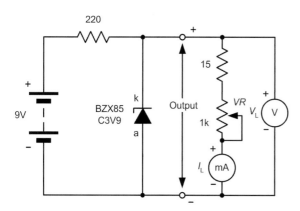

Figure 6.32   Simple zener diode voltage regulator

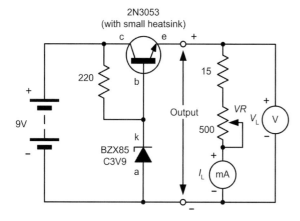

Figure 6.33   Transistor voltage regulator

Table 6.1   Table of results for the simple zener voltage regulator

Load current (mA)	Output voltage (V)
10	
20	
30	
40	
50	
60	
70	
80	
90	
100	

Table 6.2   Table of results for the transistor voltage regulator

Load current (mA)	Output voltage (V)
25	
50	
75	
100	
125	
150	
175	
200	
225	
250	

Measurements and calculations

For each circuit, record your results in a table showing corresponding values of I_L and V_L (see Tables 6.1 and 6.2).

Plot graphs showing V_L plotted against I_L for each circuit (see Figures 6.34 and 6.35). By constructing a tangent to each graph, determine the output resistance of each regulator circuit. For the simple shunt zener diode regulator the output resistance should be calculated at $I_L = 30$ mA whilst for the transistor voltage regulator the output resistance should be calculated at $I_L = 100$ mA.

Conclusion

Comment on the shape of each graph. Is this what you would expect? Compare the performance of each circuit and, in particular, the range of load currents over which effective regulation is achieved. Which of the transistors had the lowest value of output resistance? Can you suggest why this is?

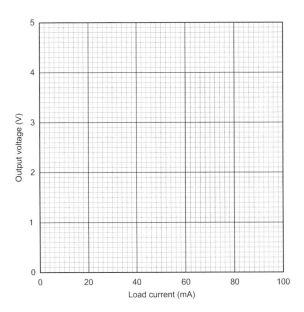

Figure 6.34 Graph layout that can be used for the load characteristic of the simple zener voltage regulator (construct a tangent to the graph at $I_L = 30$ mA in order to determine the output resistance)

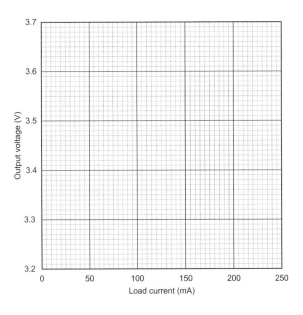

Figure 6.35 Graph layout that can be used for the load characteristic of the simple zener voltage regulator (construct a tangent to the graph at $I_L = 100$ mA in order to determine the output resistance)

Important formulae introduced in this chapter

Maximum value of series resistor for a simple shunt zener diode voltage regulator:
(page 122)

$$R_S \text{ max.} = R_L \times \left(\frac{V_{IN}}{V_{IN}} - 1 \right)$$

Minimum value of series resistor for a simple shunt zener diode voltage regulator:
(page 122)

$$R_S \text{ min.} = \frac{V_{IN}V_Z - V_Z^2}{P_Z \text{ max.}}$$

Output resistance of a power supply:
(page 123)

$$R_{out} = \frac{\text{change in output voltage}}{\text{change in output current}} = \frac{\Delta V_{out}}{\Delta I_{out}}$$

Input (line) regulation of a power supply:
(page 123)

$$\text{Regulation} = \frac{\text{change in output voltage}}{\text{change in line (input) voltage}} \times 100\%$$

Symbols introduced in this chapter

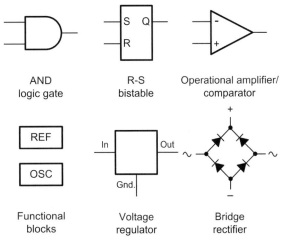

Figure 6.36 Circuit symbols introduced in this chapter

Problems

6.1   A half-wave rectifier is fitted with an *R–C* smoothing filter comprising $R = 200\ \Omega$ and $C = 50\ \mu F$. If 2 V of 400 Hz ripple appear at the input of the circuit, determine the mount of ripple appearing at the output.

6.2   The *L–C* smoothing filter fitted to a 50 Hz mains operated full-wave rectifier circuit consists of $L = 4$ H and $C = 500\ \mu F$. If 4 V of ripple appear at the input of the circuit, determine the amount of ripple appearing at the output.

6.3   If a 9 V zener diode is to be used in a simple shunt regulator circuit to supply a load having a nominal resistance of 300 Ω, determine the maximum value of series resistor for operation in conjunction with a supply of 15 V.

6.4   The circuit of a d.c. power supply is shown in Fig. 6.37. Determine the voltages that will appear at test points A, B and C.

6.5   In Fig. 6.37, determine the current flowing in *R*1 and the power dissipated in D5 when the circuit is operated without any load connected.

6.6   In Fig. 6.37, determine the effect of each of the following fault conditions:
(a) *R*1 open-circuit;
(b) D5 open-circuit;
(c) D5 short-circuit.

6.7   A 220 V AC supply feeds a 20:1 step-down transformer, the secondary of which is connected to a bridge rectifier and reservoir capacitor. Determine the approximate d.c. voltage that will appear across the reservoir capacitor under 'no-load' conditions.

6.8   The following data were obtained during a load test carried out on a d.c. power supply:
Output voltage (no-load) = 8.5 V
Output voltage (800 mA load) = 8.1V
Determine the output resistance of the power supply and estimate the output voltage at a load current of 400 mA.

6.9   The following data were obtained during a regulation test on a d.c. power supply:
Output voltage (a.c. input: 230 V) = 15 V
Output voltage (a.c. input: 190 V) = 14.6 V
Determine the regulation of the power supply and estimate the output voltage when the input voltage is 245 V.

6.10   Figure 6.38 shows a switching regulator circuit that produces an output of 9 V for an input of 4.5 V. What type of regulator is this? Between which pins of IC1 is the switching transistor connected? Which pin on IC1 is used to feed back a proportion of the output voltage to the internal comparator stage?

Answers to these problems appear on page 375.

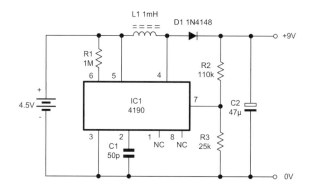

Figure 6.38   See Question 6.10

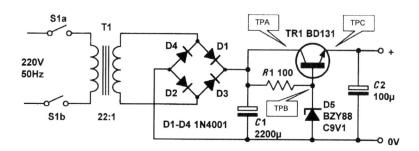

Figure 6.37   See Questions 6.4, 6.5 and 6.6

7

Amplifiers

This chapter introduces the basic concepts of amplifiers and amplification. It describes the most common types of amplifier and outlines the basic classes of operation used in both linear and non-linear amplifiers. The chapter also describes methods for predicting the performance of an amplifier based on equivalent circuits and on the use of semiconductor characteristics and load lines. Once again, we conclude with a selection of practical circuits that can be built and tested.

Types of amplifier

Many different types of amplifier are found in electronic circuits. Before we explain the operation of transistor amplifiers in detail, we shall briefly describe the main types of amplifier.

a.c. coupled amplifiers

In a.c. coupled amplifiers, stages are coupled together in such a way that d.c. levels are isolated and only the a.c. components of a signal are transferred from stage to stage.

d.c. coupled amplifiers

In d.c. (or direct) coupled amplifiers, stages are coupled together in such a way that stages are not isolated to d.c. potentials. Both a.c. and d.c. signal components are transferred from stage to stage.

Large-signal amplifiers

Large-signal amplifiers are designed to cater for appreciable voltage and/or current levels (typically from 1V to 100 V or more).

Small-signal amplifiers

Small-signal amplifiers are designed to cater for low-level signals (normally less than 1V and often much smaller). Small-signal amplifiers have to be specially designed to combat the effects of noise.

Figure 7.1   Part of a high-gain, wideband d.c. coupled amplifier using discrete components

Audio frequency amplifiers

Audio frequency amplifiers operate in the band of frequencies that is normally associated with audio signals (e.g. 20 Hz to 20 kHz).

Wideband amplifiers

Wideband amplifiers are capable of amplifying a very wide range of frequencies, typically from a few tens of hertz to several megahertz.

Radio frequency amplifiers

Radio frequency amplifiers operate in the band of frequencies that is normally associated with radio signals (e.g. from 100 kHz to over 1 GHz). Note that it is desirable for amplifiers of this type to be frequency selective and thus their frequency response may be restricted to a relatively narrow band of frequencies (see Fig. 7.9 on page 135).

Low-noise amplifiers

Low-noise amplifiers are designed so that they contribute negligible noise (signal disturbance) to the signal being amplified. These amplifiers are usually designed for use with very small signal levels (usually less than 10 mV or so).

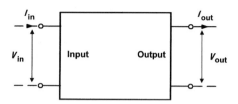

Figure 7.2   Block diagram for an amplifier showing input and output voltages and currents

Gain

One of the most important parameters of an amplifier is the amount of amplification or gain that it provides. Gain is simply the ratio of output voltage to input voltage, output current to input current, or output power to input power (see Fig. 7.2). These three ratios give, respectively, the voltage gain, current gain and power gain. Thus:

Voltage gain, $A_v = \dfrac{V_{out}}{V_{in}}$

Current gain, $A_i = \dfrac{I_{out}}{I_{in}}$

Power gain, $A_p = \dfrac{P_{out}}{P_{in}}$

Note that, since power is the product of current and voltage ($P = I\,V$), we can infer that:

$$A_p = \frac{P_{out}}{P_{in}} = \frac{I_{out} \times V_{out}}{I_{in} \times V_{in}} = \frac{I_{out}}{I_{in}} \times \frac{V_{out}}{V_{in}} = A_i \times A_v$$

Example 7.1

An amplifier produces an output voltage of 2 V for an input of 50 mV. If the input and output currents in this condition are, respectively, 4 mA and 200 mA, determine:

(a) the voltage gain;
(b) the current gain;
(c) the power gain.

Solution

(a) The voltage gain is calculated from:

$$A_v = \frac{V_{out}}{V_{in}} = \frac{2\ V}{50\ mV} = 40$$

(b) The current gain is calculated from:

$$A_i = \frac{I_{out}}{I_{in}} = \frac{200\ mA}{4\ mA} = 50$$

(c) The power gain is calculated from:

$$A_p = \frac{I_{out} \times V_{out}}{I_{in} \times V_{in}} = \frac{200\ mA \times 2\ V}{4\ mA \times 50\ mV} = \frac{0.4\ W}{200\ \mu W} = 2{,}000$$

Note that the same result is obtained from:

$$A_p = A_i \times A_v = 50 \times 40 = 2{,}000$$

Class of operation

An important requirement of most amplifiers is that the output signal should be a faithful copy of the input signal, albeit somewhat larger in amplitude. Other types of amplifier are non-linear, in which case their input and output waveforms will not necessarily be similar. In practice, the degree of **linearity** provided by an amplifier can be affected by a number of factors including the amount of bias applied (see later) and the amplitude of the input signal.

It is also worth noting that a linear amplifier will become non-linear when the applied input signal exceeds a threshold value. Beyond this value the amplifier is said to be **overdriven** and the output will become increasingly distorted if the input signal is further increased.

Amplifiers are usually designed to be operated with a particular value of bias supplied to the active devices (i.e. transistors). For linear operation, the

active device(s) must be operated in the linear part of their **transfer characteristic** (V_{out} plotted against V_{in}). In Fig. 7.3 the input and output signals for an amplifier are operating in linear mode. This form of operation is known as **Class A** and the **bias point** is adjusted to the mid-point of the linear part of the transfer characteristic. Furthermore, current will flow in the active devices used in a Class A amplifier during a complete cycle of the signal waveform. At no time does the current fall to zero.

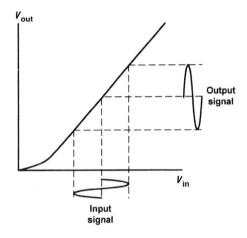

Figure 7.3   Class A (linear) operation

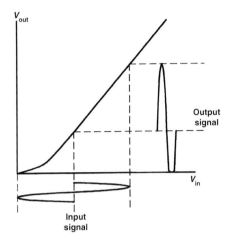

Figure 7.4   Effect of reducing bias and increasing input signal amplitude (the output waveform is no longer a faithful reproduction of the input)

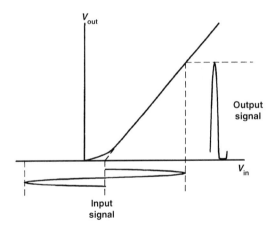

Figure 7.5   Class AB operation (bias set at projected cut-off)

Figure 7.4 shows the effect of moving the bias point down the transfer characteristic and, at the same time, increasing the amplitude of the input signal. From this, you should notice that the extreme negative portion of the output signal has become distorted. This effect arises from the non-linearity of the transfer characteristic that occurs near the origin (i.e. the zero point). Despite the obvious non-linearity in the output waveform, the active device(s) will conduct current during a complete cycle of the signal waveform.

Now consider the case of reducing the bias even further while further increasing the amplitude of the input signal (see Fig. 7.5). Here the bias point has been set at the projected cut-off point. The negative portion of the output signal becomes cut off (or **clipped**) and the active device(s) will cease to conduct for this part of the cycle. This mode of operation is known as **Class AB**.

Now let's consider what will happen if no bias at all is applied to the amplifier (see Fig. 7.6). The output signal will only comprise a series of positive half-cycles and the active device(s) will only be conducting during half-cycles of the waveform (i.e. they will only be operating 50% of the time).

This mode of operation is known as **Class B** and is commonly used in high-efficiency push-pull power amplifiers where the two active devices in the output stage operate on alternate half-cycles of the waveform.

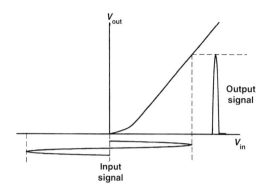

Figure 7.6   Class B operation (no bias applied)

Finally, there is one more class of operation to consider. The input and output waveforms for **Class C** operation are shown in Fig. 7.7. Here the bias point is set at beyond the cut-off (zero) point and a very large input signal is applied. The output waveform will then comprise a series of quite sharp positive-going pulses. These pulses of current or

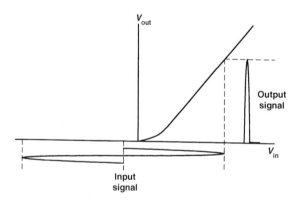

Figure 7.7   Class C operation (bias is set beyond cut-off)

voltage can be applied to a tuned circuit load in order to recreate a sinusoidal signal. In effect, the pulses will excite the tuned circuit and its inherent flywheel action will produce a sinusoidal output waveform. This mode of operation is only used in RF power amplifiers that must operate at very high levels of efficiency. Table 7.1 summarizes the classes of operation used in amplifiers.

Input and output resistance

Input resistance is the ratio of input voltage to input current and it is expressed in ohms. The input of an amplifier is normally purely resistive (i.e. any reactive component is negligible) in the middle of its working frequency range (i.e. the **mid-band**). In some cases, the reactance of the input may become appreciable (e.g. if a large value of stray capacitance appears in parallel with the input resistance). In such cases we would refer to **input impedance** rather than input resistance.

Output resistance is the ratio of open-circuit output voltage to short-circuit output current and is measured in ohms. Note that this resistance is internal to the amplifier and should not be confused with the resistance of a load connected externally.

As with input resistance, the output of an amplifier is normally purely resistive and we can safely ignore any reactive component. If this is not the case, we would once again need to refer to **output impedance** rather than output resistance.

Figure 7.8 shows how the input and output resistances are 'seen' looking into the input and output terminals, respectively. We shall be returning to this equivalent circuit a little later in this chapter. Finally, it's important to note that, although these resistance are meaningful in terms of the signals present, they cannot be measured using a conventional meter!

Table 7.1   Classes of operation

Class of operation	Bias point	Conduction angle (typical)	Efficiency (typical)	Application
A	Mid-point	360°	5% to 20%	Linear audio amplifiers
AB	Projected cut-off	210°	20% to 40%	Push-pull audio amplifiers
B	At cut-off	180°	40% to 70%	Push-pull audio amplifiers
C	Beyond cut-off	120°	70% to 90%	Radio frequency power amplifiers

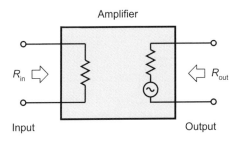

Figure 7.8 Input and output resistances 'seen' looking into the input and output terminals, respectively

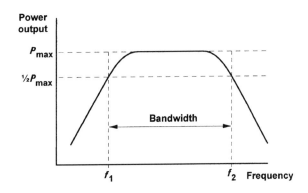

Figure 7.10 Frequency response and bandwidth (output power plotted against frequency)

Frequency response

The frequency response characteristics for various types of amplifier are shown in Fig. 7.9. Note that, for response curves of this type, frequency is almost invariably plotted on a **logarithmic scale**.

The frequency response of an amplifier is usually specified in terms of the upper and lower **cut-off frequencies** of the amplifier. These frequencies are those at which the output power has dropped to 50% (otherwise known as the **−3dB points**) or where the voltage gain has dropped to 70.7% of its mid-band value.

Figures 7.10 and 7.11, respectively, show how the bandwidth can be expressed in terms of either power or voltage (the cut-off frequencies, f_1 and f_2, and bandwidth are identical).

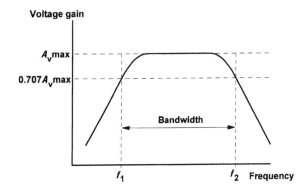

Figure 7.11 Frequency response and bandwidth (output voltage plotted against frequency)

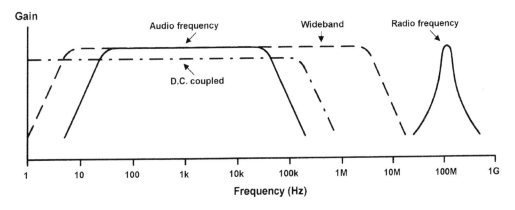

Figure 7.9 Frequency response and bandwidth (output power plotted against frequency)

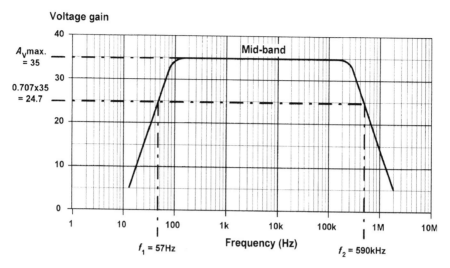

Figure 7.12   See Example 7.2

Example 7.2

Determine the mid-band voltage gain and upper and lower cut-off frequencies for the amplifier whose frequency response is shown in Fig. 7.12.

Solution

The mid-band voltage gain corresponds with the flat part of the frequency response characteristic. At the point the voltage gain reaches a maximum of 35 (see Fig. 7.12).

The voltage gain at the two cut-off frequencies can be calculated from:

$$A_v \text{ cut-off} = 0.707 \times A_v \text{ max} = 0.707 \times 35 = 24.7$$

This value of gain intercepts the frequency response graph at $f_1 = 57$ Hz and $f_2 = 590$ kHz (see Fig. 7.12).

Bandwidth

The bandwidth of an amplifier is usually taken as the difference between the upper and lower cut-off frequencies (i.e. $f_2 - f_1$ in Figs 7.10 and 7.11). The bandwidth of an amplifier must be sufficient to accommodate the range of frequencies present within the signals that it is to be presented with.

Many signals contain **harmonic** components (i.e. signals at 2f, 3f, 4f, etc. where f is the frequency of the **fundamental** signal). To reproduce a square wave, for example, requires an amplifier with a very wide bandwidth (note that a square wave comprises an infinite series of harmonics). Clearly it is not possible to *perfectly* reproduce such a wave but it does explain why it can be desirable for an amplifier's bandwidth to greatly exceed the highest signal frequency that it is required to handle!

Phase shift

Phase shift is the phase angle between the input and output signal voltages measured in degrees. The measurement is usually carried out in the mid-band where, for most amplifiers, the phase shift remains relatively constant. Note also that conventional single-stage transistor amplifiers provide phase shifts of either 180° or 360°.

Negative feedback

Many practical amplifiers use negative feedback in order to precisely control the gain, reduce distortion and improve bandwidth. The gain can be reduced to a manageable value by feeding back a small

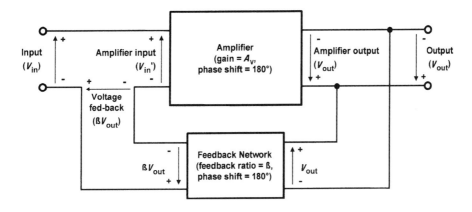

Figure 7.13   Amplifier with negative feedback applied

proportion of the output. The amount of feedback determines the overall (or **closed-loop**) gain. Because this form of feedback has the effect of reducing the overall gain of the circuit, this form of feedback is known as **negative feedback**. An alternative form of feedback, where the output is fed back in such a way as to reinforce the input (rather than to subtract from it) is known as **positive feedback**. This form of feedback is used in oscillator circuits (see Chapter 9).

Figure 7.13 shows the block diagram of an amplifier stage with negative feedback applied. In this circuit, the proportion of the output voltage fed back to the input is given by β and the overall voltage gain will be given by:

Overall gain, $G = \dfrac{V_{out}}{V_{in}}$

Now $V_{in}' = V_{in} - \beta V_{out}$ (by applying Kirchhoff's Voltage Law) (note that the amplifier's input voltage has been *reduced* by applying negative feedback) thus:

$V_{in} = V_{in}' + \beta V_{out}$

and

$V_{out} = A_v \times V_{in}$ (Note that A_v is the **internal gain** of the amplifier)

Hence:

Overall gain, $G = \dfrac{A_v \times V'_{in}}{V'_{in} + \beta V_{out}} = \dfrac{A_v \times V'_{in}}{V'_{in} + \beta \left(A_v \times V'_{in} \right)}$

Thus:

$$G = \frac{A_v}{1 + \beta A_v}$$

Hence, the overall gain with negative feedback applied will be less than the gain without feedback. Furthermore, if A_v is very large (as is the case with an operational amplifier, see Chapter 8) the overall gain with negative feedback applied will be given by:

$G = 1/\beta$ (when A_v is very large)

Note, also, that the **loop gain** of a feedback amplifier is defined as the product of β and A_v.

Example 7.3

An amplifier with negative feedback applied has an open-loop voltage gain of 50 and one-tenth of its output is fed back to the input (i.e. $\beta = 0.1$). Determine the overall voltage gain with negative feedback applied.

Solution

With negative feedback applied the overall voltage gain will be given by:

$$G = \frac{A_v}{1+\beta A_v} = \frac{50}{1+(0.1\times50)} = \frac{50}{6} = 8.33$$

Example 7.4

If, in Example 7.3, the amplifier's open-loop voltage gain increases by 20%, determine the percentage increase in overall voltage gain.

Solution

The new value of voltage gain will be given by:

$$A_v = A_v + 0.2A_v = 1.2 \times 50 = 60$$

The overall voltage gain with negative feedback will then be:

$$G = \frac{A_v}{1+\beta A_v} = \frac{60}{1+(0.1\times60)} = \frac{60}{7} = 7.14$$

The increase in overall voltage gain, expressed as a percentage, will thus be:

$$\frac{8.57-8.33}{8.33}\times100\% = 2.88\%$$

Note that this example illustrates one of the important benefits of negative feedback in stabilizing the overall gain of an amplifier stage.

Example 7.5

An integrated circuit that produces an open-loop gain of 100 is to be used as the basis of an amplifier stage having a precise voltage gain of 20. Determine the amount of feedback required.

Solution

Re-arranging the formula, $G = \dfrac{A_v}{1+\beta A_v}$

to make β the subject gives:

$$\beta = \frac{1}{G} - \frac{1}{A_v}$$

thus

$$\beta = \frac{1}{20} - \frac{1}{100} = 0.05 - 0.01 = 0.04$$

Transistor amplifiers

Regardless of what type of transistor is employed, three basic circuit configurations are used. These three circuit configurations depend upon which one of the three transistor connections is made common to both the input and the output. In the case of bipolar transistors, the configurations are known as **common emitter**, **common collector** (or **emitter follower**) and **common base**. Where field effect

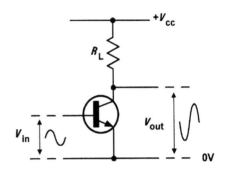

Figure 7.14 Common-emitter configuration

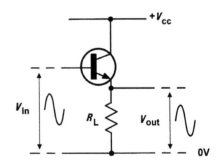

Figure 7.15 Common-collector (emitter follower) configuration

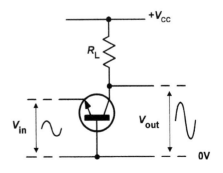

Figure 7.16 Common-base configuration

transistors are used, the corresponding configurations are **common source**, **common drain** (or **source follower**) and **common gate**.

The three basic circuit configurations (Figs 7.14 to 7.19) exhibit quite different performance characteristics, as shown in Tables 7.2 and 7.3 (typical values are given in brackets).

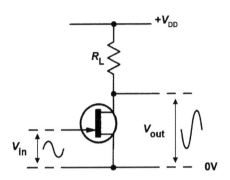

Figure 7.17 Common-source configuration

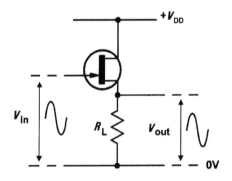

Figure 7.18 Common-drain (source follower) configuration

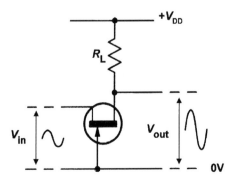

Figure 7.19 Common-gate configuration

Equivalent circuits

One method of determining the behaviour of an amplifier stage is to make use of an equivalent circuit. Figure 7.20 shows the basic equivalent circuit of an amplifier. The output circuit is reduced to its Thévenin equivalent (see Chapter 3) comprising a voltage generator ($A_v \times V_{in}$) and a series input resistance, resistance (R_{out}). This simple model allows us to forget the complex circuitry that might exist within the amplifier box!

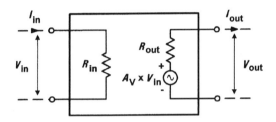

Figure 7.20 Common-source configuration

In practice, we use a slightly more complex equivalent circuit model in the analysis of a transistor amplifier. The most frequently used equivalent circuit is that which is based on **hybrid parameters** (or **h-parameters**). In this form of analysis, a transistor is replaced by four components; h_i, h_r, h_f and h_o (see Table 7.4). In order to indicate which one of the operating modes is used we add a further subscript letter to each h-parameter; e for common emitter, b for common base and c for common collector (see Table 7.5). However, to keep things simple, we will only consider common-emitter operation here.

Common-emitter input resistance (h_{ie})

The input resistance of a transistor is the resistance that is effectively 'seen' between its input terminals. As such, it is the ratio of the voltage between the input terminals to the current flowing into the input. In the case of a transistor operating in common-emitter mode, the input voltage is the voltage developed between the base and emitter, V_{be}, while the input current is the current supplied

Table 7.2 BJT amplifier circuit configurations

Parameter	Mode of operation		
	Common emitter (Fig. 7.14)	Common collector (Fig. 7.15)	Common base (Fig. 7.16)
Voltage gain	medium/high (40)	unity (1)	high (200)
Current gain	high (200)	high (200)	unity (1)
Power gain	very high (8,000)	high (200)	high (200)
Input resistance	medium (2.5 kΩ)	high (100 kΩ)	low (200 Ω)
Output resistance	medium/high (20 kΩ)	low (100 Ω)	high (100 kΩ)
Phase shift	180°	0°	0°
Typical applications	General-purpose AF and RF amplifiers	Impedance matching; input and output stages	RF and VHF/UHF amplifiers

Table 7.3 JFET amplifier circuit configurations

Parameter	Mode of operation		
	Common source (Fig. 7.17)	Common drain (Fig. 7.18)	Common gate (Fig. 7.19)
Voltage gain	medium (40)	unity (1)	high (250)
Current gain	very high (200,000)	very high (200,000)	unity (1)
Power gain	very high (800,000)	very high (200,000)	high (250)
Input resistance	very high (1 MΩ)	very high (10 MΩ)	low (500 Ω)
Output resistance	medium/high (50 kΩ)	low (200 Ω)	high (150 kΩ)
Phase shift	180°	0°	0°
Typical applications	General-purpose AF and RF amplifiers	Impedance matching; input and output stages	RF and VHF/UHF amplifiers

Table 7.4 General hybrid parameters

h_i input resistance, $\dfrac{\Delta V_{in}}{\Delta I_{in}}$

h_r reverse voltage transfer ratio, $\dfrac{\Delta V_{in}}{\Delta V_{out}}$

h_f forward current transfer ratio, $\dfrac{\Delta I_{out}}{\Delta I_{in}}$

h_o output conductance, $\dfrac{\Delta I_{out}}{\Delta V_{out}}$

Table 7.5 Common emitter mode h-parameters

h_{ie} input resistance, $\dfrac{\Delta V_{be}}{\Delta I_b}$

h_{re} reverse voltage transfer ratio, $\dfrac{\Delta V_{be}}{\Delta V_{ce}}$

h_{fe} forward current transfer ratio, $\dfrac{\Delta I_c}{\Delta I_b}$

h_{oe} output conductance, $\dfrac{\Delta I_c}{\Delta V_{ce}}$

to the base, I_b.

Figure 7.21 shows the current and voltage at the input of a common-emitter amplifier stage while Fig. 7.22 shows how the small-signal input resistance, R_{in}, appears between the base and emitter. Note that R_{in} is not a discrete component it is inside the transistor. From the foregoing we can deduce that:

$$R_{in} = \frac{V_{be}}{I_b}$$

(note that this is similar to the expression for h_{ie}).

The transistor's input characteristic can be used to predict the input resistance of a transistor amplifier stage. Since the input characteristic is non-linear (recall that very little happens until the base-emitter voltage exceeds 0.6 V), the value of input resistance will be very much dependent on the exact point on the graph at which the transistor is being operated. Furthermore, we might expect quite different values of resistance according to whether we are dealing with larger d.c. values or smaller incremental changes (a.c. values). Since this can be a rather difficult concept, it is worth expanding on it.

Figure 7.23 shows a typical input characteristic in which the transistor is operated with a base current, I_B, of 50 μA. This current produces a base-emitter voltage, V_{BE}, of 0.65 V. The input resistance corresponding to these steady (d.c.) values will be given by:

$$R_{in} = \frac{V_{BE}}{I_B} = \frac{0.65\ V}{50\ \mu A} = 13\ k\Omega$$

Now, suppose that we apply a steady bias current of, say, 70 μA and superimpose on this a signal that varies above and below this value, swinging through a total change of 100 μA (i.e. from 20 μA to 120 μA). Figure 7.24 shows that this produces a base-emitter voltage change of 0.05 V. The input resistance seen by this small-signal input current is given by:

$$R_{in} = \frac{\text{change in } V_{be}}{\text{change in } I_b} = \frac{\Delta V_{be}}{\Delta I_b} = \frac{0.05\ V}{100\ \mu A} = 500\ \Omega$$

In other words:

$h_{ie} = 500\ \Omega$ (since $h_{ie} = \dfrac{\Delta V_{be}}{\Delta I_b}$)

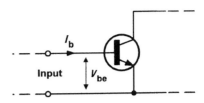

Figure 7.21 Voltage and current at the input of a common-emitter amplifier

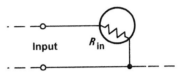

Figure 7.22 Input resistance of a common-emitter amplifier stage

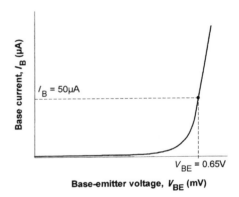

Figure 7.23 Using the input characteristic to determine the large-signal (static) input resistance of a transistor connected in common-emitter mode

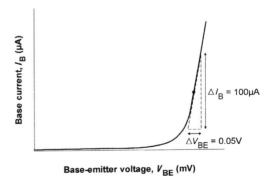

Figure 7.24 Using the input characteristic to determine the small-signal input resistance of a transistor connected in common-emitter mode

It is worth comparing this value with the steady (d.c.) value. The appreciable difference is entirely attributable to the shape of the input characteristic!

Common-emitter current gain (h_{ie})

The current gain produced by a transistor is the ratio of output current to input current. In the case of a transistor operating in common-emitter mode, the input current is the base current, I_b, while the output current is the collector current, I_c.

Figure 7.25 shows the small-signal input and output currents and voltages for a common-emitter amplifier stage. The magnitude of the current produced at the output of the transistor is equal to the current gain, A_i, multiplied by the applied base current, I_b. Since the output current is the current flowing in the collector, I_c, we can deduce that:

$$I_c = A_i \times I_b$$

where $A_i = h_{fe}$ (the common-emitter current gain).

Figure 7.26 shows how this current source appears between the collector and emitter. Once again, the current source is not a discrete component—it appears *inside* the transistor.

The transistor's transfer characteristic can be used to predict the current gain of a transistor amplifier stage. Since the transfer characteristic is linear, the current gain remains reasonably constant over a range of collector current. Figure 7.27 shows a typical transfer characteristic in which the transistor is operated with a base current, I_B, of 240 μA. This current produces a collector current, I_C, of 12 mA. The current gain corresponding to these steady (d.c.) values will be given by:

$$A_i = \frac{I_C}{I_B} = \frac{2.5 \text{ mA}}{50 \text{ μA}} = 50$$

(note that this is similar to the expression for h_{fe}).

Now, suppose that we apply a steady bias current of, say, 240 μA and superimpose on this a signal that varies above and below this value, swinging through a total change of 220 μA (i.e. from 120 μA to 360 μA). Figure 7.28 shows that this produces a collector current swing of 10 mA.

The small-signal a.c. current gain is given by:

$$A_i = \frac{\text{change in } I_c}{\text{change in } I_b} = \frac{\Delta I_c}{\Delta I_b} = \frac{10 \text{ mA}}{220 \text{ μA}} = 45.45$$

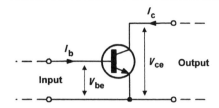

Figure 7.25 Input and output currents and voltages in a common-emitter amplifier stage

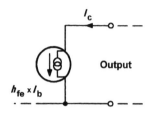

Figure 7.26 Equivalent output current source in a common-emitter amplifier stage

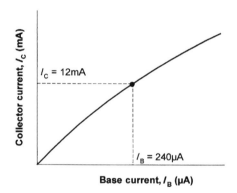

Figure 7.27 Using the transfer characteristic to determine the large-signal (static) current gain of a transistor connected in common-emitter mode

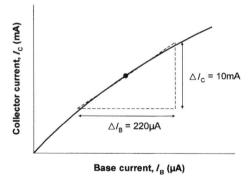

Figure 7.28 Using the transfer characteristic to determine the small-signal current gain of a transistor connected in common-emitter mode

Once again, it is worth comparing this value with the steady-state value (h_{FE}). Since the transfer characteristic is reasonably linear, the values are quite close (45.45 compared with 50). However, if the transfer characteristic was *perfectly* linear the value of h_{fe} would be *exactly* the same as that for h_{FE}.

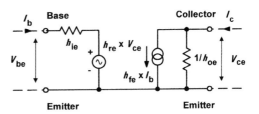

Figure 7.29 *h*-parameter equivalent circuit for a transistor amplifier

h-parameter equivalent circuit for a transistor in common-emitter mode

A complete *h*-parameter equivalent circuit for a transistor operating in common-emitter mode is shown in Fig. 7.29. We have already shown how the two most important parameters, h_{ie} and h_{fe}, can be found from the transistor's characteristic curves.

The remaining parameters, h_{re} and h_{oe}, can, in many applications, be ignored. A typical set of *h*-parameters for a BFY50 transistor is shown in Table 7.6. Note how small h_{re} and h_{oe} are for a real transistor!

Table 7.6 *h*-parameters for a BFY50 transistor

h_{ie} (Ω)	h_{re}	h_{fe}	h_{oe} (S)
250	0.85×10^{-4}	80	35×10^{-6}

Measured at $I_C = 1$ mA, $V_{CE} = 5$ V

Example 7.6

A BFY50 transistor is used in a common-emitter amplifier stage with $R_L = 10$ kΩ and $I_C = 1$ mA. Determine the output voltage produced by an input signal of 10 mV. (You may ignore the effect of h_{re} and any bias components that may be present externally.)

Solution

The equivalent circuit (with h_{re} replaced by a short circuit) is shown in Fig. 7.30. The load effectively appears between the collector and emitter while the input signal appears between base and emitter. First we need to find the value of input current, I_b, from:

$$I_b = \frac{V_{in}}{h_{ie}} = \frac{10 \text{ mV}}{250 \text{ }\Omega} = 40 \text{ }\mu A$$

Next we find the value of current generated, I_f, from:

$$I_f = h_{fe} \times I_b = 80 \times 40 \text{ }\mu A = 320 \text{ }\mu A$$

This value of current is shared between the internal resistance between collector and emitter (i.e. $1/h_{oe}$) and the external load, R_L. To determine the value of

collector current, we can apply the current divider theorem (Chapter 3):

$$I_c = I_f \times \frac{\frac{1}{h_{oe}}}{\frac{1}{h_{oe}} + R_L} = 320 \text{ }\mu A \times \frac{\frac{1}{(80 \times 10^{-6})}}{\frac{1}{(80 \times 10^{-6})} + 10 \text{ k}\Omega}$$

Thus:

$$I_c = 320 \text{ }\mu A \times \frac{12.5 \text{ k}\Omega}{12.5 \text{ k}\Omega + 10 \text{ k}\Omega}$$

from which:

Ic = 320 μA × 0.555 = 177.6 μA

Finally, we can determine the output voltage from:

$V_{out} = I_c \times R_L = 177.6 \text{ }\mu A \times 10 \text{ k}\Omega = 1.776$ V

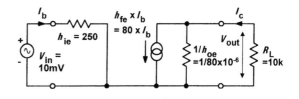

Figure 7.30 See Example 7.6

Voltage gain

We can use hybrid parameters to determine the voltage gain of a transistor stage. We have already shown how the voltage gain of an amplifier stage is given by:

$$A_v = \frac{V_{out}}{V_{in}}$$

In the case of a common-emitter amplifier stage, $V_{out} = Vi_{ce}$ and $V_{in} = V_{be}$. If we assume that h_{oe} and h_{re} are negligible, then:

$$V_{out} = V_{ce} = I_c \times R_L = I_f \times R_L = h_{fe} \times I_b \times R_L$$

and

$$V_{in} = V_{be} = I_b \times R_{in} = I_b \times h_{ie}$$

Thus:

$$A_v = \frac{V_{out}}{V_{in}} = \frac{h_{fe} \times I_b \times R_L}{I_b \times h_{ie}} = \frac{h_{fe} \times R_L}{h_{ie}}$$

Example 7.7

A transistor has $h_{fe} = 150$ and $h_{ie} = 1.5$ kΩ. Assuming that hre and hoe are both negligible, determine the value if load resistance required to produce a voltage gain of 200.

Solution

Re-arranging $A_v = \dfrac{h_{fe} \times R_L}{h_{ie}}$ to make R_L the subject gives:

$$R_L = \frac{A_v \times h_{ie}}{h_{fe}}$$

For a voltage gain of 200 the value of load resistance can be determined from:

$$R_L = \frac{200 \times 1.5 \text{ k}\Omega}{150} = 2 \text{ k}\Omega$$

Bias

We stated earlier that the optimum value of bias for a Class A (linear) amplifier is that value which ensures that the active devices are operated at the midpoint of their transfer characteristics. In

practice, this means that a static value of collector current will flow even when there is no signal present. Furthermore, the collector current will flow throughout the complete cycle of an input signal (i.e. conduction will take place over an angle of 360°). At no stage will the transistor be **saturated** nor should it be **cut-off** (i.e. the state in which *no* collector current flows).

In order to ensure that a static value of collector current flows in a transistor, a small current must therefore be applied to the base of the transistor. This current can be derived from the same voltage rail that supplies the collector circuit (via the load). Figure 7.31 shows a simple Class A common-emitter amplifier circuit in which the base bias resistor, $R1$, and collector load resistor, $R2$, are connected to a common positive supply rail.

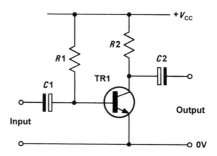

Figure 7.31   Basic Class-A common-emitter amplifier

The signal is applied to the base terminal of the transistor via a coupling capacitor, $C1$. This capacitor removes the d.c. component of any signal applied to the input terminals and ensures that the base bias current delivered by $R1$ is unaffected by any device connected to the input. $C2$ couples the signal out of the stage and also prevents d.c. current flowing appearing at the output terminals.

In order to stabilize the operating conditions for the stage and compensate for variations in transistor parameters, base bias current for the transistor can be derived from the voltage at the collector (see Fig. 7.32). This voltage is dependent on the collector current which, in turn, depends upon the base current. A negative feedback loop thus exists in which there is a degree of self-regulation. If the collector current increases, the collector voltage will fall and the base current will be reduced. The

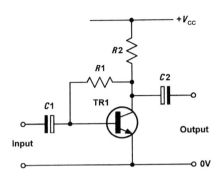

Figure 7.32 An improvement on the circuit shown in Fig. 7.31 (using negative feedback to bias the transistor)

reduction in base current will produce a corresponding reduction in collector current to offset the original change. Conversely, if the collector current falls, the collector voltage will rise and the base current will increase. This, in turn, will produce a corresponding increase in collector current to offset the original change.

The negative feedback path in Fig. 7.32 provides feedback that involves an a.c. (signal) component as well as the d.c. bias. As a result of the a.c. feedback, there is a slight reduction in signal gain. The signal gain can be increased by removing the a.c. signal component from the feedback path so that only the d.c. bias component is present. This can be achieved with the aid of a bypass capacitor as shown in Fig. 7.33. The value of bypass capacitor, $C1$, is chosen so that the component exhibits a very low reactance at the lowest signal frequency when compared with the series base bias

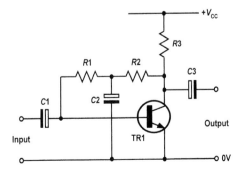

Figure 7.33 Improved version of Fig. 7.31

resistance, $R1$. The result of this potential divider arrangement is that the a.c. signal component is effectively bypassed to ground.

Figure 7.34 shows an improved form of transistor amplifier in which d.c. negative feedback is used to stabilize the stage and compensate for variations in transistor parameters, component values and temperature changes. $R1$ and $R2$ form a potential divider that determines the d.c. base potential, V_B. The base-emitter voltage (V_{BE}) is the difference between the potentials present at the base (V_B) and emitter (V_E). The potential at the emitter is governed by the emitter current (I_E). If this current increases, the emitter voltage (V_E) will increase and, as a consequence V_{BE} will fall. This, in turn, produces a reduction in emitter current which largely offsets the original change. Conversely, if the emitter current decreases, the emitter voltage (V_E) will decrease and V_{BE} will increase (remember that V_B remains constant). The increase in bias results in an increase in emitter current compensating for the original change.

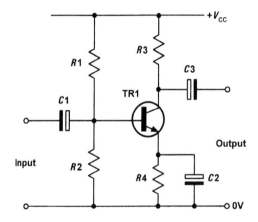

Figure 7.34 A common-emitter amplifier stage with effective bias stabilization

Example 7.8

Determine the static value of current gain and collector voltage in the circuit shown in Fig. 7.35.

Solution

Since 2 V appears across $R4$, we can determine the emitter current easily from:

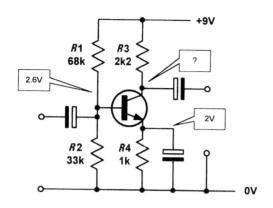

Figure 7.35 See Example 7.8

$$I_E = \frac{V_E}{R_4} = \frac{2\ V}{1\ k\Omega} = 2\ mA$$

Next we should determine the base current. This is a little more difficult. The base current is derived from the potential divider formed by $R1$ and $R2$. The potential at the junction of $R1$ and $R2$ is 2.6 V hence we can determine the currents through $R1$ and $R2$, the difference between these currents will be equal to the base current.

The current in $R2$ will be given by:

$$I_{R2} = \frac{V_B}{R2} = \frac{2.6\ V}{33\ k\Omega} = 79\ \mu A$$

The current in $R1$ will be given by:

$$I_{R1} = \frac{9\ V - V_B}{R1} = \frac{6.4\ V}{68\ k\Omega} = 94.1\ \mu A$$

Hence the base current is found from:

$$I_B = 94.1\ \mu A - 79\ \mu A = 15.1\ \mu A$$

Next we can determine the collector current from:

$$h_{FE} = \frac{I_C}{I_B} = \frac{2.0151\ mA}{15.1\ \mu A} = 133.45$$

Finally we can determine the collector voltage by subtracting the voltage dropped across $R3$ from the 9 V supply.

The voltage dropped across $R3$ will be:

$$V_{R4} = I_C \times R4 = 2.0151\ mA \times 2.2\ k\Omega = 4.43\ V$$

Hence $V_C = 9\ V - 4.43\ V = 4.57\ V$

Predicting amplifier performance

The a.c. performance of an amplifier stage can be predicted using a load line superimposed on the relevant set of output characteristics. For a bipolar transistor operating in common-emitter mode the required characteristics are I_C plotted against V_{CE} One end of the load line corresponds to the supply voltage (V_{CC}) while the other end corresponds to the value of collector or drain current that would flow with the device totally saturated. In this condition:

$$I_C = \frac{V_{CC}}{R_L}$$

where R_E is the value of collector or drain load resistance.

Figure 7.36 shows a load line superimposed on a set of output characteristics for a bipolar transistor operating in common-emitter mode. The quiescent point (or operating point) is the point on the load line that corresponds to the conditions that exist when no-signal is applied to the stage. In Fig. 7.36, the base bias current is set at 20 μA so that the quiescent point effectively sits roughly halfway along the load line. This position ensures that the collector voltage can swing both positively (above) and negatively (below) its quiescent value (V_{CQ}).

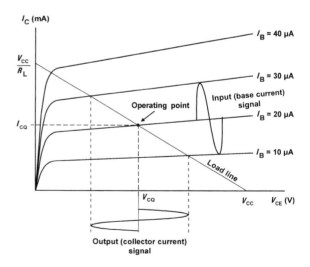

Figure 7.36 Operating point and quiescent values shown on the load line for a bipolar transistor operating in common-emitter mode

The effect of superimposing an alternating base current (of 20 μA peak-peak) to the d.c. bias current (of 20 μA) can be clearly seen. The corresponding collector current signal can be determined by simply moving up and down the load line.

Example 7.9

The characteristic curves shown in Fig. 7.37 relate to a transistor operating in common-emitter mode. If the transistor is operated with $I_B = 30$ μA, a load resistor of 1.2 kΩ and an 18 V supply, determine the quiescent values of collector voltage and current (V_{CQ} and I_{CQ}). Also determine the peak-peak output voltage that would be produced by an input signal of 40 μA peak-peak.

Solution

First we need to construct the load line. The two ends of the load line will correspond to V_{CC} (18 V) on the collector-emitter voltage axis and (18V/1.2 kΩ or 15 mA) on the collector current axis. Next we locate the **operating point** (or **quiescent point**) from the point of intersection of the $I_B = 30$ μA characteristic and the load line.

Having located the operating point we can read off the quiescent (no-signal) values of collector emitter voltage (V_{CQ}) and collector current (I_{CQ}). Hence:

$$V_{CQ} = 9.2 \text{ V and } I_{CQ} = 7.3 \text{ mA}$$

Next we can determine the maximum and minimum values of collector-emitter voltage by locating the appropriate intercept points on Fig. 7.37.

Note that the maximum and minimum values of base current will be (30 μA + 20 μA) on positive peaks of the signal and (30 μA − 20 μA) on negative peaks of the signal. The maximum and minimum values of V_{CE} are, respectively, 14.8 V and 3.3 V. Hence the output voltage swing will be (14.8 V − 3.3 V) or 11.5 V peak-peak.

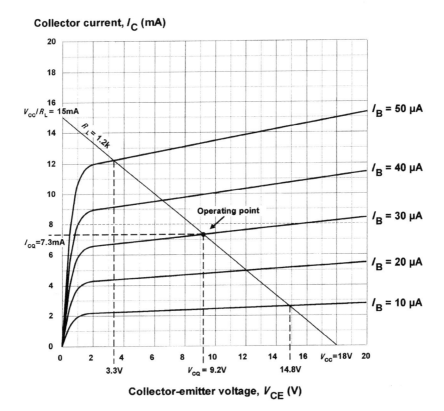

Figure 7.37   See Example 7.9

Practical amplifier circuits

The simple common-emitter amplifier stage shown in Fig. 7.38 provides a modest voltage gain (80 to 120 typical) with an input resistance of approximately 1.5 kΩ and an output resistance of around 20 kΩ. The frequency response extends from a few hertz to several hundred kilohertz. The improved arrangement shown in Fig. 7.38 provides a voltage gain of around 150 to 200 by eliminating the signal frequency negative feedback that occurs through $R1$ in Fig. 7.39.

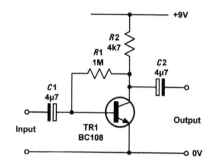

Figure 7.38 A practical common-emitter amplifier stage

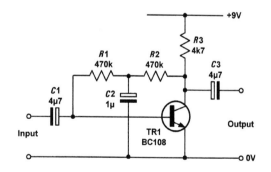

Figure 7.39 An improved version of the common-emitter amplifier stage

Figure 7.40 shows a practical common-emitter amplifier with bias stabilization. This stage provides a gain of 150 to well over 200 (depending upon the current gain, h_{ie}, of the individual transistor used). The circuit will operate with supply voltages of between 6 V and 18 V.

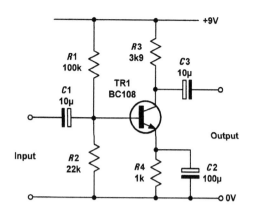

Figure 7.40 Operating point and quiescent values shown on the load line for a bipolar transistor operating in common-emitter mode

Two practical emitter-follower circuits are shown in Figs 7.41 and 7.42. These circuits offer a voltage gain of unity (1) but are ideal for matching a high resistance source to a low resistance load. It is important to note that the input resistance varies with the load connected to the output of the circuit (it is typically in the range 50 kΩ to 150 kΩ). The input resistance can be calculated by multiplying h_{fe} by the effective resistance of $R2$ in parallel with the load connected to the output terminals.

Figure 7.42 is an improved version of Fig. 7.41 in which the base current is derived from the potential divider formed by $R1$ and $R2$. Note, however, that the input resistance is reduced since $R1$ and $R2$ effectively appear in parallel with the input. The input resistance of the stage is thus typically in the region of 40 kΩ to 70 kΩ.

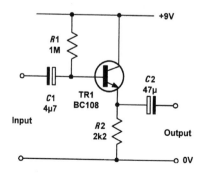

Figure 7.41 A practical emitter-follower stage

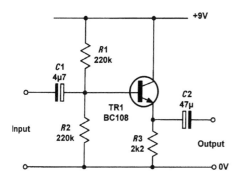

Figure 7.42   An improved emitter-follower stage

Multi-stage amplifiers

In order to provide sufficiently large values of gain, it is frequently necessary to use a number of interconnected stages within an amplifier. The overall gain of an amplifier with several stages (i.e. a multi-stage amplifier) is simply the product of the individual voltage gains. Hence:

$$A_V = A_{V1} \times A_{V2} \times A_{V3} \text{ etc.}$$

Note, however, that the bandwidth of a multi-stage amplifier will be less than the bandwidth of each individual stage. In other words, an increase in gain can only be achieved at the expense of a reduction in bandwidth.

Signals can be coupled between the individual stages of a multi-stage amplifier using one of a number of different methods shown in Fig. 7.43.

The most commonly used method is that of **R–C coupling** as shown in Fig. 7.43(a). In this coupling method, the stages are coupled together using capacitors having a low reactance at the signal frequency and resistors (which also provide a means of connecting the supply). Figure 7.44 shows a practical example of this coupling method.

A similar coupling method, known as **L–C coupling**, is shown in Fig. 7.43(b). In this method, the inductors have a high reactance at the signal frequency. This type of coupling is generally only used in RF and high-frequency amplifiers.

Two further methods, **transformer coupling** and **direct coupling** are shown in Figures 7.43(c) and 7.43(d) respectively. The latter method is used where d.c. levels present on signals must be preserved.

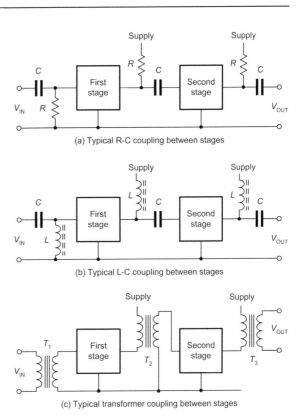

(a) Typical R-C coupling between stages

(b) Typical L-C coupling between stages

(c) Typical transformer coupling between stages

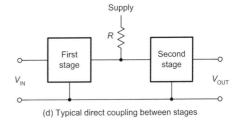

(d) Typical direct coupling between stages

Figure 7.43   Different methods used for inter-stage coupling

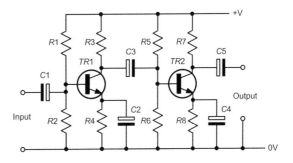

Figure 7.44   A typical two-stage high-gain R–C coupled common-emitter amplifier

Power amplifiers

The term 'power amplifier' can be applied to any amplifier that is designed to deliver an appreciable level of power. There are several important considerations for amplifiers of this type, including the ability to deliver current (as well as voltage) to a load, and also the need to operate with a reasonable degree of efficiency (recall that conventional Class A amplifiers are inefficient).

In order to deliver sufficient current to the load, power amplifiers must have a very low value of output impedance. Thus the final stage (or **output stage**) is usually based on a device operating in emitter-follower configuration. In order to operate at a reasonable level of efficiency, the output stage must operate in Class AB or Class B mode (see page 133). One means of satisfying both of these requirements is with the use of a symmetrical output stage based on complementary NPN and PNP devices.

A simple **complementary output stage** is shown in Fig. 7.45. TR1 is a suitably rated NPN device whilst TR2 is an identically rated PNP device. Both TR1 and TR2 operate as emitter followers (i.e. common-collector mode) with the output taken from the two emitters, coupled via C2 to the load. In order to bias TR1 and TR2 into Class AB mode two silicon diodes (D1 and D2) are used to provide a constant voltage drop of approximately 1.2 V between the two bases. This voltage drop is required between the bases of TR1 and TR2 in order to bring them to conduction. Since D1 and D2 are both in forward conduction (with current supplied via $R1$ and $R2$) they have little effect on the input signal (apart from shifting the d.c. level).

Figure 7.46 shows an improvement of the basic complementary output stage with the addition of a driver stage (TR1) and a means of adjusting the bias (i.e. operating point) of the two output transistors. VR1 is typically adjusted in order to produce an output stage collector current of between 15 mA and 50 mA (required for Class AB operation). With VR1 set to minimum resistance, the output stage will operate in Class B (this will produce significantly more cross-over distortion because the two devices may both be cut-off for a brief period of each cycle). Figure 7.47 shows how negative feedback bias (via $R2$) can be added in order to stabilize the output stage. Practical power amplifiers are shown in Figs 7.48 and 7.49.

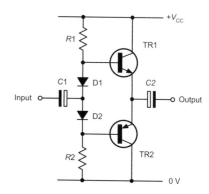

Figure 7.45 A complementary output stage

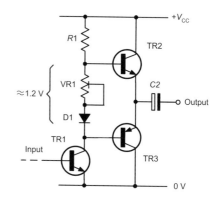

Figure 7.46 A complementary output stage with adjustable bias for the output transistors

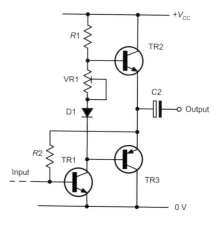

Figure 7.47 A complementary output stage with stabilized bias for the driver stage

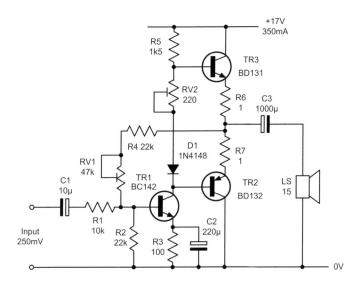

Figure 7.48 A simple power amplifier based on a Class AB complementary output stage and Class A driver stage. This amplifier provides an output of 3 W into an 8 ohm load and has a frequency response extending from 20 Hz to 50 kHz. Total harmonic distortion (THD) is less than 0.2% at 250 mW output. An input of 2 V peak-peak is required for full output.

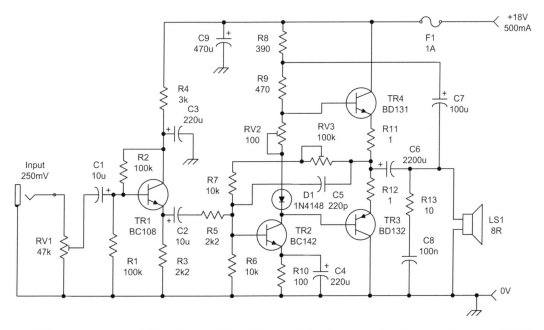

Figure 7.49 An improved 3 W audio amplifier. This amplifier has a nominal input resistance of 50 kΩ and a frequency response extending from 30 Hz to 30 kHz at the −3dB power points. Capacitor C5 is added to 'roll-off' the high frequency response (without this component the high-frequency response extends to around 100 kHz). RV2 is adjusted for a quiescent (no-signal) collector current for TR3 (and TR4) of 25 mA. RV3 is adjusted for a d.c. voltage of exactly 9 V at the junction of the two low-value thermal compensating resistors, R11 and R12. C8 and R13 form a Zobel network to provide frequency compensation of the output load.

Practical investigation

Objective

To measure the voltage gain and low-frequency response of a simple common-emitter amplifier stage.

Components and test equipment

Breadboard, d.c. voltmeter (preferably digital), AF signal generator (with variable frequency sine wave output), two AF voltmeters (or a dual beam oscilloscope), 9 V d.c. power supply (or battery), BC108 (or similar) NPN transistor, two 470 nF and two 10 μF capacitors, resistors of 1 MΩ and 2.2 kΩ 5% 0.25 W, test leads, connecting wire.

Procedure

Connect the circuit shown in Fig. 7.50. Without the signal generator connected, measure and record the no-signal d.c. collector, base and emitter voltages for TR1 (the collector voltage should be in the range 3 V to 6 V). Connect the signal generator and set it to produce a sine wave output at 1 kHz. Increase the output voltage from the signal generator until the input voltmeter reads exactly 10 mV. Measure and record the output signal voltage (see Table 7.7).

Repeat at frequencies over the range 10 Hz to 10 kHz, at each stage recording the output voltage produced in the table. Replace $C1$ and $C2$ with 10 μF capacitors and repeat the measurements.

Measurements and calculations

Use the measured value of output voltage at 1 kHz in order to determine the voltage gain of the stage.

Graphs

For each set of measurements (470 nF and 10 μF for $C1$ and $C2$) plot a graph showing the frequency response of the amplifier stage. In each case, use the graph to determine the low-frequency cut-off.

Conclusion

Comment on the no-signal voltages measured at the collector, base and emitter. Are these what you would expect? Comment on the value of voltage gain that you have obtained. Is this what you would expect? Comment on the shape of each frequency response graph. Explain why there is a difference in cut-off frequency.

Table 7.7   Results

Frequency (Hz)	Output voltage (V)
10	
20	
50	
100	
200	
500	
1 k	
2 k	
5 k	
10 k	

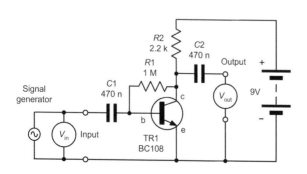

Figure 7.50   Transistor test circuit

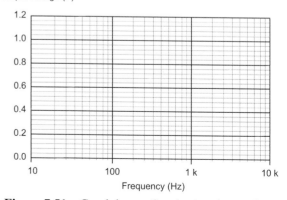

Figure 7.51   Graph layout for plotting the results

Important formulae introduced in this chapter

Voltage gain:
(page 132)

$$A_v = \frac{V_{out}}{V_{in}}$$

Current gain:
(page 132)

$$A_i = \frac{I_{out}}{I_{in}}$$

Power gain:
(page 132)

$$A_p = \frac{P_{out}}{P_{in}}$$

$$A_p = A_i \times A_v$$

Gain with negative feedback applied:
(page 137)

$$G = \frac{A_v}{1 + \beta A_v}$$

Gain (when A_v is very large):
(page 137)

$$G = 1/\beta$$

Loop gain:
(page 137)

$$G_{loop} = \beta \times A_v.$$

Input resistance (common-emitter):
(page 140)

$$h_{ie} = \frac{\Delta V_{be}}{\Delta I_b}$$

Forward current transfer ratio (common-emitter):
(page 140)

$$h_{fe} = \frac{\Delta I_c}{\Delta I_b}$$

Output conductance (common-emitter):
(page 140)

$$h_{oe} = \frac{\Delta I_c}{\Delta V_{ce}}$$

Reverse voltage transfer ratio (common-emitter):
(page 140)

$$h_{re} = \frac{\Delta V_{be}}{\Delta V_{ce}}$$

Voltage gain (common-emitter) assuming h_{re} and h_{oe} can be neglected:
(page 142)

$$A_v = \frac{h_{fe} \times R_L}{h_{ie}}$$

Symbols introduced in this chapter

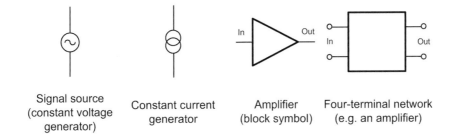

| Signal source (constant voltage generator) | Constant current generator | Amplifier (block symbol) | Four-terminal network (e.g. an amplifier) |

Figure 7.52 Circuit symbols introduced in this chapter

Problems

7.1 The following measurements were made during a test on an amplifier:

$V_{in} = 250$ mV, $I_{in} = 2.5$ mA,
$V_{out} = 10$ V, $I_{out} = 400$ mA

Determine:
(a) the voltage gain;
(b) the current gain;
(c) the power gain;
(d) the input resistance.

7.2 An amplifier has a power gain of 25 and identical input and output resistances of 600 Ω. Determine the input voltage required to produce an output of 10 V.

7.3 Determine the mid-band voltage gain and upper and lower cut-off frequencies for the amplifier whose frequency response curve is shown in Fig. 7.53. Also determine the voltage gain at frequencies of:
(a) 10 Hz
(b) 1 MHz

7.4 An amplifier with negative feedback applied has an open-loop voltage gain of 250, and 5% of its output is fed back to the input. Determine the overall voltage gain with negative feedback applied. If the open-loop voltage gain increases by 20% determine the new value of overall voltage gain.

7.5 An amplifier produces an open-loop gain of 180. Determine the amount of feedback required if it is to be operated with a precise voltage gain of 50.

7.6 A transistor has the following parameters:

$h_{ie} = 800$ Ω
$h_{re} = $ negligible
$h_{fe} = 120$
$h_{oe} = 50$ μS.

If the transistor is to be used as the basis of a common-emitter amplifier stage with $R_L = 12$ kΩ, determine the output voltage when an input signal of 2 mV is applied.

7.7 Determine the unknown current and voltages in Fig. 7.54.

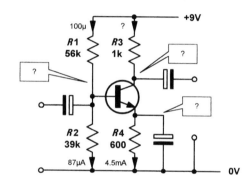

Figure 7.54 See Question 7.7

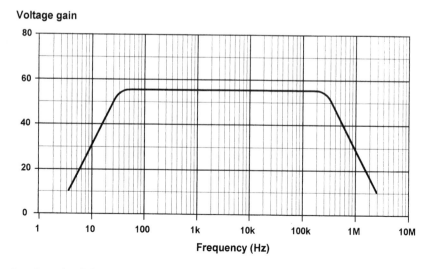

Figure 7.53 See Question 7.3

7.8 The output characteristics of a bipolar transistor are shown in Fig. 7.55. If this transistor is used in an amplifier circuit operating from a 12 V supply with a base bias current of 60 μA and a load resistor of 1 kΩ, determine the quiescent values of collector-emitter voltage and collector current. Also determine the peak-peak output voltage produced when an 80 μA peak-peak signal current is applied to the base of the transistor.

7.9 Figure 7.56 shows a simple audio power amplifier in which all of the semiconductor devices are silicon and all three transistors have an h_{FE} of 100. If $RV1$ is adjusted to produce 4.5 V at Test Point D, determine the base, emitter and collector currents and voltages for each transistor and the voltages that will appear at Test Points A to C.

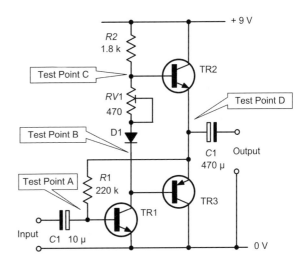

Figure 7.56 See Questions 7.9, 7.12 and 7.13

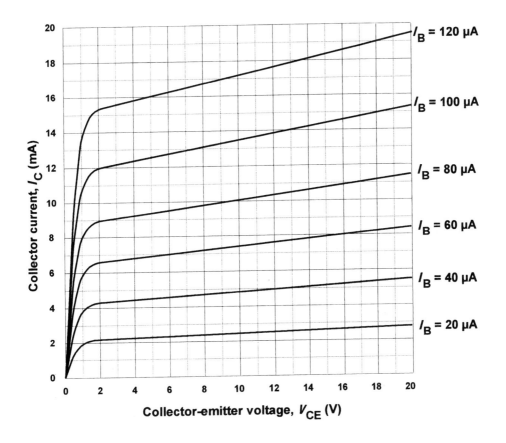

Figure 7.55 See Question 7.8

7.10   The output characteristics of a junction gate field effect transistor are shown in Fig. 7.57. If this JFET is used in an amplifier circuit operating from an 18V supply with a gate-source bias voltage of −3 V and a load resistor of 900 Ω, determine the quiescent values of drain-source voltage and drain current. Also determine the peak-peak output voltage when an input voltage of 2 V peak-peak is applied to the gate. Also determine the voltage gain of the stage.

7.11   A multi-stage amplifier consists of two R–C coupled common-emitter stages. If each stage has a voltage gain of 50, determine the overall voltage gain. Draw a circuit diagram of the amplifier and label your drawing clearly.

7.12   The following RMS voltage measurements were made during a signal test on the simple power amplifier shown in Fig. 7.56 when connected to a 15 Ω load:

$V_{in} = 50$ mV
$V_{out} = 2$ V

Determine:
(a) the voltage gain
(b) the output power
(c) the output current.

7.13   If the power amplifier shown in Fig. 7.56 produces a maximum RMS output power of 0.25 W, determine its overall efficiency if the supply current is 75 mA. Also determine the power dissipated in each of the output transistors in this condition.

Answers to these problems appear on page 375.

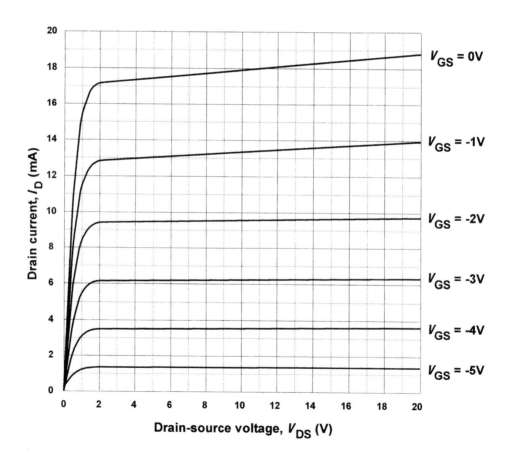

Figure 7.57   See Question 7.10

8

Operational amplifiers

Operational amplifiers are analogue integrated circuits designed for linear amplification that offer near-ideal characteristics (virtually infinite voltage gain and input resistance coupled with low output resistance and wide bandwidth).

Operational amplifiers can be thought of as universal 'gain blocks' to which external components are added in order to define their function within a circuit. By adding two resistors, we can produce an amplifier having a precisely defined gain. Alternatively, with two resistors and two capacitors we can produce a simple band-pass filter. From this you might begin to suspect that operational amplifiers are really easy to use. The good news is that they are!

Figure 8.1 A typical operational amplifier. This device is supplied in an 8-pin dual-in-line (DIL) package. It has a JFET input stage and produces a typical open-loop voltage gain of 200,000

Symbols and connections

The symbol for an operational amplifier is shown in Fig. 8.2. There are a few things to note about this. The device has two inputs and one output and no common connection. Furthermore, we often don't show the supply connections—it is often clearer to leave them out of the circuit altogether!

In Fig. 8.2, one of the inputs is marked '−' and the other is marked '+'. These polarity markings have nothing to do with the supply connections—they indicate the overall phase shift between each input and the output. The '+' sign indicates zero phase shift whilst the '−' sign indicates 180° phase shift. Since 180° phase shift produces an inverted waveform, the '−' input is often referred to as the **inverting input**. Similarly, the '+' input is known as the **non-inverting** input.

Most (but not all) operational amplifiers require a symmetrical supply (of typically ±6 V to ±15 V) which allows the output voltage to swing both positive (above 0 V) and negative (below 0 V). Figure 8.3 shows how the supply connections would appear if we decided to include them. Note that we usually have two separate supplies; a positive supply and an equal, but opposite, negative

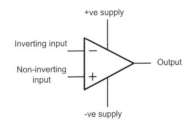

Figure 8.2 Symbol for an operational amplifier

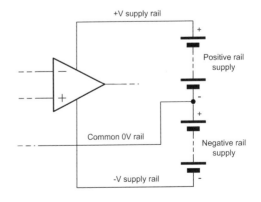

Figure 8.3 Supply connections for an operational amplifier

supply. The common connection to these two supplies (i.e., the 0 V supply connection) acts as the **common rail** in our circuit. The input and output voltages are usually measured relative to this rail.

Operational amplifier parameters

Before we take a look at some of the characteristics of 'ideal' and 'real' operational amplifiers it is important to define some of the terms and parameters that we apply to these devices.

Open-loop voltage gain

The open-loop voltage gain of an operational amplifier is defined as the ratio of output voltage to input voltage measured with no feedback applied. In practice, this value is exceptionally high (typically greater than 100,000) but is liable to considerable variation from one device to another.

Open-loop voltage gain may thus be thought of as the 'internal' voltage gain of the device, thus:

$$A_{V(OL)} = \frac{V_{OUT}}{V_{IN}}$$

where $A_{V(OL)}$ is the open-loop voltage gain, V_{OUT} and V_{IN} are the output and input voltages respectively under open-loop conditions.

In linear voltage amplifying applications, a large amount of negative feedback will normally be applied and the open-loop voltage gain can be thought of as the internal voltage gain provided by the device.

The open-loop voltage gain is often expressed in **decibels (dB)** rather than as a ratio. In this case:

$$A_{V(OL)} = 20 \log_{10} \frac{V_{OUT}}{V_{IN}}$$

Most operational amplifiers have open-loop voltage gains of 90 dB, or more.

Closed-loop voltage gain

The closed-loop voltage gain of an operational amplifier is defined as the ratio of output voltage to input voltage measured with a small proportion of the output fed back to the input (i.e. with feedback

applied). The effect of providing negative feedback is to reduce the loop voltage gain to a value that is both predictable and manageable. Practical closed-loop voltage gains range from one to several thousand but note that high values of voltage gain may make unacceptable restrictions on bandwidth, see later.

Closed-loop voltage gain is once again the ratio of output voltage to input voltage but with negative feedback is applied, hence:

$$A_{V(CL)} = \frac{V_{OUT}}{V_{IN}}$$

where $A_{V(CL)}$ is the open-loop voltage gain, V_{OUT} and V_{IN} are the output and input voltages respectively under closed-loop conditions. The closed-loop voltage gain is normally very much less than the open-loop voltage gain.

Example 8.1

An operational amplifier operating with negative feedback produces an output voltage of 2 V when supplied with an input of 400 µV. Determine the value of closed-loop voltage gain.

Solution

Now:

$$A_{V(CL)} = \frac{V_{OUT}}{V_{IN}}$$

Thus:

$$A_{V(CL)} = \frac{2}{400 \times 10^{-6}} = \frac{2 \times 10^6}{400} = 5,000$$

Expressed in decibels (rather than as a ratio) this is:

$$A_{V(CL)} = 20 \log_{10}(5,000) = 20 \times 3.7 = 74 \text{ dB}$$

Input resistance

The input resistance of an operational amplifier is defined as the ratio of input voltage to input current expressed in ohms. It is often expedient to assume that the input of an operational amplifier is purely resistive though this is not the case at high frequencies where shunt capacitive reactance may become significant. The input resistance of operational amplifiers is very much dependent on the semiconductor technology employed. In practice values range from about 2 MΩ for common bipolar types to over 10^{12} Ω for FET and CMOS devices.

Input resistance is the ratio of input voltage to input current:

$$R_{IN} = \frac{V_{IN}}{I_{IN}}$$

where R_{IN} is the input resistance (in ohms), V_{IN} is the input voltage (in volts) and I_{IN} is the input current (in amps). Note that we usually assume that the input of an operational amplifier is purely resistive though this may not be the case at high frequencies where shunt capacitive reactance may become significant.

The input resistance of operational amplifiers is very much dependent on the semiconductor technology employed. In practice, values range from about 2 MΩ for bipolar operational amplifiers to over 10^{12} Ω for CMOS devices.

Example 8.2

An operational amplifier has an input resistance of 2 MΩ Determine the input current when an input voltage of 5 mV is present.

Solution

Now:

$$R_{IN} = \frac{V_{IN}}{I_{IN}}$$

thus

$$I_{IN} = \frac{V_{IN}}{R_{IN}} = \frac{5 \times 10^{-3}}{2 \times 10^{6}} = 2.5 \times 10^{-9} \, A = 2.5 \text{ nA}$$

Output resistance

The output resistance of an operational amplifier is defined as the ratio of open-circuit output voltage to short-circuit output current expressed in ohms. Typical values of output resistance range from less than 10 Ω to around 100 Ω depending upon the configuration and amount of feedback employed.

Output resistance is the ratio of open-circuit

output voltage to short-circuit output current, hence:

$$R_{OUT} = \frac{V_{OUT(OC)}}{I_{OUT(SC)}}$$

where R_{OUT} is the output resistance (in ohms), $V_{OUT(OC)}$ is the open-circuit output voltage (in volts) and $I_{OUT(SC)}$ is the short-circuit output current (in amps).

Input offset voltage

An ideal operational amplifier would provide zero output voltage when 0 V difference is applied to its inputs. In practice, due to imperfect internal balance, there may be some small voltage present at the output. The voltage that must be applied differentially to the operational amplifier input in order to make the output voltage exactly zero is known as the input offset voltage.

Input offset voltage may be minimized by applying relatively large amounts of negative feedback or by using the offset null facility provided by a number of operational amplifier devices. Typical values of input offset voltage range from 1 mV to 15 mV. Where AC rather than DC coupling is employed, offset voltage is not normally a problem and can be happily ignored.

Full-power bandwidth

The full-power bandwidth for an operational amplifier is equivalent to the frequency at which the maximum undistorted peak output voltage swing falls to 0.707 of its low frequency (d.c.) value (the sinusoidal input voltage remaining constant). Typical full-power bandwidths range from 10 kHz to over 1 MHz for some high-speed devices.

Slew rate

Slew rate is the rate of change of output voltage with time, when a rectangular step input voltage is applied (as shown in Fig. 8.4). The slew rate of an operational amplifier is the rate of change of output voltage with time in response to a perfect step-function input. Hence:

$$\text{Slew rate} = \frac{\Delta V_{OUT}}{\Delta t}$$

where ΔV_{OUT} is the change in output voltage (in volts) and Δt is the corresponding interval of time (in s).

Slew rate is measured in V/s (or V/µs) and typical values range from 0.2 V/µs to over 20 V/µs. Slew rate imposes a limitation on circuits in which large amplitude pulses rather than small amplitude sinusoidal signals are likely to be encountered.

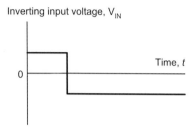

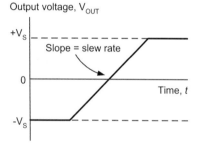

Figure 8.4   Slew rate for an operational amplifier

Operational amplifier characteristics

Having now defined the parameters that we use to describe operational amplifiers we shall now consider the desirable characteristics for an 'ideal' operational amplifier. These are:
(a) The open-loop voltage gain should be very high (ideally infinite).
(b) The input resistance should be very high (ideally infinite).
(c) The output resistance should be very low (ideally zero).
(d) Full-power bandwidth should be as wide as possible.
(e) Slew rate should be as large as possible.
(f) Input offset should be as small as possible.

Table 8.1   Comparison of operational amplifier parameters for 'ideal' and 'real' devices

Parameter	Ideal	Real
Voltage Gain	Infinite	100,000
Input Resistance	Infinite	100 MΩ
Output resistance	Zero	20 Ω
Bandwidth	Infinite	2 MHz
Slew-rate	Infinite	10 V/µs
Input offset	Zero	Less than 5 mV

The characteristics of most modern integrated circuit operational amplifiers (i.e. 'real' operational amplifiers) come very close to those of an 'ideal' operational amplifier, as witnessed by the data shown in Table 8.1.

Example 8.3

A perfect rectangular pulse is applied to the input of an operational amplifier. If it takes 4 µs for the output voltage to change from −5 V to +5 V, determine the slew rate of the device.

Solution

The slew rate can be determined from:

$$\text{Slew rate} = \frac{\Delta V_{OUT}}{\Delta t} = \frac{10 \text{ V}}{4 \text{ µs}} = 2.5 \text{ V/µs}$$

Example 8.4

A wideband operational amplifier has a slew rate of 15 V/µs. If the amplifier is used in a circuit with a voltage gain of 20 and a perfect step input of 100 mV is applied to its input, determine the time taken for the output to change level.

Solution

The output voltage change will be 20 × 100 = 2,000 mV (or 2 V). Re-arranging the formula for slew rate gives:

$$\Delta t = \frac{\Delta V_{OUT}}{\text{Slew rate}} = \frac{2 \text{ V}}{15 \text{ V/µs}} = 0.133 \text{ µs}$$

Table 8.2 Some common examples of integrated circuit operational amplifiers

Device	Type	Open-loop voltage gain (dB)	Input bias current	Slew rate (V/μs)	Application
AD548	Bipolar	100 min.	0.01 nA	1.8	Instrumentation amplifier
AD711	FET	100	25 pA	20	Wideband amplifier
CA3140	CMOS	100	5 pA	9	Low-noise wideband amplifier
LF347	FET	110	50 pA	13	Wideband amplifier
LM301	Bipolar	88	70 nA	0.4	General-purpose operational amplifier
LM348	Bipolar	96	30 nA	0.6	General-purpose operational amplifier
TL071	FET	106	30 pA	13	Wideband amplifier
741	Bipolar	106	80 nA	0.5	General-purpose operational amplifier

Operational amplifier applications

Table 8.2 shows abbreviated data for some common types of integrated circuit operational amplifier together with some typical applications.

Example 8.5

Which of the operational amplifiers in the table would be most suitable for each of the following applications:

(a) amplifying the low-level output from a piezoelectric vibration sensor

(b) a high-gain amplifier that can be used to faithfully amplify very small signals

(c) a low-frequency amplifier for audio signals.

Solution

(a) AD548 (this operational amplifier is designed for use in instrumentation applications and it offers a very low input offset current which is important when the input is derived from a piezoelectric transducer)

(b) CA3140 (this is a low-noise operational amplifier that also offers high gain and fast slew rate)

(c) LM348 or LM741 (both are general purpose operational amplifiers and are ideal for non-critical applications such as audio amplifiers).

Gain and bandwidth

It is important to note that, since the product of gain and bandwidth is a constant for any particular operational amplifier. Hence, an increase in gain can only be achieved at the expense of bandwidth, and vice versa.

Figure 8.5 shows the relationship between voltage gain and bandwidth for a typical operational amplifier (note that the axes use logarithmic, rather than linear scales). The open-loop voltage gain (i.e. that obtained with no feedback applied) is 100,000 (or 100 dB) and the bandwidth obtained in this condition is a mere 10

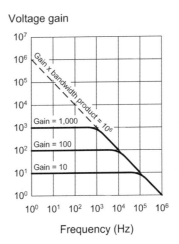

Figure 8.5 Frequency response curves for an operational amplifier

Hz. The effect of applying increasing amounts of negative feedback (and consequently reducing the gain to a more manageable amount) is that the bandwidth increases in direct proportion.

The frequency response curves in Fig. 8.5 show the effect on the bandwidth of making the closed-loop gains equal to 10,000, 1,000, 100, and 10. Table 8.3 summarizes these results. You should also note that the (gain × bandwidth) product for this amplifier is 1×10^6 Hz (i.e. 1 MHz).

We can determine the bandwidth of the amplifier when the closed-loop voltage gain is set to 46 dB by constructing a line and noting the intercept point on the response curve. This shows that the bandwidth will be 10 kHz. Note that, for this operational amplifier, the (gain × bandwidth) product is 2×10^6 Hz (or 2 MHz).

Table 8.3 Corresponding values of voltage gain and bandwidth for an operational amplifier with a gain × bandwidth product of 1×10^6

Voltage gain (A_V)	Bandwidth
1	DC to 1 MHz
10	DC to 100 kHz
100	DC to 10 kHz
1,000	DC to 1 kHz
10,000	DC to 100 Hz
100,000	DC to 10 Hz

Inverting amplifier with feedback

Figure 8.6 shows the circuit of an inverting amplifier with negative feedback applied. For the sake of our explanation we will assume that the operational amplifier is 'ideal'. Now consider what happens when a small positive input voltage is applied. This voltage (V_{IN}) produces a current (I_{IN}) flowing in the input resistor $R1$.

Since the operational amplifier is 'ideal' we will assume that:
(a) the input resistance (i.e. the resistance that appears between the inverting and non-inverting input terminals, R_{IC}) is infinite

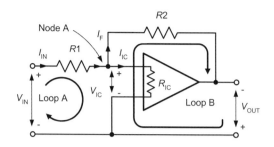

Figure 8.6   Operational amplifier with negative feedback applied

(b) the open-loop voltage gain (i.e., the ratio of V_{OUT} to V_{IN} with no feedback applied) is infinite.

As a consequence of (a) and (b):
(i) the voltage appearing between the inverting and non-inverting inputs (V_{IC}) will be zero, and
(ii) the current flowing into the chip (I_{IC}) will be zero (recall that $I_{IC} = V_{IC}/R_{IC}$ and R_{IC} is infinite).

Applying Kirchhoff's Current Law at node A gives:

$$I_{IN} = I_{IC} + I_F \text{ but } I_{IC} = 0 \text{ thus } I_{IN} = I_F \qquad (1)$$

(this shows that the current in the feedback resistor, $R2$, is the same as the input current, I_{IN}).

Applying Kirchhoff's Voltage Law to loop A gives:

$$V_{IN} = (I_{IN} \times R1) + V_{IC}$$

but $V_{IC} = 0$ thus $V_{IN} = I_{IN} \times R1$ \qquad (2)

Using Kirchhoff's Voltage Law in loop B gives:

$$V_{OUT} = -V_{IC} + (I_F \times R2)$$

but $V_{IC} = 0$ thus $V_{OUT} = I_F \times R2$ \qquad (3)

Combining (1) and (3) gives:

$$V_{OUT} = I_{IN} \times R2 \qquad (4)$$

The voltage gain of the stage is given by:

$$A_v = \frac{V_{OUT}}{V_{IN}} \qquad (5)$$

Combining (4) and (2) with (5) gives:

$$A_v = \frac{I_{IN} \times R2}{I_{IN} \times R1} = \frac{R2}{R1}$$

To preserve symmetry and minimize offset voltage, a third resistor is often included in series with the non-inverting input. The value of this resistor should be equivalent to the parallel combination of $R1$ and $R2$. Hence:

$$R3 = \frac{R1 \times R2}{R1 + R2}$$

From this point onwards (and to help you remember the function of the resistors) we shall refer to the input resistance as R_{IN} and the feedback resistance as R_F (instead of the more general and less meaningful $R1$ and $R2$, respectively).

Operational amplifier configurations

The three basic configurations for operational voltage amplifiers, together with the expressions for their voltage gain, are shown in Fig. 8.7. Supply rails have been omitted from these diagrams for clarity but are assumed to be symmetrical about 0V.

All of the amplifier circuits described previously have used direct coupling and thus have frequency response characteristics that extend to d.c. This, of course, is undesirable for many applications, particularly where a wanted a.c. signal may be superimposed on an unwanted d.c. voltage level or when the bandwidth of the amplifier greatly exceeds that of the signal that it is required to amplify. In such cases, capacitors of appropriate value may be inserted in series with the input resistor, R_{IN}, and in parallel with the feedback resistor, R_F, as shown in Fig. 8.8.

The value of the input and feedback capacitors, C_{IN} and C_F respectively, are chosen so as to roll-off the frequency response of the amplifier at the desired lower and upper cut-off frequencies, respectively. The effect of these two capacitors on an operational amplifier's frequency response is shown in Fig. 8.9.

By selecting appropriate values of capacitor, the frequency response of an inverting operational voltage amplifier may be very easily tailored to suit a particular set of requirements.

The lower cut-off frequency is determined by the value of the input capacitance, C_{IN}, and input resistance, R_{IN}. The lower cut-off frequency is given by:

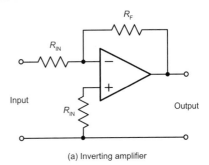

(a) Inverting amplifier

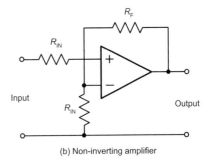

(b) Non-inverting amplifier

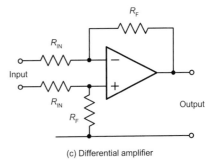

(c) Differential amplifier

Figure 8.7   The three basic configurations for operational voltage amplifiers

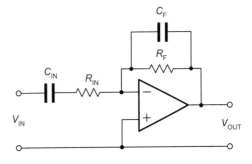

Figure 8.8   Adding capacitors to modify the frequency response of an inverting operational amplifier

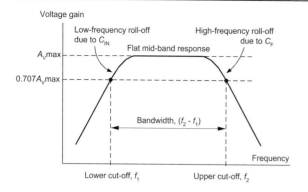

Figure 8.9　Effect of adding capacitors, C_{IN} and C_F, to modify the frequency response of an operational amplifier

$$f_1 = \frac{1}{2\pi C_{IN} R_{IN}} = \frac{0.159}{C_{IN} R_{IN}}$$

where f_1 is the lower cut-off frequency in Hz, C_{IN} is in Farads and R_{IN} is in ohms.

Provided the upper frequency response it not limited by the gain × bandwidth product, the upper cut-off frequency will be determined by the feedback capacitance, C_F, and feedback resistance, R_F, such that:

$$f_2 = \frac{1}{2\pi C_F R_F} = \frac{0.159}{C_F R_F}$$

where f_2 is the upper cut-off frequency in Hz, C_F is in Farads and R_2 is in ohms.

Example 8.6

An inverting operational amplifier is to operate according to the following specification:

Voltage gain = 100
Input resistance (at mid-band) = 10 kΩ
Lower cut-off frequency = 250 Hz
Upper cut-off frequency = 15 kHz

Devise a circuit to satisfy the above specification using an operational amplifier.

Solution

To make things a little easier, we can break the problem down into manageable parts. We shall base our circuit on a single operational amplifier configured as an inverting amplifier with capacitors to define the upper and lower cut-off frequencies, as shown in the previous figure.

The nominal input resistance is the same as the value for R_{IN}. Thus:

$$R_{IN} = 10 \text{ k}\Omega$$

To determine the value of R_F we can make use of the formula for mid-band voltage gain:

$$A_v = \frac{R2}{R1}$$

thus $R2 = A_v \times R1 = 100 \times 10 \text{ k}\Omega = 100\text{k}\Omega$

To determine the value of C_{IN} we will use the formula for the low-frequency cut-off:

$$f_1 = \frac{0.159}{C_{IN} R_{IN}}$$

from which:

$$C_{IN} = \frac{0.159}{f_1 R_{IN}} = \frac{0.159}{250 \times 10 \times 10^3}$$

hence:

$$C_{IN} = \frac{0.159}{2.5 \times 10^6} = 63 \times 10^{-9} \text{ F} = 63 \text{ nF}$$

Finally, to determine the value of C_F we will use the formula for high-frequency cut-off:

$$f_2 = \frac{0.159}{C_F R_F}$$

from which:

$$C_F = \frac{0.159}{f_2 R_{IN}} = \frac{0.159}{15 \times 10^3 \times 100 \times 10^3}$$

hence:

$$C_F = \frac{0.159}{1.5 \times 10^9} = 0.106 \times 10^{-9} \text{ F} = 106 \text{ pF}$$

For most applications the nearest preferred values (68 nF for C_{IN} and 100 pF for C_F) would be perfectly adequate. The complete circuit of the operational amplifier stage is shown in Fig. 8.10.

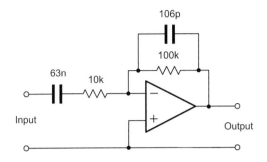

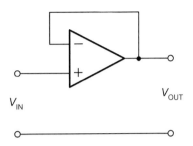

Figure 8.10   See Example 8.6. This operational amplifier has a mid-band voltage gain of 10 over the frequency range 250 Hz to 15 kHz

Figure 8.11   A voltage follower

Operational amplifier circuits

As well as their application as a general-purpose amplifying device, operational amplifiers have a number of other uses, including voltage followers, differentiators, integrators, comparators, and summing amplifiers. We shall conclude this section by taking a brief look at each of these applications.

Voltage followers

A voltage follower using an operational amplifier is shown in Fig. 8.11. This circuit is essentially an inverting amplifier in which 100% of the output is fed back to the input. The result is an amplifier that has a voltage gain of 1 (i.e. unity), a very high input resistance and a very high output resistance. This stage is often referred to as a buffer and is used for matching a high-impedance circuit to a low-impedance circuit.

Typical input and output waveforms for a voltage follower are shown in Fig. 8.12. Notice how the input and output waveforms are both in-phase (they rise and fall together) and that they are identical in amplitude.

Differentiators

A differentiator using an operational amplifier is shown in Fig. 8.13. A differentiator produces an output voltage that is equivalent to the rate of change of its input. This may sound a little complex but it simply means that, if the input voltage

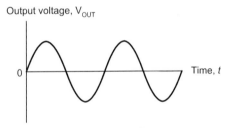

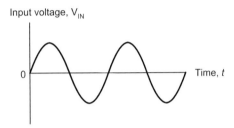

Figure 8.12 Typical input and output waveforms for a voltage follower

remains constant (i.e. if it isn't changing) the output also remains constant. The faster the input voltage changes the greater will the output be. In mathematics this is equivalent to the differential function.

Typical input and output waveforms for a differentiator are shown in Fig. 8.14. Notice how the square wave input is converted to a train of short duration pulses at the output. Note also that the output waveform is inverted because the signal has been applied to the inverting input of the operational amplifier.

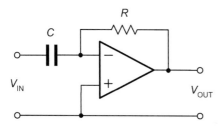

Figure 8.13 A differentiator

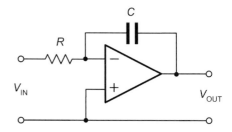

Figure 8.15 An integrator

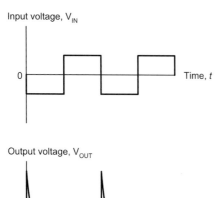

Figure 8.14 Typical input and output waveforms for a differentiator

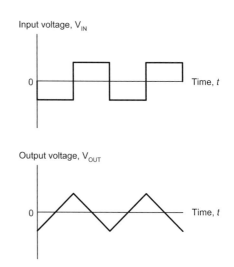

Figure 8.16 Typical input and output waveforms for an integrator

Integrators

An integrator using an operational amplifier is shown in Fig. 8.15. This circuit provides the opposite function to that of a differentiator (see earlier) in that its output is equivalent to the area under the graph of the input function rather than its rate of change. If the input voltage remains constant (and is other than 0V) the output voltage will ramp up or down according to the polarity of the input. The longer the input voltage remains at a particular value the larger the value of output voltage (of either polarity) will be produced.

Typical input and output waveforms for an integrator are shown in Fig. 8.16. Notice how the square wave input is converted to a wave that has a triangular shape. Once again, note that the output waveform is inverted.

Comparators

A comparator using an operational amplifier is shown in Fig. 8.17. Since no negative feedback has been applied, this circuit uses the maximum gain of the operational amplifier. The output voltage produced by the operational amplifier will thus rise to the maximum possible value (equal to the positive supply rail voltage) whenever the voltage present at the non-inverting input exceeds that present at the inverting input. Conversely, the output voltage produced by the operational amplifier will fall to the minimum possible value (equal to the negative supply rail voltage) whenever the voltage present at the inverting input exceeds that present at the non-inverting input.

Typical input and output waveforms for a

comparator are shown in Fig. 8.18. Notice how the output is either +15V or –15V depending on the relative polarity of the two input. A typical application for a comparator is that of comparing a signal voltage with a reference voltage. The output will go high (or low) in order to signal the result of the comparison.

Summing amplifiers

A summing amplifier using an operational amplifier is shown in Fig. 8.19. This circuit produces an output that is the sum of its two input voltages. However, since the operational amplifier is connected in inverting mode, the output voltage is given by:

$$V_{OUT} = -(V_1 + V_2)$$

where V_1 and V_2 are the input voltages (note that all of the resistors used in the circuit have the same value). Typical input and output waveforms for a summing amplifier are shown in Fig. 8.20. A typical application is that of 'mixing' two input signals to produce an output voltage that is the sum of the two.

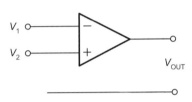

Figure 8.17 A comparator

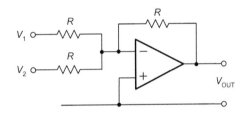

Figure 8.19 A summing amplifier

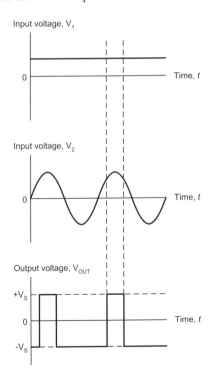

Figure 8.18 Typical input and output waveforms for a comparator

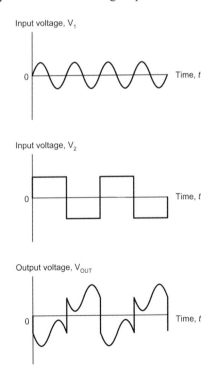

Figure 8.20 Typical input and output waveforms for a summing amplifier

Positive versus negative feedback

We have already shown how negative feedback can be applied to an operational amplifier in order to produce an exact value of gain. Negative feedback is frequently used in order to stabilize the gain of an amplifier and also to increase the frequency response (recall that, for an amplifier the product of gain and bandwidth is a constant). Positive feedback, on the other hand, results in an increase in gain and a reduction in bandwidth. Furthermore, the usual result of applying positive feedback is that an amplifier becomes unstable and oscillates (i.e. it generates an output without an input being present!). For this reason, positive feedback is only used in amplifiers when the voltage gain is less than unity.

The important thing to remember from all of this is that, when negative feedback is applied to an amplifier the overall gain is reduced and the bandwidth is increased (note that the gain × bandwidth product remains constant). When positive feedback is applied to an amplifier the overall gain increases and the bandwidth is reduced. In most cases this will result in instability and oscillation.

Multi-stage amplifiers

Multi-stage amplifiers can easily be produced using operational amplifiers. Coupling methods can be broadly similar to those described earlier in Chapter 7 (see page 149). As an example, Fig. 8.21 shows a two-stage amplifier in which each stage has a tailored frequency response. Note how $C1$ and $C3$ provide d.c. isolation between the stages as well as helping to determine the low-frequency roll-off.

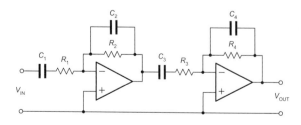

Figure 8.21   A multi-stage amplifier (both stages have tailored frequency response)

Practical investigation

Objective

To measure the voltage gain and frequency response of an inverting operational amplifier.

Components and test equipment

Breadboard, AF signal generator (with variable frequency sine wave output), two AF voltmeters (or a dual beam oscilloscope), ±9 V d.c. power supply (or two 9 V batteries), TL081 (or similar operational amplifier), 22 pF, 2.2 nF, 47 nF and 220 nF capacitors, resistors of 10 kΩ and 100 kΩ 5% 0.25 W, test leads, connecting wire.

Procedure

Connect the circuit shown in Fig. 8.22 with C_{IN} = 47 nF and C_F = 2.2 nF, set the signal generator to produce an output of 100 mV at 1 kHz. Measure and record the output voltage produced and repeat this measurement for frequencies over the range 10 Hz to 100 kHz, see Table 8.4.

Replace C_{IN} and C_F with 220 nF and 22 pF capacitors and repeat the measurements, this time over the extended frequency range from 1 Hz to 1 MHz, recording your results in Table 8.5.

Measurements and graphs

Use the measured value of output voltage at 1 kHz for both sets of measurements, in order to determine the mid-band voltage gain of the stage.

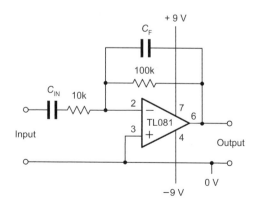

Figure 8.22   See Practical investigation

For each set of measurements plot graphs showing the frequency response of the amplifier stage (see Fig. 8.23). In each case, use the graph to determine the lower and upper cut-off frequencies.

Calculations

For each circuit calculate:
(a) the mid-band voltage gain
(b) the lower cut-off frequency
(c) the upper cut-off frequency.
Compare the calculated values with the measured values.

Table 8.4 Results (C_{IN} = 47 nF, C_F = 2.2 nF)

Frequency (Hz)	Output voltage (V)
10	
20	
40	
100	
200	
400	
1 k	
2 k	
4 k	
10 k	

Conclusion

Comment on the performance of the amplifier stage. Is this what you would expect? Do the measured values agree with those obtained by calculation? If not, suggest reasons for any differences. Suggest typical applications for the circuit.

Table 8.5 Results (C_{IN} = 220 nF, C_F = 22 pF)

Frequency (Hz)	Output voltage (V)
4	
10	
20	
40	
100	
200	
400	
1k	
10 k	
20 k	
40 k	
100 k	
200 k	
400 k	

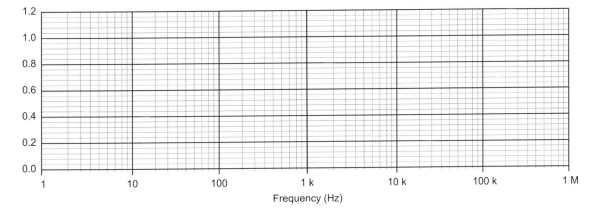

Figure 8.23 Graph layout for plotting the results

Important formulae introduced in this chapter

Open-loop voltage gain
(page 158):

$$A_{V(OL)} = \frac{V_{OUT}}{V_{IN}}$$

$$A_{V(OL)} = 20 \log_{10} \frac{V_{OUT}}{V_{IN}} \ \text{dB}$$

Closed-loop voltage gain:
(page 158)

$$A_{V(CL)} = \frac{V_{OUT}}{V_{IN}}$$

$$A_{V(CL)} = 20 \log_{10} \frac{V_{OUT}}{V_{IN}} \ \text{dB}$$

Input resistance:
(page 140)

$$R_{IN} = \frac{V_{IN}}{I_{IN}}$$

Output resistance:
(page 159)

$$R_{OUT} = \frac{V_{OUT(OC)}}{I_{OUT(SC)}}$$

Slew rate
(page 159)

$$\text{Slew rate} = \frac{\Delta V_{OUT}}{\Delta t}$$

Upper cut-off frequency:
(page 164)

$$f_2 = \frac{0.159}{C_F R_F}$$

Lower cut-off frequency:
(page 164)

$$f_1 = \frac{0.159}{C_{IN} R_{IN}}$$

Output voltage produced by a summing amplifier:
(page 167)

$$V_{OUT} = -(V_1 + V_2)$$

Symbol introduced in this chapter

Operational
amplifier

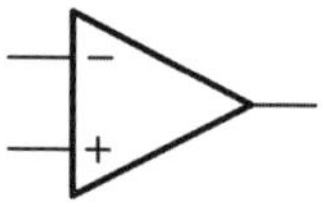

Figure 8.24 Symbol introduced in this chapter

Problems

8.1 Sketch the circuit symbol for an operational amplifier. Label each of the connections.

8.2 List four characteristics associated with an 'ideal' operational amplifier.

8.3 An operational amplifier with negative feedback applied produces an output of 1.5 V when an input of 7.5 mV is present. Determine the closed-loop voltage gain.

8.4 Sketch the circuit of an inverting amplifier based on an operational amplifier. Label your circuit and identify the components that determine the closed-loop voltage gain.

8.5 Sketch the circuit of each of the following based on the use of operational amplifiers:
(a) a comparator
(b) a differentiator
(c) an integrator.

8.6 An inverting amplifier is to be constructed having a mid-band voltage gain of 40, an input resistance of 5 kΩ and a frequency response extending from 20 Hz to 20 kHz. Devise a circuit and specify all component values required.

8.7 A summing amplifier with two inputs has R_F = 10 kΩ, and R_{IN} (for both inputs) of 2 kΩ. Determine the output voltage when one input is at −2 V and the other is +0.5 V.

8.8 During measurements on an operational amplifier under open-loop conditions, an output voltage of 12 V is produced by an input voltage of 1 mV. Determine the open-loop voltage gain expressed in dB.

8.9 With the aid of a sketch, explain what is meant by the term 'slew rate'. Why is this important?

Answers to these problems appear on page 375.

9

Oscillators

This chapter describes circuits that generate sine wave, square wave, and triangular waveforms. These oscillator circuits form the basis of clocks and timing arrangements as well as signal and function generators.

Positive feedback

In Chapter 7, we showed how negative feedback can be applied to an amplifier to form the basis of a stage which has a precisely controlled gain. An alternative form of feedback, where the output is fed back in such a way as to reinforce the input (rather than to subtract from it), is known as positive feedback.

Figure 9.1 shows the block diagram of an amplifier stage with positive feedback applied. Note that the amplifier provides a phase shift of 180° and the feedback network provides a further 180°. Thus the overall phase shift is 0°. The overall voltage gain, G, is given by:

$$\text{Overall gain, } G = \frac{V_{\text{out}}}{V_{\text{in}}}$$

By applying Kirchhoff's Voltage Law

$$V_{\text{in}}' = V_{\text{in}} + \beta V_{\text{out}}$$

thus

$$V_{\text{in}} = V_{\text{in}}' - \beta V_{\text{out}}$$

and

$$V_{\text{out}} = A_{\text{v}} \times V_{\text{in}}$$

where A_{v} is the internal gain of the amplifier. Hence:

$$\text{Overall gain, } G = \frac{A_{\text{v}} \times V_{\text{in}}'}{V_{\text{in}}' - \beta V_{\text{out}}} = \frac{A_{\text{v}} \times V_{\text{in}}'}{V_{\text{in}}' - \beta (A_{\text{v}} \times V_{\text{in}}')}$$

Thus, $G = \dfrac{A_{\text{v}}}{1 - \beta A_{\text{v}}}$

Now consider what will happen when the loop gain, βA_{v}, approaches unity (i.e., when the loop gain is just less than 1). The denominator $(1 - \beta A_{\text{v}})$ will become close to zero. This will have the effect of *increasing* the overall gain, i.e. the overall gain with positive feedback applied will be *greater* than the gain without feedback.

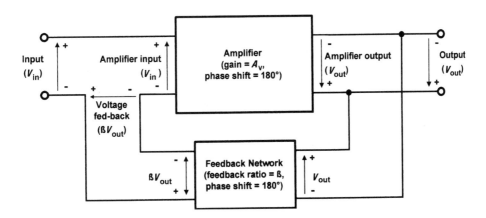

Figure 9.1 Amplifier with positive feedback applied

It is worth illustrating this difficult concept using some practical figures. Assume that you have an amplifier with a gain of 9 and one-tenth of the output is fed back to the input (i.e. $\beta = 0.1$). In this case the loop gain ($\beta \times A_v$) is 0.9.

With negative feedback applied (see Chapter 7) the overall voltage gain will be:

$$G = \frac{A_v}{1 + \beta A_v} = \frac{9}{1 + (0.1 \times 9)} = \frac{9}{1 + 0.9} = \frac{9}{1.9} = 4.7$$

With positive feedback applied the overall voltage gain will be:

$$G = \frac{A_v}{1 - \beta A_v} = \frac{10}{1 - (0.1 \times 9)} = \frac{10}{1 - 0.9} = \frac{10}{0.1} = 90$$

Now assume that you have an amplifier with a gain of 10 and, once again, one-tenth of the output is fed back to the input (i.e. $\beta = 0.1$). In this example the loop gain ($\beta \times A_v$) is exactly 1.

With negative feedback applied (see Chapter 7) the overall voltage gain will be:

$$G = \frac{A_v}{1 + \beta A_v} = \frac{10}{1 + (0.1 \times 10)} = \frac{10}{1 + 1} = \frac{10}{2} = 5$$

With positive feedback applied the overall voltage gain will be:

$$G = \frac{A_v}{1 - \beta A_v} = \frac{10}{1 - (0.1 \times 10)} = \frac{10}{1 - 1} = \frac{10}{0} = \infty$$

This simple example shows that a loop gain of unity (or larger) will result in infinite gain and an amplifier which is unstable. In fact, the amplifier will oscillate since any disturbance will be amplified and result in an output.

Clearly, as far as an amplifier is concerned, positive feedback may have an undesirable effect—instead of reducing the overall gain the effect is that of reinforcing any signal present and the output can build up into continuous oscillation if the loop gain is 1 or greater. To put this another way, oscillator circuits can simply be thought of as amplifiers that generate an output signal without the need for an input!

Conditions for oscillation

From the foregoing we can deduce that the

conditions for oscillation are:

(a) the feedback must be positive (i.e. the signal fed back must arrive back in-phase with the signal at the input);

(b) the overall loop voltage gain must be greater than 1 (i.e. the amplifier's gain must be sufficient to overcome the losses associated with any frequency selective feedback network).

Hence, to create an oscillator we simply need an amplifier with sufficient gain to overcome the losses of the network that provide positive feedback. Assuming that the amplifier provides 180° phase shift, the frequency of oscillation will be that at which there is 180° phase shift in the feedback network.

A number of circuits can be used to provide 180° phase shift, one of the simplest being a three-stage C–R ladder network that we shall meet next. Alternatively, if the amplifier produces 0° phase shift, the circuit will oscillate at the frequency at which the feedback network produces 0° phase shift. In both cases, the essential point is that the feedback should be positive so that the output signal arrives back at the input in such a sense as to reinforce the original signal.

Ladder network oscillator

A simple phase-shift oscillator based on a three-stage C–R ladder network is shown in Fig. 9.2. TR1 operates as a conventional common-emitter amplifier stage with R1 and R2 providing base bias potential and R3 and C1 providing emitter stabilization.

The total phase shift provided by the C–R ladder network (connected between collector and base) is 180° at the frequency of oscillation. The transistor provides the other 180° phase shift in order to realize an overall phase shift of 360° or 0° (note that these are the same).

The frequency of oscillation of the circuit shown in Fig. 9.2 is given by:

$$f = \frac{1}{2\pi \times \sqrt{6}CR}$$

The loss associated with the ladder network is 29, thus the amplifier must provide a gain of *at least* 29 in order for the circuit to oscillate. In practice this is easily achieved with a single transistor.

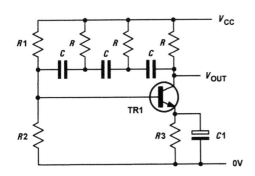

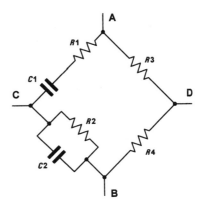

Figure 9.2 Sine wave oscillator based on a three-stage *C–R* ladder network

Figure 9.3 A Wien bridge network

Example 9.1

Determine the frequency of oscillation of a three-stage ladder network oscillator in which $C = 10$ nF and $R = 10$ kΩ.

Solution

Using

$$f = \frac{1}{2\pi \times \sqrt{6}CR}$$

gives

$$f = \frac{1}{6.28 \times 2.45 \times 10 \times 10^{-9} \times 10 \times 10^{3}}$$

from which

$$f = \frac{1}{6.28 \times 2.45 \times 10^{-4}} = \frac{10^{4}}{15.386} = 647 \text{ Hz}$$

Wien bridge oscillator

An alternative approach to providing the phase shift required is the use of a Wien bridge network (Fig. 9.3). Like the *C–R* ladder, this network provides a phase shift which varies with frequency. The input signal is applied to A and B while the output is taken from C and D. At one particular frequency, the phase shift produced by the network will be exactly zero (i.e. the input and output signals will be in-phase). If we connect the network to an amplifier producing 0° phase shift which has sufficient gain to overcome the losses of the Wien bridge, oscillation will result.

The minimum amplifier gain required to sustain oscillation is given by:

$$A_v = 1 + \frac{C1}{C2} + \frac{R2}{R1}$$

In most cases, $C1 = C2$ and $R1 = R2$, hence the minimum amplifier gain will be 3.

The frequency at which the phase shift will be zero is given by:

$$f = \frac{1}{2\pi \times \sqrt{C1C2R1R2}}$$

When $R1 = R2$ and $C1 = C2$ the frequency at which the phase shift will be zero will be given by:

$$f = \frac{1}{2\pi \times \sqrt{C^2 R^2}} = \frac{1}{2\pi CR}$$

where $R = R1 = R2$ and $C = C1 = C2$.

Example 9.2

Figure 9.4 shows the circuit of a Wien bridge oscillator based on an operational amplifier. If $C1 = C2 = 100$ nF, determine the output frequencies produced by this arrangement (a) when $R1 = R2 = 1$ kΩ and (b) when $R1 = R2 = 6$ kΩ.

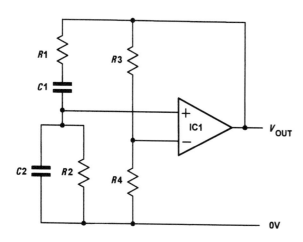

Figure 9.4 Sine wave oscillator based on a Wien bridge network (see Example 9.2)

Solution

(a)   When $R1 = R2 = 1$ kΩ

$$f = \frac{1}{2\pi CR}$$

where $R = R1 = R1$ and $C = C1 = C2$.

Thus

$$f = \frac{1}{6.28 \times 100 \times 10^{-9} \times 1 \times 10^{3}}$$

$$f = \frac{10^{4}}{6.28} = 1.59 \text{ kHz}$$

(b)   When $R1 = R1 = 6$ kΩ

$$f = \frac{1}{2\pi CR}$$

where $R = R1 = R1$ and $C = C1 = C2$.

Thus

$$f = \frac{1}{6.28 \times 100 \times 10^{-9} \times 6 \times 10^{3}}$$

$$f = \frac{10^{4}}{37.68} = 265 \text{ Hz}$$

Multivibrators

There are many occasions when we require a square wave output from an oscillator rather than a sine wave output. Multivibrators are a family of oscillator circuits that produce output waveforms consisting of one or more rectangular pulses. The term 'multivibrator' simply originates from the fact that this type of waveform is rich in harmonics (i.e. 'multiple vibrations').

Multivibrators use regenerative (i.e. positive) feedback; the active devices present within the oscillator circuit being operated as switches, being alternately cut off and driven into saturation. The principal types of multivibrator are:

(a) **astable multivibrators** that provide a continuous train of pulses (these are sometimes also referred to as free-running multivibrators)

(b) **monostable multivibrators** that produce a single output pulse (they have one stable state and are thus sometimes also referred to as 'one-shot')

(c) **bistable multivibrators** that have two stable states and require a trigger pulse or control signal to change from one state to another.

Figure 9.5 This high-speed bistable multivibrator uses two general-purpose silicon transistors and works at frequencies of up to 1 MHz triggered from an external signal

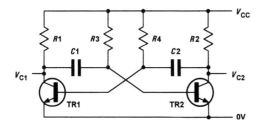

Figure 9.6 Astable multivibrator using BJTs

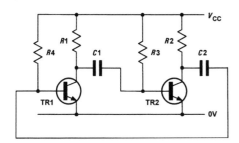

Figure 9.7 Circuit of Fig. 9.6 redrawn to show two common-emitter amplifier stages with positive feedback

The astable multivibrator

Figure 9.6 shows a classic form of astable multivibrator based on two transistors. Figure 9.7 shows how this circuit can be redrawn in an arrangement that more closely resembles a two-stage common-emitter amplifier with its output connected back to its input. In Fig. 9.5, the values of the base resistors, $R3$ and $R4$, axe such that the sufficient base current will be available to completely saturate the respective transistor. The values of the collector load resistors, $R1$ and $R2$, are very much smaller than $R3$ and $R4$. When power is first applied to the circuit, assume that TR2 saturates before TR1 when the power is first applied (in practice one transistor would always saturate before the other due to variations in component tolerances and transistor parameters).

As TR2 saturates, its collector voltage will fall rapidly from $+V_{CC}$ to 0 V. This drop in voltage will be transferred to the base of TR1 via $C1$. This negative-going voltage will ensure that TR1 is initially placed in the non-conducting state. As long as TR1 remains cut off, TR2 will continue to be saturated. During this time, $C1$ will charge via $R4$

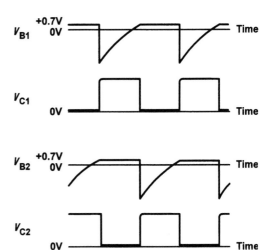

Figure 9.8 Waveforms for the BJT multivibrator shown in Fig. 9.6

and TR1's base voltage will rise exponentially from $-V_{CC}$ towards $+V_{CC}$. However, TR1's base voltage will not rise much above 0 V because, as soon as it reaches $+0.7$ V (sufficient to cause base current to flow), TR1 will begin to conduct. As TR1 begins to turn on, its collector voltage will rapidly fall from $+V_{CC}$ 0 V. This fall in voltage is transferred to the base of TR2 via $C1$ and, as a consequence, TR2 will turn off. $C1$ will then charge via $R3$ and TR2's base voltage will rise exponentially from $-V_{CC}$ towards $+V_{CC}$. As before, TR2's base voltage will not rise much above 0 V because, as soon as it reaches $+0.7$ V (sufficient to cause base current to flow), TR2 will start to conduct. The cycle is then repeated indefinitely.

The time for which the collector voltage of TR2 is low and TR1 is high ($T1$) will be determined by the time constant, $R4 \times C1$. Similarly, the time for which the collector voltage of TR1 is low and TR2 is high ($T2$) will be determined by the time constant, $R3 \times C1$.

The following approximate relationships apply:

$$T1 = 0.7\ C2\ R4 \quad \text{and} \quad T2 = 0.7\ C1\ R3$$

Since one complete cycle of the output occurs in a time, $T = T1 + T2$, the periodic time of the output is given by:

$$T = 0.7\ (C2\ R4 + C1\ R3)$$

Finally, we often require a symmetrical **square wave** output where $T1 = T2$. To obtain such an output, we should make $R3 = R4$ and $C1 = C1$, in which case the periodic time of the output will be given by:

$$T = 1.4 \, C \, R$$

where $C = C1 = C2$ and $R = R3 = R4$. Waveforms for the astable oscillator are shown in Fig. 9.8.

Example 9.3

The astable multivibrator in Fig. 9.6 is required to produce a square wave output at 1 kHz. Determine suitable values for $R3$ and $R4$ if $C1$ and $C2$ are both 10 nF.

Solution

Since a square wave is required and $C1$ and $C2$ have identical values, R3 must be made equal to R4. Now:

$$T = \frac{1}{f} = \frac{1}{1 \times 10^3} = 1 \times 10^{-3} \text{ s}$$

Re-arranging $T = 1.4CR$ to make R the subject gives:

$$R = \frac{T}{1.4C} = \frac{1 \times 10^{-3}}{1.4 \times 10 \times 10^{-9}} = \frac{1 \times 10^6}{14} = 0.071 \times 10^6$$

hence

$$R = 71 \times 10^3 = 71 \text{ k}\Omega$$

Other forms of astable oscillator

Figure 9.9 shows the circuit diagram of an alternative form of astable oscillator which produces a triangular output waveform. Operational amplifier IC1 forms an integrating stage while IC2 is connected with positive feedback to ensure that oscillation takes place.

Assume that the output from IC2 is initially at, or near, $+V_{CC}$ and capacitor, C, is uncharged. The voltage at the output of IC2 will be passed, via R, to IC1. Capacitor, C, will start to charge and the output voltage of IC1 will begin to fall.

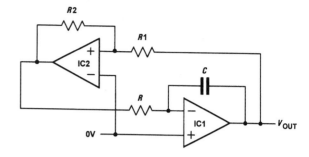

Figure 9.9 Astable oscillator using operational amplifiers

Eventually, the output voltage will have *fallen* to a value that causes the polarity of the voltage at the non-inverting input of IC2 to change from positive to negative. At this point, the output of IC2 will rapidly fall to $-V_{CC}$. Again, this voltage will be passed, via R, to IC1. Capacitor, C, will then start to charge in the other direction and the output voltage of IC1 will begin to rise.

Some time later, the output voltage will have *risen* to a value that causes the polarity of the non-inverting input of IC2 to revert to its original (positive) state and the cycle will continue indefinitely.

The upper threshold voltage (i.e. the maximum positive value for V_{out}) will be given by:

$$V_{UT} = V_{CC} \times \left(\frac{R1}{R2} \right)$$

The lower threshold voltage (i.e. the maximum negative value for V_{out}) will be given by:

$$V_{LT} = -V_{CC} \times \left(\frac{R1}{R2} \right)$$

Single-stage astable oscillator

A simple form of astable oscillator that produces a square wave output can be built using just one operational amplifier, as shown in Fig. 9.10. The circuit employs positive feedback with the output fed back to the non-inverting input via the potential divider formed by $R1$ and $R2$. This circuit can make a very simple square wave source with a

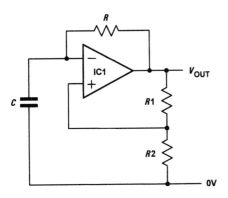

Figure 9.10 Single-stage astable oscillator using an operational amplifier

frequency that can be made adjustable by replacing R with a variable or preset resistor.

Assume that C is initially uncharged and the voltage at the inverting input is slightly less than the voltage at the non-inverting input. The output voltage will rise rapidly to $+V_{CC}$ and the voltage at the inverting input will begin to rise exponentially as capacitor C charges through R.

Eventually, the voltage at the inverting input will have reached a value that causes the voltage at the inverting input to exceed that present at the non-inverting input. At this point, the output voltage will rapidly fall to $-V_{CC}$. Capacitor, C, will then start to charge in the other direction and the voltage at the inverting input will begin to fall exponentially.

Eventually, the voltage at the inverting input will have reached a value that causes the voltage at the inverting input to be less than that present at the non-inverting input. At this point, the output voltage will rise rapidly to $+V_{CC}$ once again and the cycle will continue indefinitely.

The upper threshold voltage (i.e. the maximum positive value for the voltage at the inverting input) will be given by:

$$V_{UT} = V_{CC} \times \left(\frac{R2}{R1 + R2} \right)$$

The lower threshold voltage (i.e. the maximum negative value for the voltage at the inverting input) will be given by:

$$V_{LT} = -V_{CC} \times \left(\frac{R2}{R1 + R2} \right)$$

Finally, the time for one complete cycle of the output waveform produced by the astable oscillator is given by:

$$T = 2CR \ln \left(1 + 2 \left(\frac{R2}{R1} \right) \right)$$

Crystal controlled oscillators

A requirement of some oscillators is that they accurately maintain an exact frequency of oscillation. In such cases, a quartz crystal can be used as the frequency determining element. The quartz crystal (a thin slice of quartz in a hermetically sealed enclosure, see Fig. 9.11) vibrates whenever a potential difference is applied across its faces (this phenomenon is known as the piezoelectric effect). The frequency of oscillation is determined by the crystal's 'cut' and physical size.

Most quartz crystals can be expected to stabilize the frequency of oscillation of a circuit to within a few parts in a million. Crystals can be manufactured for operation in **fundamental mode** over a frequency range extending from 100 kHz to around 20 MHz and for **overtone** operation from 20 MHz to well over 100 MHz. Figure 9.12 shows a simple crystal oscillator circuit in which the crystal provides feedback from the drain to the source of a junction gate FET.

Figure 9.11 A quartz crystal (this crystal is cut to be resonant at 4 MHz and is supplied in an HC18 wire-ended package)

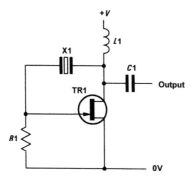

Figure 9.12 A simple JFET oscillator

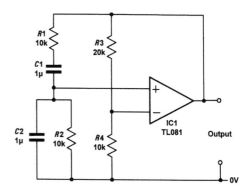

Figure 9.14 Practical sine wave oscillator based on a Wien bridge

Practical oscillator circuits

Figure 9.13 shows a practical sine wave oscillator based on a three-stage C–R ladder network. The circuit provides an output of approximately 1V pk-pk at 1.97 kHz.

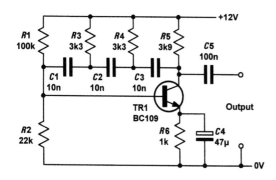

Figure 9.13 A practical sine wave oscillator based on a phase shift ladder network

A practical Wien bridge oscillator is shown in Fig. 9.14. This circuit produces a sine wave output at 16 Hz. The output frequency can easily be varied by making R1 and R2 a 10 kΩ dual-gang potentiometer and connecting a fixed resistor of 680 Ω in series with each. In order to adjust the loop gain for an optimum sine wave output it may be necessary to make R3/R4 adjustable. One way of doing this is to replace both components with a 10 kΩ multi-turn potentiometer with the sliding contact taken to the inverting input of IC1.

An astable multivibrator is shown in Fig. 9.15. This circuit produces a square wave output of 5 V pk-pk at approximately 690 Hz.

A triangle wave generator is shown in Fig. 9.16. This circuit produces a symmetrical triangular output waveform at approximately 8 Hz. If desired, a simultaneous square wave output can be derived from the output of IC2. The circuit requires symmetrical supply voltage rails (not shown in Fig. 9.14) of between ±9V and ±15 V.

Figure 9.17 shows a single-stage astable oscillator. This circuit produces a square wave output at approximately 13 Hz.

Finally, Fig. 9.18 shows a high-frequency crystal oscillator that produces an output of approximately 1V pk-pk at 4 MHz. The precise frequency of operation depends upon the quartz crystal employed (the circuit will operate with fundamental mode crystals in the range 2 MHz to about 12 MHz).

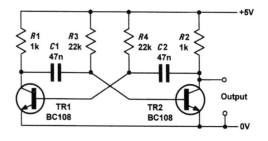

Figure 9.15 A practical square wave oscillator based on an astable multivibrator

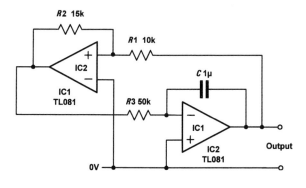

Figure 9.16 A practical triangle wave generator

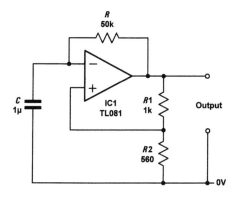

Figure 9.17 A single-stage astable oscillator that produces a square wave output

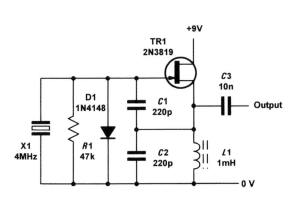

Figure 9.18 A practical high-frequency crystal oscillator

Practical investigation

Objective

To investigate a simple operational amplifier astable oscillator.

Components and test equipment

Breadboard, oscilloscope, ±9 V d.c. power supply (or two 9 V batteries), 741CN (or similar operational amplifier), 10 n, 22 n, 47 n and 100 n capacitors, resistors of 100 kΩ, 1 kΩ and 680 Ω 5% 0.25 W, test leads, connecting wire.

Procedure

Connect the circuit shown in Fig. 9.19 with $C = 47$ nF. Set the oscilloscope timebase to the 2 ms/cm range and Y-attenuator to 1 V/cm. Adjust the oscilloscope so that it triggers on a positive edge and display the output waveform produced by the oscillator. Make a sketch of the waveform using the graph layout shown in Fig. 9.20.

Measure and record (using Table 9.1) the time for one complete cycle of the output. Repeat this measurement with $C = 10$ nF, 22 nF and 100 nF.

Calculations

For each value of C, calculate the periodic time of the oscillator's output and compare this with the measured values.

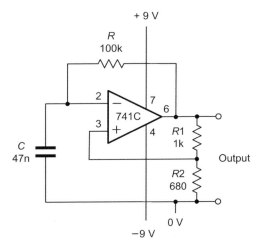

Figure 9.19 Astable oscillator circuit used in the Practical investigation

Conclusion

Comment on the performance of the astable oscillator. Is this what you would expect? Do the measured values agree with those obtained by calculation? If not, suggest reasons for any differences. Suggest typical applications for the circuit.

Table 9.1   Table of results and calculated values

C	Measured periodic time	Calculated periodic time
10 nF		
22 nF		
47 nF		
100 nF		

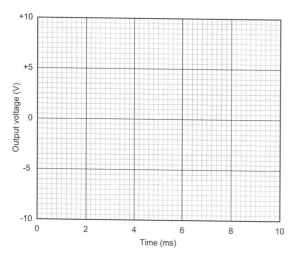

Figure 9.20   Graph layout for sketching the output waveform produced by the astable oscillator

Symbol introduced in this chapter

Quartz crystal
(or piezo resonator)

Figure 9.21   Symbol introduced in this chapter

Important formulae introduced in this chapter

Gain with positive feedback
(page 171):

$$G = \frac{A_v}{1 - \beta A_v}$$

Loop gain:
(page 171)

$$L = \beta A_v$$

Output frequency of a three-stage C–R ladder network oscillator:
(page 172)

$$f = \frac{1}{2\pi \times \sqrt{6}CR}$$

Output frequency of a Wien bridge oscillator:
(page 173)

$$f = \frac{1}{2\pi CR}$$

Time for which a multivibrator output is 'high':
(page 175)

$$T1 = 0.7 \; C2 \; R4$$

Time for which a multivibrator output is 'low':
(page 175)

$$T2 = 0.7 \; C1 \; R3$$

Periodic time for the output of a square wave mutivibrator:
(pages 175 and 176)

$$T = 0.7 \; (C2 \; R4 + C1 \; R3)$$

when $C = C1 = C2$ and $R = R3 = R4$

$$T = 1.4 \; C \; R$$

Periodic time for the output of a single-stage astable oscillator:
(page 177)

$$T = 2CR \ln\left(1 + 2\left(\frac{R2}{R1}\right)\right)$$

Problems

9.1 An amplifier with a gain of 8 has 10% of its output fed back to the input. Determine the gain of the stage (a) with negative feedback, (b) with positive feedback.

9.2 A phase-shift oscillator is to operate with an output at 1 kHz. If the oscillator is based on a three-stage ladder network, determine the required values of resistance if three capacitors of 10 nF are to be used.

9.3 A Wien bridge oscillator is based on the circuit shown in Fig. 9.4 but $R1$ and $R2$ are replaced by a dual-gang potentiometer. If $C1 = C2 = 22$ nF determine the values of $R1$ and $R2$ required to produce an output at exactly 400 Hz.

9.4 Determine the peak-peak voltage developed across $C1$ in the oscillator circuit shown in Fig. 9.22.

9.5 Determine the periodic time and frequency of the output signal produced by the oscillator circuit shown in Fig. 9.22.

9.6 An astable multivibrator circuit is required to produce an asymmetrical rectangular output which has a period of 4 ms and is to be 'high' for 1 ms and 'low' for 3 ms. If the timing capacitors are both to be 100 nF, determine the values of the two timing resistors required.

9.7 Explain, briefly, how the astable multivibrator shown in Fig. 9.23 operates. Illustrate your answer using a waveform sketch.

9.8 Determine the output frequency of the signal produced by the circuit shown in Fig. 9.23.

9.9 Explain, briefly, how the Wien bridge oscillator shown in Fig. 9.24 operates. What factors affect the choice of values for $R3$ and $R4$?

9.10 Determine the output frequency of the signal produced by the circuit shown in Fig. 9.24.

9.11 Sketch the circuit of an oscillator that will produce a triangular waveform output. Explain briefly how the circuit operates and suggest a means of varying the output frequency over a limited range.

9.12 Distinguish between the following types of mulitivibrator circuit:

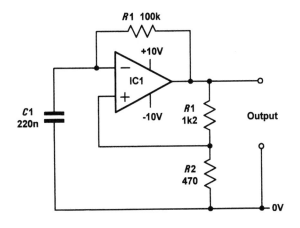

Figure 9.22   See Questions 9.4 and 9.5.

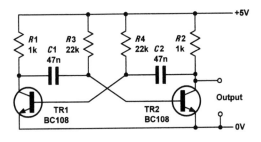

Figure 9.23   See Questions 9.7 and 9.8.

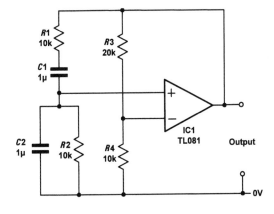

Figure 9.24   See Questions 9.9 and 9.10.

(a) astable multivibrators, (b) monostable multivibrators, (c) bistable multivibrators.

9.13 Derive an expression (in terms of $R3$ and $R4$) for the minimum value of voltage gain required to produce oscillation in the circuit shown in Fig. 9.25.

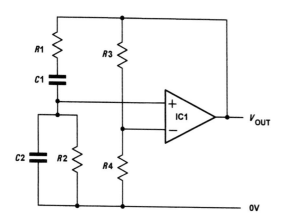

Figure 9.25   See Question 9.13

9.14 Design an oscillator circuit that will generate the output waveform shown in Fig. 9.26. Sketch a circuit diagram for the oscillator and specify all component values (including supply voltage). Give reasons for your choice of oscillator circuit.

9.15 Design an oscillator circuit that will generate the output waveform shown in Fig. 9.27. Sketch a circuit diagram for the oscillator and specify all component values (including supply voltage). Give reasons for your choice of oscillator circuit.

9.16 Design an oscillator circuit that will generate the output waveform shown in Fig. 9.28. Sketch a circuit diagram for the oscillator and specify all component values (including supply voltage). Give reasons for your choice of oscillator circuit.

9.17 Briefly explain the term 'piezoelectric effect'.

9.18 Sketch the circuit diagram of simple single-stage crystal oscillator and explain the advantages of using a quartz crystal as the frequency determining element.

Answers to these problems appear on page 375.

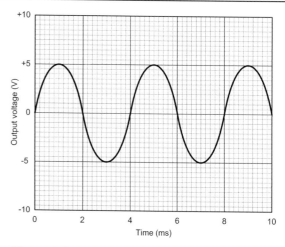

Figure 9.26   See Question 9.14

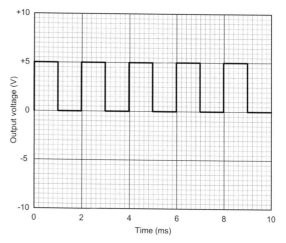

Figure 9.27   See Question 9.15

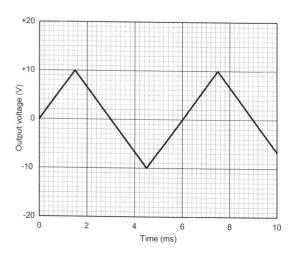

Figure 9.28   See Question 9.16

10

Logic circuits

This chapter introduces electronic circuits and devices that are associated with digital rather than analogue circuitry. These logic circuits are used extensively in digital systems and form the basis of clocks, counters, shift registers and timers.

The chapter starts by introducing the basic logic functions (AND, OR, NAND, NOR, etc.) together with the symbols and truth tables that describe the operation of the most common logic gates. We then show how these gates can be used in simple combinational logic circuits before moving on to introduce bistable devices, counters and shift registers. The chapter concludes with a brief introduction to the two principal technologies used in modern digital logic circuits, TTL and CMOS.

Logic functions

Electronic logic circuits can be used to make simple decisions like:

If dark then put on the light.

and

If temperature is less then 20°C then connect the supply to the heater.

They can also be used to make more complex decisions like:

If 'hour' is greater than 11 and '24 hour clock' is not selected then display message 'pm'.

All of these logical statements are similar in form. The first two are essentially:

If {condition} *then* {action}.

while the third is a compound statement of the form:

If {condition 1} *and not* {condition 2} *then* {action}.

Both of these statements can be readily implemented using straightforward electronic circuits. Because this circuitry is based on discrete states and since the behaviour of the circuits can be described by a set of logical statements, it is referred to as **digital logic**.

Switch and lamp logic

In the simple circuit shown in Fig. 10.1 a battery is connected to a lamp via a switch. There are two possible states for the switch, open and closed but the lamp will only operate when the switch is closed. We can summarize this using Table 10.1.

Since the switch can be only in one of the two states (i.e. open or closed) at any given time, the open and closed conditions are mutually exclusive. Furthermore, since the switch cannot exist in any other state than completely open or completely closed (i.e. there is no intermediate or half-open state) the circuit uses binary or 'two-state' logic. The logical state of the switch can be represented by the **binary digits**, 0 and 1. For example, if logical 0 is synonymous with open (or 'off') and logical 1 is equivalent to closed (or 'on'), then:

Switch open (off) = 0
Switch closed (on) = 1

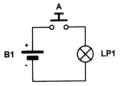

Figure 10.1 Simple switch and lamp circuit

Table 10.1 Simple switching logic

Condition	Switch	Comment
1	Open	No light produced
2	Closed	Light produced

A	LP1
0	0
1	1

Figure 10.2 Truth table for the switch and lamp

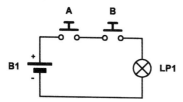

Figure 10.3 AND switch and lamp logic

Table 10.2 Simple AND switching logic

Condition	Switch A	Switch B	Comment
1	Open	Open	No light produced
2	Open	Closed	No light produced
3	Closed	Open	No light produced
4	Closed	Closed	Light produced

A	B	Y
0	0	0
0	1	0
1	0	0
1	1	1

Figure 10.4 Truth table for the switch and lamp

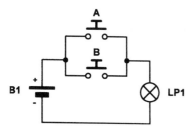

Figure 10.5 OR switch and lamp logic

We can now rewrite the **truth table** in terms of the binary states as shown in Fig. 10.2 where:

No light (off) = 0
Light (on) = 1

AND logic

Now consider the circuit with two switches shown in Fig. 10.3. Here the lamp will only operate when switch A is closed *and* switch B is closed. However, let's look at the operation of the circuit in a little more detail.

Since there are two switches (A and B) and there are two possible states for each switch (open or closed), there is a total of four possible conditions for the circuit. We can summarize these conditions in Table 10.2.

Since each switch can only be in one of the two states (i.e. open or closed) at any given time, the open and closed conditions are mutually exclusive. Furthermore, since the switches cannot exist in any other state than completely open or completely closed (i.e. there are no intermediate states) the circuit uses **binary logic**. We can thus represent the logical states of the two switches by the binary digits, 0 and 1.

Once again, if we adopt the convention that an open switch can be represented by 0 and a closed switch by 1, we can rewrite the truth table in terms of the binary states shown in Fig. 10.4 where:

No light (off) = 0
Light (on) = 1

OR logic

Figure 10.5 shows another circuit with two switches. This circuit differs from that shown in Fig. 10.3 by virtue of the fact that the two switches are connected in parallel rather than in series. In this case the lamp will operate when either of the two switches is closed. As before, there is a total of four possible conditions for the circuit. We can summarize these conditions in Table 10.3.

Once again, adopting the convention that an open switch can be represented by 0 and a closed switch by 1, we can rewrite the truth table in terms of the binary states as shown in Fig. 10.6.

Table 10.3 Simple OR switching logic

Condition	Switch A	Switch B	Comment
1	Open	Open	No light produced
2	Open	Closed	Light produced
3	Closed	Open	Light produced
4	Closed	Closed	Light produced

A	B	Y
0	0	0
0	1	1
1	0	1
1	1	1

Figure 10.6 Truth table for OR logic

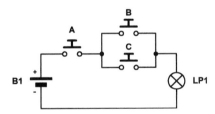

Figure 10.7 See Example 10.1

A	B	C	Y
0	0	0	0
0	0	1	0
0	1	0	0
0	1	1	0
1	0	0	0
1	0	1	1
1	1	0	1
1	1	1	1

Figure 10.8 See Example 10.1

Example 10.1

Figure 10.7 shows a simple switching circuit. Describe the logical state of switches A, B, and C in order to operate the lamp. Illustrate your answer with a truth table.

Solution

In order to operate the lamp, switch A *and* either switch B *or* switch C must be operated. The truth table is shown in Fig. 10.8.

Logic gates

Logic gates are circuits designed to produce the basic logic functions, AND, OR, etc. These circuits are designed to be interconnected into larger, more complex, logic circuit arrangements. Since these circuits form the basic building blocks of all digital systems, we have summarized the action of each of the gates in the next section. For each gate we have included its British Standard (BS) symbol together with its American Standard (MIL/ANSI) symbol. We have also included the truth tables and Boolean expressions (using '+' to denote OR, '·' to denote AND, and '‾' to denote NOT). Note that, while inverters and buffers each have only one input, exclusive-OR gates have two inputs and the other basic gates (AND, OR, NAND and NOR) are commonly available with up to eight inputs.

Buffers (Fig. 10.9)

Buffers do not affect the logical state of a digital signal (i.e. a logic 1 input results in a logic 1 output whereas a logic 0 input results in a logic 0 output). Buffers are normally used to provide extra current drive at the output but can also be used to regularize the logic levels present at an interface. The Boolean expression for the output, Y, of a buffer with an input, X, is:

$Y = X$

Inverters (Fig. 10.10)

Inverters are used to complement the logical state (i.e. a logic 1 input results in a logic 0 output and vice versa). Inverters also provide extra current drive and, like buffers, are used in interfacing applications where they provide a means of regularizing logic levels present at the input or output of a digital system. The Boolean expression for the output, Y, of a buffer with an input, X, is:

$$Y = \overline{X}$$

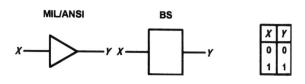

Figure 10.9 Symbols and truth table for a buffer

AND gates (Fig. 10.11)

AND gates will only produce a logic 1 output when all inputs are simultaneously at logic 1. Any other input combination results in a logic 0 output. The Boolean expression for the output, Y, of an AND gate with inputs, A and B, is:

$$Y = A \cdot B$$

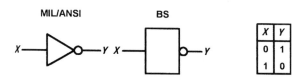

Figure 10.10   Symbols and truth table for an inverter

OR gates (Fig. 10.12)

OR gates will produce a logic 1 output whenever any one, or more, inputs are at logic 1. Putting this another way, an OR gate will only produce a logic 0 output whenever all of its inputs are simultaneously at logic 0. The Boolean expression for the output, Y, of an OR gate with inputs, A and B, is:

$$Y = A + B$$

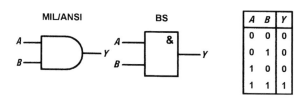

Figure 10.11 Symbols and truth table for an AND gate

NAND gates (Fig. 10.13)

NAND (i.e. NOT-AND) gates will only produce a logic 0 output when all inputs are simultaneously at logic 1. Any other input combination will produce a logic 1 output. A NAND gate, therefore, is nothing more than an AND gate with its output inverted! The circle shown at the output denotes this inversion. The Boolean expression for the output, Y, of a NAND gate with inputs, A and B, is:

$$Y = \overline{A \cdot B}$$

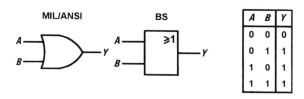

Figure 10.12 Symbols and truth table for an OR gate

NOR gates (Fig. 10.14)

NOR (i.e. NOT-OR) gates will only produce a logic 1 output when all inputs are simultaneously at logic 0. Any other input combination will produce a

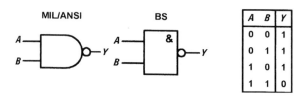

Figure 10.13 Symbols and truth table for a NAND gate

logic 0 output. A NOR gate, therefore, is simply an OR gate with its output inverted. A circle is again used to indicate inversion. The Boolean expression for the output, Y, of a NOR gate with inputs, A and B, is:

$$Y = \overline{A+B}$$

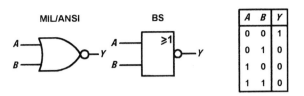

Figure 10.14 Symbols and truth table for a NOR gate

Exclusive-OR gates (Fig. 10.15)

Exclusive-OR gates will produce a logic 1 output whenever either one of the inputs is at logic 1 and the other is at logic 0. Exclusive-OR gates produce a logic 0 output whenever both inputs have the same logical state (i.e. when both are at logic 0 or both are at logic 1). The Boolean expression for the output, Y, of an exclusive-OR gate with inputs, A and B, is:

$$Y = A \cdot \overline{B} + B \cdot \overline{A}$$

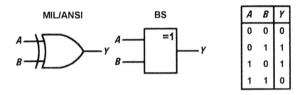

Figure 10.15 Symbols and truth table for an exclusive-OR gate

Combinational logic

By using a standard range of logic levels (i.e. voltage levels used to represent the logic 1 and logic 0 states) logic circuits can be combined together in order to solve complex logic functions.

Example 10.2

A logic circuit is to be constructed that will produce a logic 1 output whenever two, or more, of its three inputs are at logic 1.

Solution

This circuit could be more aptly referred to as a **majority vote** circuit. Its truth table is shown in Fig. 10.16. Figure 10.17 shows the logic circuitry required.

Example 10.3

Show how an arrangement of basic logic gates (AND, OR and NOT) can be used to produce the exclusive-OR function.

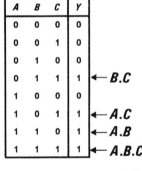

$$Y = (B.C) + (A.C) + (A.B) + A.B.C$$

Figure 10.16 See Example 10.2

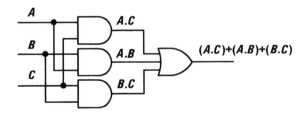

Figure 10.17 See Example 10.2

Solution

In order to solve this problem, consider the Boolean expression for the exclusive-OR function:

$$Y = A \cdot \overline{B} + B. \overline{A}$$

This expression takes the form:

$$Y = P + Q \quad \text{where } P = A \cdot \overline{B} \text{ and } Q = B \cdot \overline{A}$$

$A \cdot \overline{B}$ and $Q = B \cdot \overline{A}$ can be obtained using two two-input AND gates and the result (i.e. P and Q) can then be applied to an OR gate with two inputs. $\overline{A}$ and $\overline{B}$ can be produced using inverters. The complete solution is shown in Fig. 10.18.

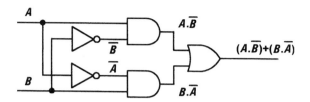

Figure 10.18 See Example 10.3

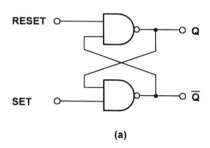

(a)

Bistables

The output of a bistable has two stables states (logic 0 or logic 1) and, once set in one or other of these states, the device will remain at a particular logic level for an indefinite period until reset. A bistable thus constitutes a simple form of 'memory cell' because it will remain in its latched state (whether **set** or **reset**) until a signal is applied to it in order to change its state (or until the supply is disconnected).

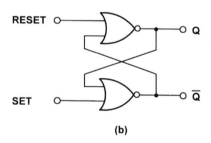

(b)

Figure 10.19 R-S bistables using cross-coupled NAND and NOR gates

R-S bistables

The simplest form of bistable is the R-S bistable. This device has two inputs, SET and RESET, and complementary outputs, Q and $\overline{Q}$. A logic 1 applied to the SET input will cause the Q output to become (or remain at) logic 1 while a logic 1 applied to the RESET input will cause the Q output to become (or remain at) logic 0. In either case, the bistable will remain in its SET or RESET state until an input is applied in such a sense as to change the state.

Two simple forms of R-S bistable based on cross-coupled logic gates are shown in Fig. 10.19. Figure 10.19(a) is based on NAND gates while Fig. 10.19(b) is based on NOR gates.

The simple cross-coupled logic gate bistable has a number of serious shortcomings (consider what

would happen if a logic 1 was simultaneously present on both the SET and RESET inputs!) and practical forms of bistable make use of much improved purpose-designed logic circuits such as D-type and J-K bistables.

D-type bistables

The D-type bistable has two inputs: D (standing variously for 'data' or 'delay') and CLOCK (CLK). The data input (logic 0 or logic 1) is clocked into the bistable such that the output state only changes when the clock changes state. Operation is thus said to be synchronous. Additional subsidiary inputs (which are invariably active low) are provided which can be used to directly set or reset the bistable. These are usually called PRESET (PR) and CLEAR (CLR). D-type bistables are used both as latches (a simple form of memory) and as binary dividers. The simple circuit arrangement in Fig. 10.20 together with the timing diagram shown in Fig. 10.21 illustrate the operation of D-type bistables.

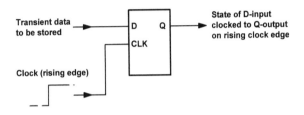

Figure 10.20 D-type bistable operation

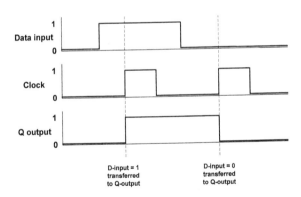

Figure 10.21 Timing diagram for the D-type bistable

J-K bistables

J-K bistables have two clocked inputs (J and K), two direct inputs (PRESET and CLEAR), a CLOCK (CK) input, and outputs (Q and $\overline{Q}$). As with R-S bistables, the two outputs are complementary (i.e. when one is 0 the other is 1, and vice versa). Similarly, the PRESET and CLEAR inputs are invariably both active low (i.e. a

0 on the PRESET input will set the Q output to 1 whereas a 0 on the CLEAR input will set the Q output to 0). Tables 10.4 and 10.5 summarize the operation of a J-K bistable respectively for the PRESET and CLEAR inputs and for clocked operation.

Table 10.4 Input and output states for a J-K bistable (PRESET and CLEAR inputs)

Inputs		Output	Comments
PRESET	CLEAR	Q_{N+1}	
0	0	?	Indeterminate
0	1	0	Q output changes to 0 (i.e. Q is reset) regardless of the clock state
1	0	1	Q output changes to 1 (i.e. Q is set) regardless of the clock state
1	1	See below	Enables clocked operation (refer to Table 10.5)

Note: The preset and clear inputs operate regardless of the clock.

Table 10.5 Input and output states for a J-K bistable (clocked operation)

Inputs		Output	Comments
J	K	Q_{N+1}	
0	0	Q_N	No change in state of the Q output on the next clock transition
0	1	0	Q output changes to 0 (i.e. Q is reset) on the next clock transition
1	0	1	Q output changes to 1 (i.e. Q is set) on the next clock transition
1	1	Q_N	Q output changes to the opposite state on the next clock transition

Note: Q_{N+1} means 'Q after the next clock transition' while Q_N means 'Q in whatever state it was before'.

J-K bistables are the most sophisticated and flexible of the bistable types and they can be configured in various ways including binary dividers, shift registers, and latches.

Figure 10.22 shows the arrangement of a four-stage binary counter based on J-K bistables. The **timing diagram** for this circuit is shown in Fig. 10.23. Each stage successively divides the clock input signal by a factor of two. Note that a logic 1 input is transferred to the respective Q-output on the falling edge of the clock pulse and all J and K inputs must be taken to logic 1 to enable binary counting.

Figure 10.24 shows the arrangement of a four-stage shift register based on J-K bistables. The timing diagram for this circuit is shown in Fig. 10.25. Note that each stage successively feeds data to the next stage. Note that all data transfer occurs on the falling edge of the clock pulse.

Example 10.4

A logic arrangement has to be designed so that it produces the pulse train shown in Fig. 10.27. Devise a logic circuit arrangement that will

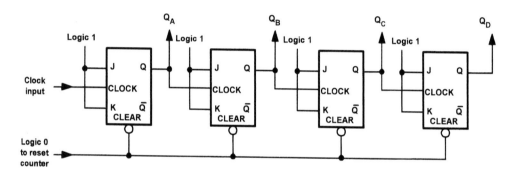

Figure 10.22 Four-stage binary counter using J-K bistables

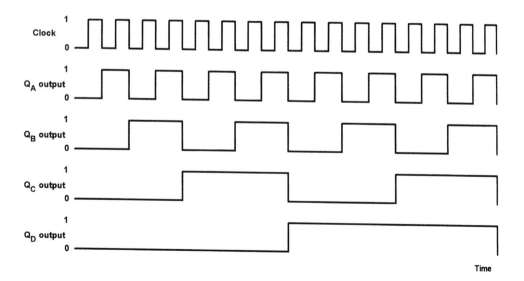

Figure 10.23 Timing diagram for the four-stage binary counter shown in Fig. 10.22

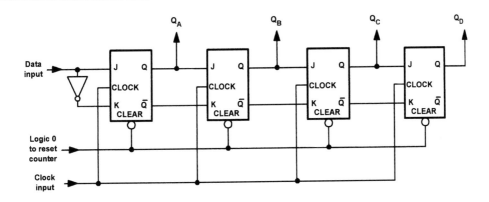

Figure 10.24 Four-stage shift register using J-K bistables

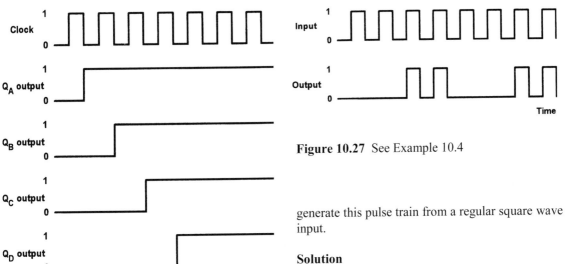

Figure 10.25 Timing diagram for the four-stage shift register shown in Fig. 10.24

Figure 10.27 See Example 10.4

generate this pulse train from a regular square wave input.

Solution

A two-stage binary divider (based on J-K bistables) can be used together with a two-input AND gate as shown in Fig. 10.26. The waveforms for this logic arrangement are shown in Fig. 10.28.

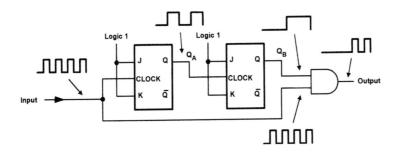

Figure 10.26 See Example 10.4

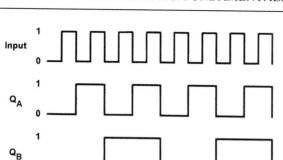

Figure 10.28 Waveforms for the logic arrangement shown in Fig. 10.26

Integrated circuit logic devices

The task of realizing a complex logic circuit is made simple with the aid of digital integrated circuits. Such devices are classified according to the semiconductor technology used in their fabrication (the logic family to which a device belongs is largely instrumental in determining its operational characteristics, such as power consumption, speed, and immunity to noise).

The relative size of a digital integrated circuit (in terms of the number of active devices that it contains) is often referred to as its scale of integration and the terminology in Table 10.5 is commonly used.

Table 10.5 Scale of integration

Scale of integration	Abbreviation	Number of logic gates*
Small	SSI	1 to 10
Medium	MSI	10 to 100
Large	LSI	100 to 1,000
Very large	VLSI	1,000 to 10,000
Super large	SLSI	10,000 to 100,000

* or active circuitry of equivalent complexity

The two basic logic families are CMOS (complementary metal oxide semiconductor) and TTL (transistor transistor logic). Each of these families is then further sub-divided. Representative circuits for a two-input AND gate in both technologies are shown in Figs 10.29 and 10.30.

The most common family of TTL logic devices is known as the 74-series. Devices from this family are coded with the prefix number 74. Variants within the family are identified by letters which follow the initial 74 prefix, as shown in Table 10.6.

The most common family of CMOS devices is known as the 4000-series. Variants within the family are identified by the suffix letters given in Table 10.7.

Table 10.6 TTL device coding—infix letters

Infix	Meaning
None	Standard TTL device
ALS	Advanced low-power Schottky
C	CMOS version of a TTL device
F	'Fast' (a high-speed version)
H	High-speed version
S	Schottky input configuration (improved speed and noise immunity)
HC	High-speed CMOS version (CMOS compatible inputs)
HCT	High-speed CMOS version (TTL compatible inputs)
LS	Low-power Schottky

Table 10.7 CMOS device coding—the most common variants of the 4000 family are identified using these suffix letters

Infix	Meaning
None	Standard CMOS device
A	Standard (unbuffered) CMOS device
B, BE	Improved (buffered) CMOS device
UB, UBE	Improved (unbuffered) CMOS device

Example 10.5

Identify each of the following integrated circuits:

(i) 4001UBE;
(ii) 74LS14.

Solution

Integrated circuit (i) is an improved (unbuffered) version of the CMOS 4001 device. Integrated circuit (ii) is a low-power Schottky version of the TTL 7414 device.

Date codes

It is also worth noting that the vast majority of logic devices and other digital integrated circuits are marked with a four digit date code. The code often appears alongside or below the device code. The first two digits of this code give the year of manufacture while the last two digits specify the week of manufacture.

Example 10.6

An integrated circuit marked '4050B 9832'. What type of device is it and when was it manufactured?

Solution

The device is a buffered CMOS 4050 manufactured in the 32nd week of 1998.

Logic levels

Logic levels are simply the range of voltages used to represent the logic states 0 and 1. The logic levels for CMOS differ markedly from those associated with TTL. In particular, CMOS logic levels are relative to the supply voltage used while the logic levels associated with TTL devices tend to be absolute (see Table 10.8).

Noise margin

The noise margin of a logic device is a measure of its ability to reject noise and spurious signals; the

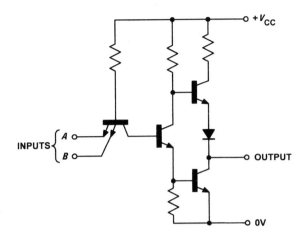

Figure 10.29 Two-input TTL NAND gate

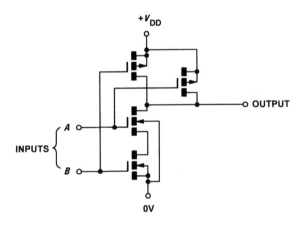

Figure 10.30 Two-input CMOS NAND gate

Table 10.8 Logic levels for CMOS and TTL logic devices

Condition	CMOS	TTL
Logic 0	Less than $1/3 V_{DD}$	More than 2 V
Logic 1	More than $2/3 V_{DD}$	Less than 0.8 V
Indeterminate	Between $1/3 V_{DD}$ and $2/3 V_{DD}$	Between 0.8 V and 2 V

Note: V_{DD} is the positive supply associated with CMOS devices

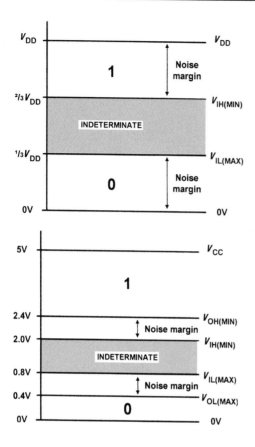

Figure 10.31 Logic levels and noise margins for TTL and CMOS devices

larger the noise margin the better is its ability to perform in an environment in which noise is present. Noise margin is defined as the difference between the minimum values of high state output and high state input voltage and the maximum values of low state output and low state input voltage. Hence:

Noise margin $= V_{OH(MIN)} - V_{IH(MIN)}$

or

Noise margin $= V_{OL(MAX)} - V_{OH(MAX)}$

where $V_{OH(MIN)}$ is the minimum value of high state (logic 1) output voltage, $V_{IH(MIN)}$ is the minimum value of high state (logic 1) input voltage, $V_{OL(MAX)}$ is the maximum value of low state (logic 0) output voltage, and $V_{OH(MAX)}$ is the maximum value of low state (logic 0) input voltage.

The noise margin for standard 7400 series TTL is typically 400 mV while that for CMOS is $1/3V_{DD}$, as shown in Fig. 10.31.

Table 10.9 compares the more important characteristics of common members of the TTL family with their buffered CMOS logic counterparts. Finally, Fig. 10.32 shows the packages and pin connections for two common logic devices, the 74LS00 (quad two-input NAND gate) and the 4001UBE (quad two-input NOR gate).

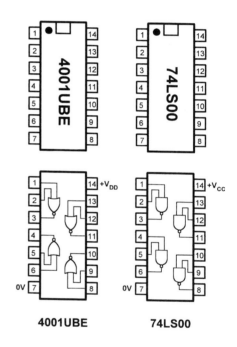

Figure 10.32 Packages and pin connections for two common logic devices

Example 10.7

Show how a 4001UBE device (see Fig. 10.32) can be connected to form a simple cross-coupled bistable. Sketch a circuit diagram showing pin connections and include LEDs that will indicate the output state of the bistable.

Solution

See Practical investigation on Page 195. Note that only two of the four logic gates have been used.

Table 10.9 Characteristics of common logic families

Characteristic	Logic family			
	74	74LS	74HC	40BE
Maximum supply voltage (V)	5.25	5.25	5.5	18
Minimum supply voltage (V)	4.75	4.75	4.5	3
Static power dissipation (mW per gate at 100 kHz)	10	2	negligible	negligible
Dynamic power dissipation (mW per gate at 100 kHz)	10	2	0.2	0.1
Typical propagation delay (ns)	10	10	10	105
Maximum clock frequency (MHz)	35	40	40	12
Speed-power product (pJ at 100 kHz)	100	20	1.2	11
Minimum output current (mA at $V_O = 0.4$ V)	16	8	4	1.6
Fan-out (number of standard loads that can be driven)	40	20	10	4
Maximum input current (mA at $V_1 = 0.4$ V)	−1.6	−0.4	0.001	−0.001

Practical investigation

Objective

To investigate the operation of a simple bistable based on cross-coupled NOR gates.

Components and test equipment

Breadboard, 9 V d.c. power supply (or a 9 V battery), 4001BE quad two-input buffered CMOS NOR gate, red and green LEDs, operational amplifier), two 1 kΩ and two 47 kΩ 5% 0.25 W resistors, test leads, connecting wire.

Procedure

Connect the circuit shown in Fig. 10.33 (see also Fig. 10.34 for the corresponding breadboard layout). Note that the green LED should become illuminated when the bistable is in the SET condition (i.e. when Q is at logic 1) and the red LED should become illuminated when the bistable is in the RESET condition. Note also that the 47 kΩ resistors act as **pull-up resistors**. They are used to ensure that the respective input goes to logic 1 when the corresponding link is removed.

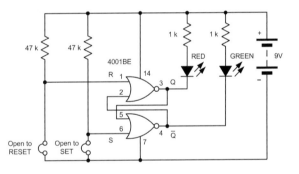

Figure 10.33 Bistable circuit used in the Practical investigation. The LEDs are used to indicate the state of the outputs

With both links in place (i.e. SET = 0 and RESET = 0) observe and record (using a truth table) the state of the outputs.

Remove the RESET link (to make RESET = 1) whilst leaving the SET link in place (to keep SET = 0). Once again, observe and record the state of the outputs. Replace the RESET link (to make RESET = 0) and check that the bistable does not change

Figure 10.34   Breadboard circuit layout

state. Now remove the SET link (to make SET = 1). Once again, observe and record the state of the outputs. Replace the SET link (to make SET = 0) and once again check that the bistable does not change state. Finally, remove both links (to make SET = 1 and RESET = 1) and observe the state of the outputs in this **disallowed state**.

Conclusion

Comment on the truth table produced. Is this what you would expect? What happened when both SET and RESET inputs were at logic 1? Suggest a typical application for the circuit.

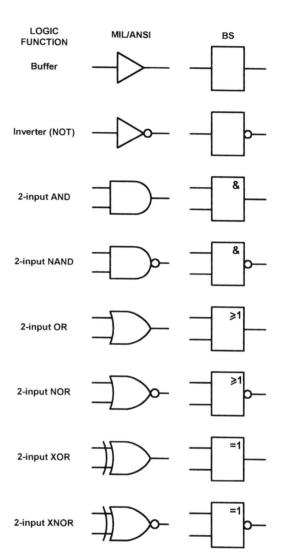

Figure 10.36   Logic gate symbols

Symbols introduced in this chapter

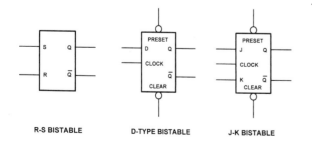

Figure 10.35   Bistable symbols

Important formulae introduced in this chapter

Noise margin:
(page 194)

Noise margin = $V_{OH(MIN)} - V_{IH(MIN)}$

or

Noise margin = $V_{OL(MAX)} - V_{OH(MAX)}$

Problems

10.1 Show how a four-input AND gate can be made from three two-input AND gates.

10.2 Show how a four-input OR gate can be made from three two-input OR gates.

10.3 Construct the truth table for the logic gate arrangement shown in Fig. 10.37.

10.4 Using only two-input NAND gates, show how each of the following logical functions can be satisfied:
(a) two-input AND;
(b) two-input OR;
(c) four-input AND.
In each case, use the minimum number of gates. (Hint: a two-input NAND gate can be made into an inverter by connecting its two inputs together)

10.5 The rocket motor of an air-launched missile will operate if, and only if, the following conditions are satisfied:
(i) 'launch' signal is at logic 1;
(ii) 'unsafe height' signal is at logic 0;
(iii) 'target lock' signal is at logic 1.
Devise a suitable logic arrangement that will satisfy this requirement. Simplify your answer using the minimum number of logic gates.

10.6 An automatic sheet metal guillotine will operate if the following conditions are satisfied:
(i) 'guard lowered' signal is at logic 1;
(ii) 'feed jam' signal is at logic 0;
(iii) 'manual start' signal is at logic 1.
The sheet metal guillotine will also operate if the following conditions are satisfied:
(i) 'manual start' signal is at logic 1;
(ii) 'test key' signal is at logic 1.
Devise a suitable logic arrangement that will satisfy this requirement. Use the minimum number of logic gates.

10.7 Devise a logic arrangement using no more than four two-input gates that will satisfy the truth table shown in Fig. 10.38.

10.8 Devise a logic arrangement that will produce the output waveform from the three input waveforms shown in Fig. 10.39.

10.9 A logic device is marked '74LS90 2798'. To which family and sub-family of logic does it belong and when was the device manufactured?

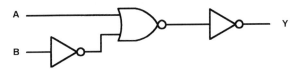

Figure 10.37   See Questions 10.3 and 10.22

A	B	C	Y
0	0	0	0
0	0	1	1
0	1	0	0
0	1	1	1
1	0	0	0
1	0	1	1
1	1	0	1
1	1	1	1

Figure 10.38   See Question 10.7

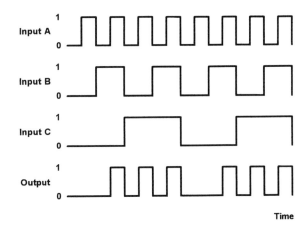

Figure 10.39   See Question 10.8.

10.10 A logic family recognizes a logic 1 input as being associated with any voltage between 2.0 V and 5.5 V. The same family produces an output in the range 2.6 V to 5.0 V corresponding to a logic 1 output. Determine the noise margin.

10.11 Sketch the circuit of a bistable using:
(a) two NAND gates;
(b) two NOR gates.
Label the inputs and outputs on your diagram.

10.12 Sketch the symbol of each of the following types of bistable:
(a) an R-S bistable;
(b) a D-type bistable;
(c) a J-K bistable.
Label your drawings clearly.

10.13 With the aid of a diagram, explain how a three-stage binary counter can be built using J-K bistables.

10.14 With the aid of a diagram, explain how a three-stage shift register can be built using J-K bistables.

10.15 Identify each of the logic devices shown in Fig. 10.40.

10.16 Explain, in relation to the scale of integration, what is meant by the terms (a) MSI and (b) VLSI.

10.17 Figure 10.41 shows the internal schematic for a logic device. Identify the logic family to which this device belongs and state its logic function.

10.18 Specify typical logic levels for the logic device shown in Fig. 10.41. In relation to your answer, explain the significance of the indeterminate region and explain how this effects the noise margin of the device.

10.19 Sketch the logic gate arrangement of a four input majority vote circuit. Using a truth table, briefly explain the operation of the circuit.

10.20 Show, with the aid of a logic diagram, how an exclusive-OR gate can be built using two-input NAND gates.

10.21 Specify typical values for the power dissipation, propagation delay, and maximum clock speed for (a) low-power Schottky TTL and (b) buffered CMOS.

10.22 Devise a logic gate arrangement using only two-input NAND gates that will perform the same logic function as the arrangement shown in Fig. 10.37. Simplify your answer as far as possible using the minimum number of logic gates.

Answers to these problems appear on page 375.

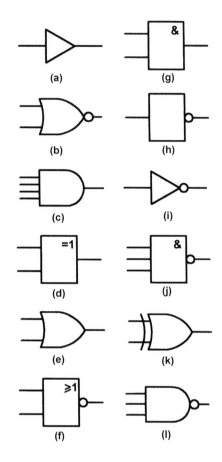

Figure 10.40   See Question 10.15

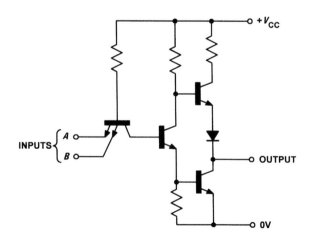

Figure 10.41   See Questions 10.17 and 10.18

11

Microprocessors

Many of today's complex electronic systems are based on the use of a microprocessor or microcontroller. Such systems comprise hardware that is controlled by software. If it is necessary to change the way that the system behaves it is the software (rather than the hardware) that is changed.

In this chapter we provide an introduction to microprocessors and explain, in simple terms, both how they operate and how they are used. We shall start by explaining some of the terminology that is used to describe different types of system that involve the use of a microprocessor or a similar device.

Microprocessor systems

Microprocessor systems are usually assembled on a single PCB comprising a microprocessor CPU together with a number of specialized support chips. These very large scale integrated (VLSI) devices provide input and output to the system, control and timing as well as storage for programs and data.

Typical applications for microprocessor systems include the control of complex industrial processes. Typical examples are based on families of chips such as the Z80CPU plus Z80PIO, Z80CTC, and Z80SIO.

Single-chip microcomputers

A single-chip microcomputer is a complete computer system (comprising CPU, RAM and ROM etc.) in a single VLSI package. A single-chip microcomputer requires very little external circuitry in order to provide all of the functions associated with a complete computer system (but usually with limited input and output capability).

Single-chip microcomputers may be programmed using in-built programmable memories or via external memory chips. Typical applications of single-chip microcomputers include computer printers, instrument controllers, and displays. A typical example is the Z84C.

Microcontrollers

A microcontroller is a single-chip microcomputer that is designed specifically for control rather than general-purpose applications. They are often used to satisfy a particular control requirement, such as controlling a motor drive. Single-chip microcomputers, on the other hand, usually perform a variety of different functions and may control several processes at the same time.

Typical applications include control of peripheral devices such as motors, drives, printers, and minor sub-system components. Typical examples are the Z86E, 8051, 68705 and 89C51.

PIC microcontrollers

A PIC microcontroller is a general-purpose microcontroller device that is normally used in a stand-alone application to perform simple logic, timing and input/output control. PIC devices provide a flexible low-cost solution that very effectively bridges the gap between single-chip computers and the use of discrete logic and timer chips, as explained in Chapter 18.

A number of PIC and microcontroller devices have been produced that incorporate a high-level language interpreter. The resident interpreter allows developers to develop their programs languages such as BASIC rather than having to resort to more complex assembly language. This feature makes PIC microcontrollers very easy to use. PIC microcontrollers are used in 'self-contained' applications involving logic, timing and simple analogue to digital and digital to analogue conversion. Typical examples are the PIC12C508 and PIC16C620.

Programmed logic devices

Whilst not an example of a microprocessor device, a programmed logic device (PLD) is a programmable chip that can carry out complex logical operations. For completeness, we have included a reference to such devices here. PLDs are capable of replacing a large number of conventional logic gates, thus minimising chip-count and reducing printed circuit board sizes. Programming is relatively straightforward and simply requires the derivation of complex logic functions using Boolean algebra (see Chapter 10) or truth tables. Typical examples are the 16L8 and 22V10.

Figure 11.2   A Z80 microprocessor

Programmable logic controllers

Programmable logic controllers (PLC) are microprocessor based systems that are used for controlling a wide variety of automatic processes, from operating an airport baggage handling system to brewing a pint of your favourite lager. PLCs are rugged and modular and they are designed specifically for operation in the process control environment.

The control program for a PLC is usually stored in one or more semiconductor memory devices. The program can be entered (or modified) by means of a simple hand-held programmer, a laptop controller, or downloaded over a local area network (LAN). PLC manufacturers include Allen Bradley, Siemens and Mitsubishi.

Microprocessor systems

The basic components of any microprocessor system (see Fig. 11.1) are:

(a) a central processing unit (CPU)
(b) a memory, comprising both 'read/write' and 'read only' devices (commonly called RAM and ROM respectively)
(c) a means of providing input and output (I/O). For example, a keypad for input and a display for output.

In a microprocessor system the functions of the CPU are provided by a single very large scale integrated (VLSI) microprocessor chip (see Fig. 11.2). This chip is equivalent to many thousands of

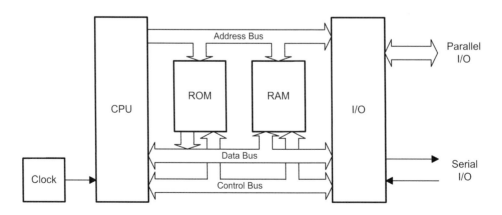

Figure 11.1   Block diagram of a microprocessor system

individual transistors. Semiconductor devices are also used to provide the read/write and read-only memory. Strictly speaking, both types of memory permit 'random access' since any item of data can be retrieved with equal ease regardless of its actual location within the memory. Despite this, the term 'RAM' has become synonymous with semiconductor read/write memory.

The basic components of the system (CPU, RAM, ROM and I/O) are linked together using a multiple-wire connecting system know as a **bus** (see Fig. 11.1). Three different buses are present, these are:

(a) the **address bus** used to specify memory locations;
   the **data bus** on which data is transferred between devices; and
(c) the **control bus** which provides timing and control signals throughout the system.

The number of individual lines present within the address bus and data bus depends upon the particular microprocessor employed. Signals on all lines, no matter whether they are used for address, data, or control, can exist in only two basic states: logic 0 (**low**) or logic 1 (**high**). Data and addresses are represented by **binary numbers** (a sequence of 1s and 0s) that appear respectively on the data and address bus.

Many microprocessors designed for control and instrumentation applications make use of an 8-bit data bus and a 16-bit address bus. Others have data and address buses which can operate with as many as 128-bits at a time.

The largest binary number that can appear on an 8-bit data bus corresponds to the condition when all eight lines are at logic 1. Therefore the largest value of data that can be present on the bus at any instant of time is equivalent to the binary number 11111111 (or 255). Similarly, most the highest address that can appear on a 16-bit address bus is 1111111111111111 (or 65,535). The full range of data values and addresses for a simple microprocessor of this type is thus:

Data	from	00000000
	to	11111111
Addresses	from	0000000000000000
	to	1111111111111111

Data representation

Binary numbers – particularly large ones – are not very convenient. To make numbers easier to handle we often convert binary numbers to **hexadecimal** (base 16). This format is easier for mere humans to comprehend and offers the advantage over denary (base 10) in that it can be converted to and from binary with ease. The first sixteen numbers in binary, denary, and hexadecimal are shown in the table below. A single hexadecimal character (in the range zero to F) is used to represent a group of four binary digits (bits). This group of four bits (or single hex. character) is sometimes called a **nibble**.

A **byte** of data comprises a group of eight bits. Thus a byte can be represented by just two hexadecimal (hex) characters. A group of sixteen bits (a word) can be represented by four hex characters, thirty-two bits (a double word by eight hex. characters, and so on).

The value of a byte expressed in binary can be easily converted to hex by arranging the bits in groups of four and converting each nibble into hexadecimal using the table shown below:

Table 11.1 Binary, denary and hexadecimal

Binary (base 2)	*Denary (base 10)*	*Hexadecimal (base 16)*
0000	0	0
0001	1	1
0010	2	2
0011	3	3
0100	4	4
0101	5	5
0110	6	6
0111	7	7
1000	8	8
1001	9	9
1010	10	A
1011	11	B
1100	12	C
1101	13	D
1110	14	E
1111	15	F

Note that, to avoid confusion about whether a number is hexadecimal or decimal, we often place a $ symbol before a hexadecimal number or add an H to the end of the number. For example, 64 means decimal 'sixty-four'; whereas, $64 means hexadecimal 'six-four', which is equivalent to decimal 100. Similarly, 7FH means hexadecimal 'seven-F' which is equivalent to decimal 127.

Example 11.1

Convert hexadecimal A3 into binary.

Solution

From Table 11.1, A = 1010 and 3 = 0101. Thus A3 in hexadecimal is equivalent to 10100101 in binary.

Example 11.2

Convert binary 11101000 binary to hexadecimal.

Solution

From Table 11.1, 1110 = E and 1000 = 8. Thus 11101000 in binary is equivalent to E8 in hexadecimal.

Data types

A byte of data can be stored at each address within the total memory space of a microprocessor system. Hence one byte can be stored at each of the 65,536 memory locations within a microprocessor system having a 16-bit address bus.

Individual bits within a byte are numbered from 0 (least significant bit) to 7 (most significant bit). In the case of 16-bit words, the bits are numbered from 0 (least significant bit) to 15 (most significant bit).

Negative (or signed) numbers can be represented using **two's complement** notation where the leading (most significant) bit indicates the sign of the number (1 = negative, 0 = positive). For example, the signed 8-bit number 10000001 represents the denary number -1.

The range of integer data values that can be represented as bytes, words and long words are shown in Table 11.2.

Table 11.2   Data types

Data type	Bits	Range of values
Unsigned byte	8	0 to 255
Signed byte	8	-128 to $+127$
Unsigned word	16	0 to 65,535
Signed word	16	$-32,768$ to $+32,767$

Data storage

The semiconductor ROM within a microprocessor system provides storage for the program code as well as any permanent data that requires storage. All of this data is referred to as non-volatile because it remains intact when the power supply is disconnected.

The semiconductor RAM within a microprocessor system provides storage for the transient data and variables that are used by programs. Part of the RAM is also be used by the microprocessor as a temporary store for data whilst carrying out its normal processing tasks.

It is important to note that any program or data stored in RAM will be lost when the power supply is switched off or disconnected. The only exception to this is CMOS RAM that is kept alive by means of a small battery. This **battery-backed memory** is used to retain important data, such as the time and date.

When expressing the amount of storage provided by a memory device we usually use Kilobytes (Kbyte). It is important to note that a Kilobyte of memory is actually 1,024 bytes (not 1,000 bytes). The reason for choosing the Kbyte rather than the kbyte (1,000 bytes) is that 1,024 happens to be the nearest power of 2 (note that $2^{10} = 1,024$).

The capacity of a semiconductor ROM is usually specified in terms of an address range and the number of bits stored at each address. For example, 2 K × 8 bits (capacity 2 Kbytes), 4 K × 8 bits (capacity 4 Kbytes), and so on. Note that it is not always necessary (or desirable) for the entire memory space of a microprocessor to be populated by memory devices.

The microprocessor

The microprocessor **central processing unit** (**CPU**) forms the heart of any microprocessor or microcomputer system computer and, consequently, its operation is crucial to the entire system.

The primary function of the microprocessor is that of fetching, decoding, and executing instructions resident in memory. As such, it must be able to transfer data from external memory into its own internal registers and vice versa. Furthermore, it must operate predictably, distinguishing, for example, between an operation contained within an instruction and any accompanying addresses of read/write memory locations. In addition, various system housekeeping tasks need to be performed including being able to suspend normal processing in order to respond to an external device that needs attention.

The main parts of a microprocessor CPU are:

(a) **registers** for temporary storage of addresses and data;
(b) an **arithmetic logic unit** (ALU) that performs arithmetic and logic operations;
(c) a unit that receives and decodes instructions; and
(d) a means of controlling and timing operations within the system.

Figure 11.3 shows the principal internal features of a typical 8-bit microprocessor. We will briefly explain each of these features in turn:

Accumulator

The accumulator functions as a source and destination register for many of the basic microprocessor operations. As a **source register** it contains the data that will be used in a particular

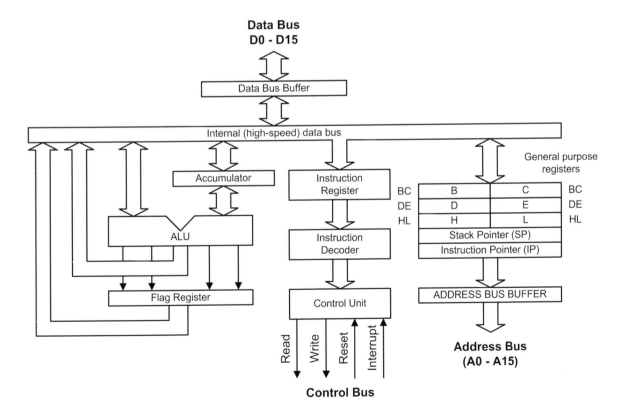

Figure 11.3   Internal architecture of a typical 8-bit microprocessor CPU

operation whilst as a **destination register** it will be used to hold the result of a particular operation. The accumulator (or **A-register**) features in a very large number of microprocessor operations, consequently more reference is made to this register than any others.

Instruction register

The instruction register provides a temporary storage location in which the current microprocessor instruction is held whilst it is being decoded. Program instructions are passed into the microprocessor, one at time, through the data bus.

On the first part of each **machine cycle**, the instruction is fetched and decoded. The instruction is executed on the second (and subsequent) machine cycles. Each machine cycle takes a finite time (usually less than a microsecond) depending upon the frequency of the microprocessor's clock.

Data bus (D0 to D7)

The external data bus provides a highway for data that links all of the system components (such as random access memory, read-only memory, and input/output devices) together. In an 8-bit system, the data bus has eight data lines, labelled D0 (the **least significant bit**) to D7 (**the most significant bit**) and data is moved around in groups of eight bits, or **bytes**. With a sixteen bit data bus the data lines are labelled D0 to D15, and so on.

Data bus buffer

The data bus buffer is a temporary register through which bytes of data pass on their way into, and out of, the microprocessor. The buffer is thus referred to as **bi-directional** with data passing out of the microprocessor on a **write operation** and into the processor during a **read operation**. The direction of data transfer is determined by the **control unit** as it responds to each individual program instruction.

Internal data bus

The internal data bus is a high-speed data highway that links all of the microprocessor's internal elements together. Data is constantly flowing backwards and forwards along the internal data bus lines.

General-purpose registers

Many microprocessor operations (for example, adding two 8-bit numbers together) require the use of more than one register. There is also a requirement for temporarily storing the partial result of an operation whilst other operations take place. Both of these needs can be met by providing a number of general-purpose registers. The use to which these registers are put is left mainly up to the programmer.

Stack pointer

When the time comes to suspend a particular task in order to briefly attend to something else, most microprocessors make use of a region of external random access memory (RAM) known as a **stack**. When the main program is interrupted, the microprocessor temporarily places in the stack the contents of its internal registers together with the address of the next instruction in the main program. When the interrupt has been attended to, the microprocessor recovers the data that has been stored temporarily in the stack together with the address of the next instruction within the main program. It is thus able to return to the main program exactly where it left off and with all the data preserved in its registers. The stack pointer is simply a register that contains the address of the last used stack location.

Program counter

Programs consist of a sequence of instructions that are executed by the microprocessor. These instructions are stored in external random access memory (RAM) or read-only memory (ROM). Instructions must be fetched and executed by the microprocessor in a strict sequence. By storing the address of the next instruction to be executed, the program counter allows the microprocessor to keep track of where it is within the program. The program counter is automatically incremented when each instruction is executed.

Address bus buffer

The address bus buffer is a temporary register through which addresses (in this case comprising 16-bits) pass on their way out of the microprocessor. In a simple microprocessor, the address buffer is unidirectional with addresses placed on the address bus during both read and write operations. The address bus lines are labelled A0 to A15, where A0 is the least-significant address bus line and A16 is the most significant address bus line. Note that a 16-bit address bus can be used to communicate with 65,536 individual memory locations. At each location a single byte of data is stored.

Control bus

The control bus is a collection of signal lines that are both used to control the transfer of data around the system and also to interact with external devices. The control signals used by microprocessors tend to differ with different types, however the following are commonly found:

READ   an output signal from the microprocessor that indicates that the current operation is a read operation

WRITE   an output signal from the microprocessor that indicates that the current operation is a write operation

RESET   a signal that resets the internal registers and initializes the program counter so that the program can be re-started from the beginning

IRQ   interrupt request from an external device attempting to gain the attention of the microprocessor (the request may be obeyed or ignored according to the state of the microprocessor at the time that the interrupt request is received).

NMI   non-maskable interrupt (i.e. an interrupt signal that cannot be ignored by the microprocessor).

Address bus (A0 to A15)

The address bus provides a highway for addresses that links with all of the system components (such as random access memory, read-only memory, and input/output devices). In a system with a 16-bit address bus, there are sixteen address lines, labelled A0 (the least significant bit) to A15 (the most significant bit). In a system with a 32-bit address bus there are 32 address lines, labelled A0 to A31, and so on.

Instruction decoder

The instruction decoder is nothing more than an arrangement of logic gates that acts on the bits stored in the instruction register and determines which instruction is currently being referenced. The instruction decoder provides output signals for the microprocessor's control unit.

Control unit

The control unit is responsible for organizing the orderly flow of data within the microprocessor as well as generating, and responding to, signals on the control bus. The control unit is also responsible for the timing of all data transfers. This process is synchronized using an internal or external clock signal (not shown in Fig. 11.3).

Arithmetic logic unit (ALU)

As its name suggests, the ALU performs arithmetic and logic operations. The ALU has two inputs (in this case these are both 8-bits wide). One of these inputs is derived from the Accumulator whilst the other is taken from the internal data bus via a temporary register (not shown in Fig. 11.3). The operations provided by the ALU usually include addition, subtraction, logical AND, logical OR, logical exclusive-OR, shift left, shift right, etc. The result of most ALU operations appears in the accumulator.

Flag register (or status register)

The result of an ALU operation is sometimes important in determining what subsequent action takes place. The flag register contains a number of individual bits that are set or reset according to the outcome of an ALU operation. These bits are referred to as flags. The following flags are available in most microprocessors:

ZERO   the zero flag is set when the result of an ALU operation is zero (i.e. a byte value of 00000000)

CARRY   the carry flag is set whenever the result of an ALU operation (such as addition) generates a carry bit (in other words, when the result cannot be contained within an 8-bit register)

INTERRUPT the interrupt flag indicates whether external interrupts are currently enabled or disabled.

Clocks

The clock used in a microprocessor system is simply an accurate and stable square wave generator. In most cases the frequency of the square wave generator is determined by a quarts crystal. A simple 4 MHz square wave clock oscillator (together with the clock waveform that is produces) is shown in Fig. 11.4. Note that one complete clock cycle is sometimes referred to as a T-state.

Microprocessors sometimes have an internal clock circuit in which case the quartz crystal (or other resonant device) is connected directly to pins on the microprocessor chip. In Fig. 11.5(a) an external clock is shown connected to a microprocessor whilst in Fig.11.5(b) and internal clock oscillator is used.

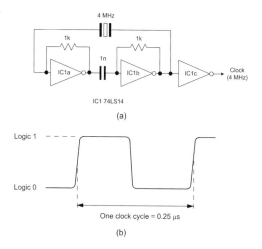

(a)

(b)

Figure 11.4 (a) A typical microprocessor clock circuit (b) waveform produced by the clock circuit

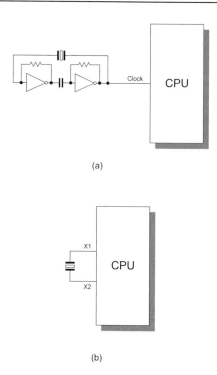

(a)

(b)

Figure 11.5 (a) An external CPU clock, and (b) an internal CPU clock

Microprocessor operation

The majority of operations performed by a microprocessor involve the movement of data. Indeed, the program code (a set of instructions stored in ROM or RAM) must itself be fetched from memory prior to execution. The microprocessor thus performs a continuous sequence of instruction fetch and execute cycles. The act of fetching an instruction code (or operand or data value) from memory involves a read operation whilst the act of moving data from the microprocessor to a memory location involves a write operation – see Fig. 11.6.

Each cycle of CPU operation is known as a machine cycle. Program instructions may require several machine cycles (typically between two and five). The first machine cycle in any cycle consists of an instruction fetch (the instruction code is read from the memory) and it is known as the M1 cycle. Subsequent cycles M2, M3, and so on, depend on the type of instruction that is being executed. This fetch-execute sequence is shown in Fig. 11.7.

Microprocessors determine the source of data (when it is being read) and the destination of data (when it is being written) by placing a unique address on the address bus. The address at which the data is to be placed (during a write operation) or from which it is to be fetched (during a read operation) can either constitute part of the memory of the system (in which case it may be within ROM

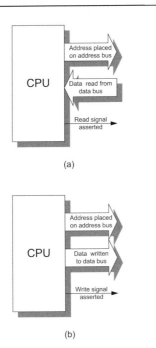

(a)

(b)

Figure 11.6   (a) Read, and (b) write operations

or RAM) or it can be considered to be associated with input/output (I/O).

Since the data bus is connected to a number of VLSI devices, an essential requirement of such chips (e.g., ROM or RAM) is that their data outputs should be capable of being isolated from the bus whenever necessary. These chips are fitted with select or enable inputs that are driven by address decoding logic (not shown in Fig. 11.2). This logic ensures that ROM, RAM and I/O devices never simultaneously attempt to place data on the bus!

The inputs of the address decoding logic are derived from one, or more, of the address bus lines. The address decoder effectively divides the available memory into blocks corresponding to a particular function (ROM, RAM, I/O, etc). Hence, where the processor is reading and writing to RAM, for example, the address decoding logic will ensure that only the RAM is selected whilst the ROM and I/O remain isolated from the data bus.

Within the CPU, data is stored in several registers. Registers themselves can be thought of as a simple pigeon-hole arrangement that can store as many bits as there are holes available. Generally, these devices can store groups of sixteen or thirty-two bits. Additionally, some registers may be configured as either one register of sixteen bits or two registers of thirty-two bits.

Some microprocessor registers are accessible to the programmer whereas others are used by the microprocessor itself. Registers may be classified as either general purpose or dedicated. In the latter case a particular function is associated with the register, such as holding the result of an operation or signalling the result of a comparison. A typical microprocessor and its register model is shown in Fig. 11.8.

The arithmetic logic unit

The ALU can perform arithmetic operations (addition and subtraction) and logic (complementation, logical AND, logical OR, etc). The ALU operates on two inputs (sixteen or thirty-two bits in length depending upon the CPU type) and it provides one output (again of sixteen or thirty-two bits). In addition, the ALU status is

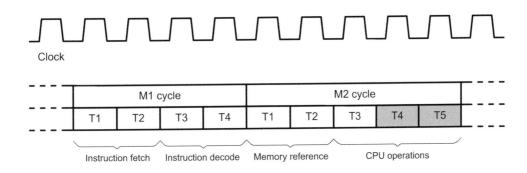

Figure 11.7   A typical timing diagram for a microprocessor's fetch-execute cycle

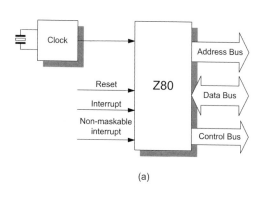

Main register set

Accumulator (A)	Flags (F)
(B)	(C)
(D)	(E)
(H)	(L)

Special purpose registers

Interrupt Vector (I)	Memory Refresh (R)
Index Register (IX)	
Index Register (IY)	
Stack Pointer (SP)	
Program Counter (PC)	

(b)

Figure 11.8 The Z80 microprocessor (showing some of its more important control signals) together with its register model

preserved in the **flag register** so that, for example, an overflow, zero or negative result can be detected.

The control unit is responsible for the movement of data within the CPU and the management of control signals, both internal and external. The control unit asserts the requisite signals to read or write data as appropriate to the current instruction.

Input and output

The transfer of data within a microprocessor system involves moving groups of 8, 16 or 32-bits using the bus architecture described earlier. Consequently it is a relatively simple matter to transfer data into and out of the system in parallel form. This process is further simplified by using a **Programmable Parallel I/O** device (a Z80PIO, 8255, or equivalent VLSI chip). This device provides registers for the temporary storage of data that not only buffer the data but also provide a degree of electrical isolation from the system data bus.

Parallel data transfer is primarily suited to high-speed operation over relatively short distances, a typical example being the linking of a microcomputer to an adjacent dot matrix printer.

There are, however, some applications in which parallel data transfer is inappropriate, the most common example being data communication by means of telephone lines. In such cases data must be sent serially (one bit after another) rather than in parallel form.

To transmit data in serial form, the parallel data from the microprocessor must be reorganized into a stream of bits. This task is greatly simplified by using an LSI interface device that contains a shift register that is loaded with parallel data from the data bus. This data is then read out as a serial bit stream by successive shifting. The reverse process, serial-to-parallel conversion, also uses a shift register. Here data is loaded in serial form, each bit shifting further into the register until it becomes full. Data is then placed simultaneously on the parallel output lines. The basic principles of parallel-to-serial and serial-to-parallel data conversion are illustrated in Fig. 11.9.

An example program

The following example program (see Table 11.3) is written in assembly code. The program transfers 8-bit data from an input port (Port A), complements

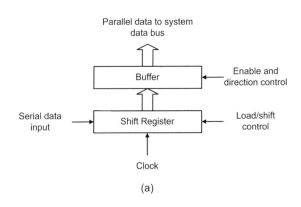

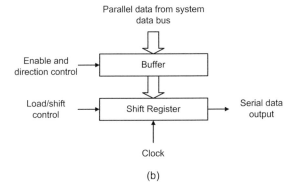

Figure 11.9 (a) Serial-to-parallel data conversion, and (b) parallel-to-serial data conversion

Table 11.3 A simple example program

Address	Data	Assembly code	Comment
2002	DB FF	IN A, (FFH)	Get a byte from Port A
2002	2F	CPL	Invert the byte
2003	D3 FE	OUT (FEH), A	Output the byte to Port B
2005	C3 00 20	JP 2000	Go round again

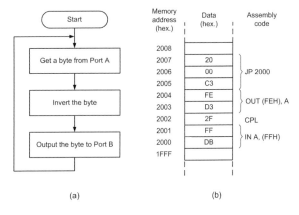

Figure 11.10 (a) Flowchart for the example program and (b) the eight bytes of program code stored in memory

(i.e. inverts) the data (by changing 0's to 1's and 1's to 0's in every bit position) and then outputs the result to an output port (Port B). The program repeats indefinitely.

Just three microprocessor instructions are required to carry out this task together with a fourth (jump) instruction that causes the three instructions to be repeated over and over again. A program of this sort is most easily written in assembly code which consists of a series of easy to remember mnemonics. The flowchart for the program is shown in Fig. 11.10(a).

The program occupies a total of eight bytes of memory, starting at a hexadecimal address of 2000 as shown in Fig. 11.10(b). You should also note that the two ports, A and B, each have unique addresses; Port A is at hexadecimal address FF whilst Port B is at hexadecimal address FE.

Interrupts

A program that simply executes a loop indefinitely has a rather limited practical application. In most microprocessor systems we want to be able to interrupt the normal sequence of program flow in order to alert the microprocessor to the need to do something. We can do this with a signal known as an **interrupt**. There are two types of interrupt; maskable and non-maskable.

When a **non-maskable interrupt** input is asserted, the processor must suspend execution of the current instruction and respond immediately to the interrupt. In the case of a **maskable interrupt**, the processor's response will depend upon whether interrupts are currently enabled or disabled (when enabled, the CPU will suspend its current task and carry out the requisite interrupt service routine).

The response to interrupts can be enabled or disabled by means of appropriate program instructions. In practice, interrupt signals may be generated from a number of sources and since each will require its own customized response a mechanism must be provided for identifying the source of the interrupt and calling the appropriate interrupt service routine. In order to assist in this task, the microprocessor may use a dedicated programmable interrupt controller chip.

A microcontroller system

Figure 11.11 shows the arrangement of a typical microcontroller system. The sensed quantities (temperature, position, etc.) are converted to corresponding electrical signals by means of a number of sensors. The outputs from the sensors (in either digital or analogue form) are passed as input signals to the microcontroller. The microcontroller also accepts inputs from the user. These user set options typically include target values for variables (such as desired room temperature), limit values (such as maximum shaft speed), or time constraints (such as 'on' time and 'off' time, delay time, etc).

The operation of the microcontroller is controlled by a sequence of software instructions known as a control program. The control program operates continuously, examining inputs from sensors, user settings, and time data before making changes to the output signals sent to one or more controlled devices.

The controlled quantities are produced by the controlled devices in response to output signals from the microcontroller. The controlled device generally converts energy from one form into energy in another form. For example, the controlled device might be an electrical heater that converts electrical energy from the AC mains supply into heat energy thus producing a given temperature (the controlled quantity).

In most real-world systems there is a requirement for the system to be automatic or self-regulating. Once set, such systems will continue to operate without continuous operator intervention. The output of a self-regulating system is fed back to its input in order to produce what is known as a closed loop system. A good example of a closed-loop system is a heating control system that is designed to maintain a constant room temperature and humidity within a building regardless of changes in the outside conditions.

In simple terms, a microcontroller must produce a specific state on each of the lines connected to its output ports in response to a particular combination of states present on each of the lines connected to its input ports (see Fig. 11.11). Microcontrollers must also have a central processing unit (CPU) capable of performing simple arithmetic, logical and timing operations.

The input port signals can be derived from a number of sources including:

- switches (including momentary action push-buttons)
- sensors (producing logic-level compatible outputs)
- keypads (both encoded and unencoded types).

The output port signals can be connected to a number of devices including:

- LED indicators (both individual and multiple bar types)
- LED seven segment displays (via a suitable interface)
- motors and actuators (both linear and rotary types) via a suitable buffer/driver or a dedicated interface)
- relays (both conventional electromagnetic types and optically couple solid-state types)
- transistor drivers and other solid-state switching devices.

Input devices

Input devices supply information to the computer system from the outside world. In an ordinary personal computer, the most obvious input device is the keyboard. Other input devices available on a PC are the mouse (pointing device), scanner and modem. Microcontrollers use much simpler input devices. These need be nothing more than individual switches or contacts that make and break but many other types of device are also used including many types of sensor that provide logic

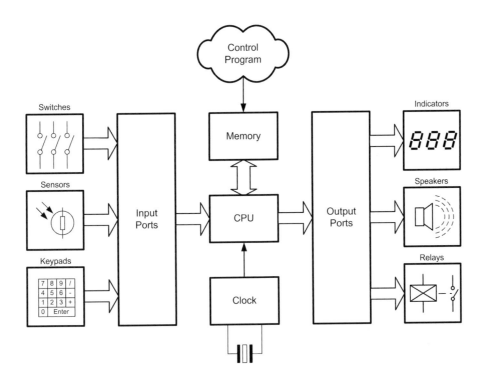

Figure 11.11 A microcontroller system with typical inputs and outputs

level outputs (such as float switches, proximity detectors, light sensors, etc).

It is important to note that, in order to be connected directly to the input port of a microcontroller, an input device must provide a logic compatible signal. This is because microcontroller inputs can only accept digital input signals with the same voltage levels as the logic power source. The 0 V ground level (often referred to as V_{SS} in the case of a CMOS microcontroller) and the positive supply (V_{DD} in the case of a CMOS microcontroller) is invariably 5 V ± 5%. A level of approximately 0V indicates a logic 0 signal and a voltage approximately equal to the positive power supply indicates a logic 1 signal.

Other input devices may sense analogue quantities (such as velocity) but use a digital code to represent their value as an input to the microcontroller system. Some microcontrollers provide an internal analogue-to-digital converter (ADC) in order to greatly simplify the connection of analogue sensors as input devices but where this

facility isn't available it will be necessary to use an external ADC which usually takes the form of a single integrated circuit. The resolution of the ADC will depend upon the number of bits used and 8, 10, and 12-bit devices are common in control applications.

Output devices

Output devices are used to communicate information or actions from a computer system to the outside world. In a personal computer system, the most common output device is the CRT (cathode ray tube) display. Other output devices include printers and modems. As with input devices, microcontroller systems often use much simpler output devices. These may be nothing more than LEDs, piezoelectric sounders, relays and motors. In order to be connected directly to the output port of a microcontroller, an output device must, once again, be able to accept a logic compatible signal.

Where analogue quantities (rather than simple digital on/off operation) are required at the output a digital-to-analogue converter (DAC) will be needed. All of the functions associated with a DAC can be provided by a single integrated circuit. As with an ADC, the output resolution of a DAC depends on the number of bits and 8, 10, and 12-bits are common in control applications.

Interface circuits

Finally, where input and output signals are not logic compatible (i.e. when they are outside the range of signals that can be connected directly to the microcontroller) some additional interface circuitry may be required in order to shift the voltage levels or to provide additional current drive. Additional circuitry may also be required when a load (such as a relay or motor) requires more current than is available from a standard logic device or output port. For example, a common range of interface circuits (solid-state relays) is available that will allow a microcontroller to be easily interfaced to an AC mains-connected load. It then becomes possible for a small microcontroller (operating from only a 5 V DC supply) to control a central heating system operating from 240 V AC mains.

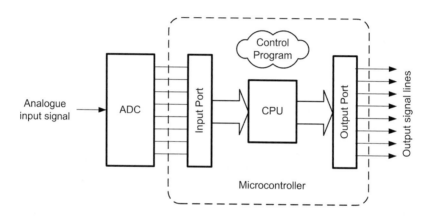

Figure 11.12   An analogue input signal can be connected to a microcontroller input port via an analogue-to-digital converter (ADC)

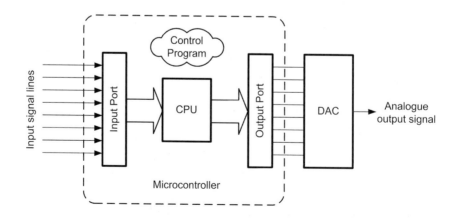

Figure 11.13   An analogue output signal can be produced by connecting a digital-to-analogue converter (DAC) to a microcontroller output power

Practical investigation

Objective

To investigate the operation of a Z80 microprocessor using a simulator to run a simple assembly language program.

Simulator

Oshonsoft Z80 Simulator or similar Integrated Development Environment (IDE). The Oshonsoft IDE can be downloaded from www.oshonsoft.com. Alternatively, a Z80 development system with resident assembler and I/O ports can be used (note that port addresses used in the Practical Investigation may need to be modified in order to agree with those available).

System configuration and required operation

For the purposes of this investigation we shall assume that the system has eight input switches connected to port address 80H and eight LED indicators connected to output port address 82H. The LEDs are to be operated from their corresponding input switches on the following basis:

Inputs (Port 80H)

Switch 'on' = logic 1
Switch 'off' = logic 0

Outputs (Port 82H)

Logic 1 = LED 'off'
Logic 0 = LED 'on'

Note that, in order to illuminate an LED the logic 1 input must be inverted in order to produce a logic 0 output. Hence it will be necessary to read the byte from port 80H, invert (i.e. complement) it, and then output it to port 82H.

The program is to continue to run, detecting the state of the input switches and turning the appropriate LEDs on, until such time as all of the input switches are set to the 'off' position. If this situation is detected the program is to halt.

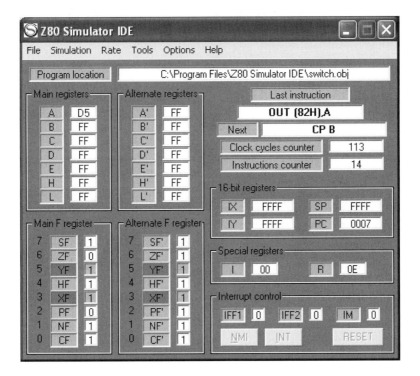

Figure 11.14 The Z80 Simulator Integrated Development Environment (IDE)

Procedure

Start the Z80 Simulator (see Fig. 11.14) then select Tools and Assembler and enter the assembly language code shown in Fig. 11.15. When complete select File and Save your assembly language source code with a suitable name (e.g. switch.asm).

Next select Tools and Assemble. Correct any errors in the assembly language source code and repeat the process until the source code assembles without error. Save the final assembly code and then quit the assembler in order to return to the main IDE screen.

Select File and Load Program and then select the object code (switch.obj) that has just been produced by the assembler. You will then need to configure the input and output ports before you run the program. You can do this by selecting Tools and Peripheral Devices (see Fig. 11.16). Configure the peripheral devices so that the input port address is 80H and the output port address to 82H. Then set some of the input bits by clicking on the indicators (which will change colour).

Next, return to the main IDE, select Tools and Simulation Log Viewer and then Rate and Slow. Finally, select Simulation and Start (or press the F1 key). You will see the program instructions and the contents of the Z80's registers displayed as the program is executed (see Fig. 11.17). If you have left the Peripheral Devices window open you will also be able to change the switch settings and observe the effect that this has. Finally, you should set all of the switches to the 'off' position and check that the program exits from the loop and reaches the HALT instruction.

A further program

If you have been able to run the simple program successfully you might like to try your hand at developing your own assembly language program using some of the assembly language instructions shown in Table 11.4. This program should shift the bits on output Port 82H to the left by the number of places indicated by data from Port 80H. The two ports can be preset before running the program.

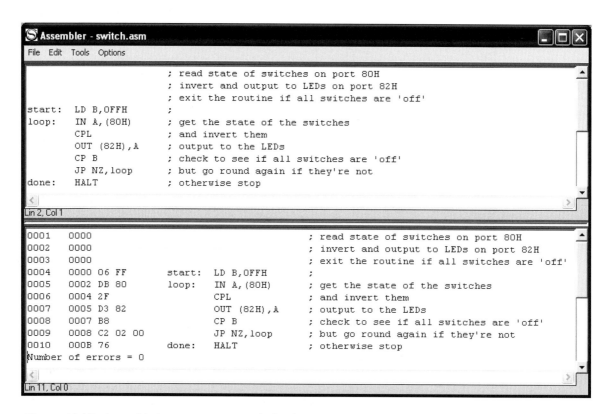

Figure 11.15 Assembly language source code for the Practical investigation

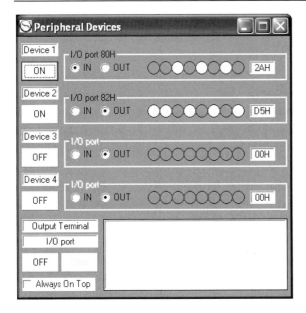

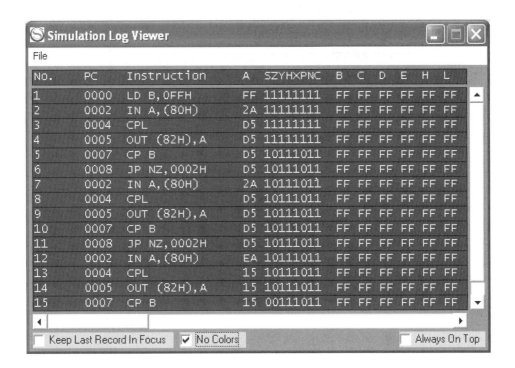

Table 11.4 Some selected Z80 assembly language instructions

Assembly code	Meaning
LD A, data	Load the Accumulator with 8-bit data
LD B, data	Load the B register with 8-bit data
LD C, data	Load the C register with 8-bit data
LD C,A	Load the C register with 8-bit data from the Accumulator
CP B	Compare the value in the B register with the value in the Accumulator
DEC C	Decrement the C register
CPL	Complement (i.e. invert) the contents of the Accumulator
SLA A	Shift the contents of the Accumulator left by one bit
JP NZ, label	Jump to the symbolic address label of the Zero flag has been set
HALT	Halt (suspend program execution)
IN A, (port)	Input the data from the specified port to the Accumulator
OUT (port), A	Output the data from the Accumulator to the specified port

Figure 11.16 Input and output data can be set and viewed (respectively) using the Peripheral Devices dialogue

Figure 11.17 Simulation log showing execution of individual program instructions

Symbols introduced in this chapter

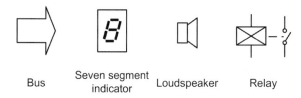

Bus   Seven segment   Loudspeaker   Relay
         indicator

Figure 11.18   Symbols introduced in this chapter

Problems

11.1   Convert 3A hexadecimal to binary.

11.2   Convert 11000010 binary to hexadecimal.

11.3   Convert 63 decimal to
       (a) binary
       (b) hexadecimal.

11.4   Which of the following numbers is the largest?
       (a) 19H
       (b) $13
       (c) 33_{10}
       (d) 11101_2.

11.5   How many unique addresses are available to a microprocessor CPU that has a 20-bit address bus?

11.6   What is the largest unsigned data value that can appear on a 10-bit data bus?

11.7   What is the largest negative data value that can be represented using signed 16-bit binary numbers?

11.8   The following fragment of assembly language code is executed using a Z80 microprocessor:
       IN A, (FEH)
       CPL
       OUT (FFH), A
       HALT
       (a) What are the addresses of the input and output ports?
       (b) If a data value of 10101111 appears at the input port what value will appear at the output port after the code has been executed?

11.9   Give examples of (a) two input devices, and (b) two output devices commonly used in microprocessor systems.

11.10 Explain the purpose of the circuit shown in Fig. 11.19 and state the function of the components marked A, B and C.

11.11 Sketch two cycles of the typical output waveform produced by the circuit shown in Fig. 11.19. Include labelled axes of time and voltage.

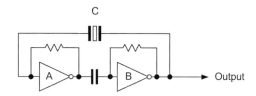

Figure 11.19   See Questions 11.10 and 11.11

11.12 Identify, and briefly explain the purpose of, the features labelled P, Q, R, S, T, U, and V in the microcomputer system shown in Fig. 11.20.

11.13 Explain the function of four common control bus signals used in a microcomputer system.

11.14 Explain the need for an ADC when a temperature sensor is to be interfaced to a microcomputer system.

Answers to these problems appear on page 376.

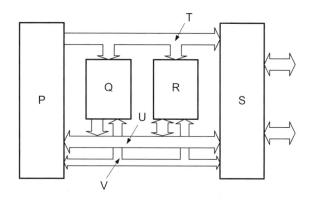

Figure 11.20   See Question 11.12

12

The 555 timer

The 555 timer is without doubt one of the most versatile integrated circuit chips ever produced. Not only is it a neat mixture of analogue and digital circuitry but its applications are virtually limitless in the world of timing and digital pulse generation. The device also makes an excellent case study for newcomers to electronics because it combines a number of important concepts and techniques.

Internal features

To begin to understand how timer circuits operate, it is worth spending a few moments studying the internal circuitry of the 555 timer, see Fig. 12.1. Essentially, the device comprises two operational amplifiers (used as comparators – see page 166) together with an R-S bistable element (see page 188). In addition, an inverting buffer (see page 185) is incorporated so that an appreciable current can be delivered to a load. The main features of the device are shown in Table 12.1.

Unlike standard TTL logic devices, the 555 timer can both **source** and **sink** current. It's worth taking a little time to explain what we mean by these two terms:

(a)   When **sourcing** current, the 555's output (pin-3) is in the **high** state and current will then flow *out of* the output pin into the load and down to 0V, as shown in Fig. 12.2(a).

(b)   When **sinking** current, the 555's output (pin-3) is in the **low** state in which case current will flow from the positive supply (+Vcc) through the load and *into* the output (pin-3), as shown in Fig. 12.2(b).

Returning to Fig. 12.1, the single transistor switch, TR1, is provided as a means of rapidly discharging an external timing capacitor. Because the series chain of resistors, $R1$, $R2$ and $R3$, all have identical values, the supply voltage (V_{CC}) is divided equally across the three resistors. Hence the voltage at the non-inverting input of IC1 is one-third of the

Table 12.1   Internal features of the 555 timer

Feature	Function
A	A potential divider comprising R_1, R_2 and R_3 connected in series. Since all three resistors have the same values the input voltage (V_{CC}) will be divided into thirds, i.e. one third of V_{CC} will appear at the junction of R_2 and R_3 whilst two thirds of V_{CC} will appear at the junction of R_1 and R_2.
B	Two operational amplifiers connected as comparators. The operational amplifiers are used to examine the voltages at the **threshold** and **trigger** inputs and compare these with the fixed voltages from the potential divider (two thirds and one third of V_{CC} respectively).
C	An R–S bistable stage. This stage can be either **set** or **reset** depending upon the output from the comparator stage. An external reset input is also provided.
D	An open-collector transistor switch. This stage is used to discharge an external capacitor by effectively shorting it out whenever the base of the transistor is driven positive.
E	An inverting power amplifier. This stage is capable of **sourcing** and **sinking** enough current (well over 100 mA in the case of a standard 555 device) to drive a small relay or another low-resistance load connected to the output.

supply voltage (V_{CC}) whilst that at the inverting input of IC2 is two-thirds of the supply voltage (V_{CC}). Hence if V_{CC} is 9 V, 3 V will appear at each resistor and the upper comparator will have 6 V applied to its inverting input whilst the lower comparator will have 3V at its non-inverting input.

The 555 family

The standard 555 timer is housed in an 8-pin dual-in-line (DIL) package and operates from supply rail voltages of between 4.5 V and 15 V. This, of

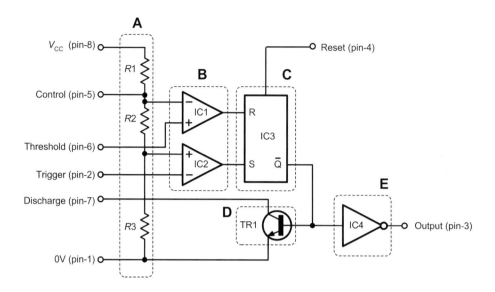

Figure 12.1   Internal arrangement of a 555 timer

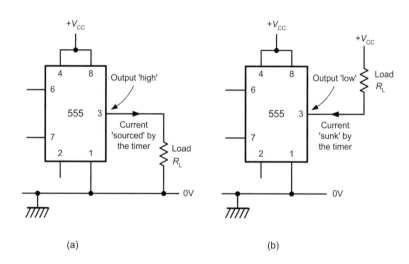

(a)   (b)

Figure 12.2   Loads connected to the output of a 555 timer: (a) current sourced by the timer when the output is high (b) current sunk by the timer when the output is low

course, encompasses the normal range for TTL devices (5 V ± 5%) and thus the device is ideally suited for use with TTL circuitry.

Several versions of the 555 timer are available, including low power (CMOS) and dual versions, as follows:

Low power (CMOS) 555

This device is a CMOS version of the 555 timer that is both pin and function compatible with its standard counterpart. By virtue of its CMOS technology the device operates over a somewhat wider range of supply voltages (2 V to 18 V) and consumes minimal operating current (120 mA) typical for an 18 V supply). Note that, by virtue of the low-power CMOS technology employed, the device does not have the same output current drive as that possessed by its standard counterparts. It can, however supply up to two standard TTL loads.

Dual 555 timer (e.g. NE556A)

This is a dual version of the standard 555 timer housed in a 14-pin DIL package. The two devices may be used entirely independently and share the same electrical characteristics as the standard 555.

Low-power (CMOS) dual 555 (e.g. ICM75561PA)

This is a dual version of the low-power CMOS 555 timer contained in a 14-pin DIL package. The two devices may again be used entirely independently and share the same electrical characteristics as the low-power CMOS 555.

Pin connecting details for the above devices can be found in Appendix 4.

Monostable pulse generator

Figure 12.3 shows a standard 555 timer operating as a **monostable pulse generator**. The monostable timing period (i.e. the time for which the output is high) is initiated by a falling edge trigger pulse applied to the **trigger** input (pin-2), see Fig. 12.4.

When this falling edge trigger pulse is received and falls below one-third of the supply voltage, the output of IC2 goes **high** and the bistable will be placed in the **set** state. The inverted Q output of the bistable then goes **low**, TR1 is placed in the **off** (non-conducting) state and the output voltage (pin-3) goes high.

The capacitor, C, then charges through the series resistor, R, until the voltage at the threshold input reaches two-thirds of the supply voltage (V_{cc}). At this point, the output of the upper comparator changes state and the bistable is **reset**. The inverted Q output then goes high, TR1 is driven into conduction and the final output goes **low**. The device then remains in the inactive state until another falling trigger pulse is received.

The trigger and output waveforms produced by the circuit of Fig. 12.3 are shown in Fig. 12.4. The waveform has the following properties:

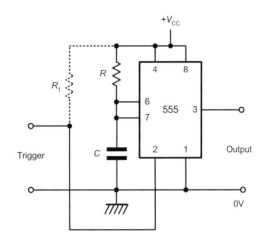

Figure 12.3   555 monostable configuration. C and R are the timing components

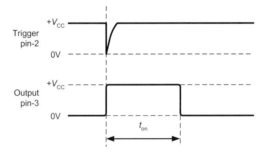

Figure 12.4   555 monostable configuration. C and R are the timing components

Time for which output is high: $t_{on} = 1.1\,CR$

Recommended trigger pulse width: $t_{tr} < \dfrac{t_{on}}{4}$

where t_{on} and t_{tr} are in seconds, C is in Farads and R is in ohms.

The period of the 555 monostable output can be changed very easily by simply altering the values of the timing resistor, R, and/or timing capacitor, C. Doubling the value of R will double the timing period. Similarly, doubling the value of C will double the timing period.

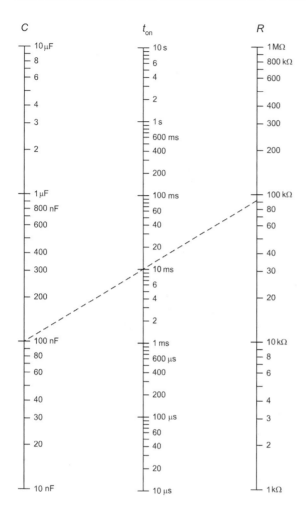

Figure 12.5   Chart for determining values of C, t_{on} and R for a 555 operating in monostable mode. The dotted line shows how a 10 ms pulse will be produced when $C = 100$ nF and $R = 91$ kΩ (see Example 12.1)

Values for C and R can be selected over quite a wide range but it is worth noting that the performance of the timer may become unpredictable if the values of these components are outside the recommended range:

$C = 470$ pF to 470 μF

$R = 1$ kΩ to 3.3 MΩ

For any particular monostable timing period, the required values for C and R can be determined from the formula shown earlier or by using the graph shown in Fig. 12.5. The output period can be easily adjusted by making R a preset resistor with a value of about twice that of the calculated value.

Example 12.1

Design a timer circuit that will produce a 10 ms pulse when a negative-going trigger pulse is applied to it.

Solution

Using the circuit shown in Fig. 12.4, the value of monostable timing period can be calculated from the formula:

$$t_{on} = 1.1\,C\,R$$

We need to choose an appropriate value for C that is in the range stated earlier. Since we require a fairly modest time period we will choose a mid-range value for C. This should help to ensure that the value of R is neither too small nor too large. A value of 100 nF should be appropriate and should also be easy to obtain. Making R the subject of the formula and substituting for $C = 100$ nF gives:

$$R = \frac{t_{on}}{1.1C} = \frac{10 \text{ ms}}{1.1 \times 100 \text{ nF}} = \frac{10 \times 10^{-3}}{110 \times 10^{-9}}$$

From which:

$$R = \frac{10}{110} \times 10^6 = 0.091 \times 10^6 \ \Omega = 91 \text{ k}\Omega$$

Alternatively, the chart shown in Fig. 12.5 can be used.

Example 12.2

Design a timer circuit that will produce a +5V output for a period of 60 s when a 'start' button is operated. The time period is to be aborted when a 'stop' button is operated.

Solution

For the purposes of this question we shall assume that the 'start' and 'stop' buttons both have normally-open (NO) actions.

The value of monostable timing period can be calculated from the formula:

$$t_{on} = 1.1\,C\,R$$

We need to choose an appropriate value for C that is in the range stated earlier. Since we require a fairly long time period we will choose a relatively large value of C in order to avoid making the value of R too high. A value of 100 μF should be appropriate and should also be easy to obtain. Making R the subject of the formula and substituting for $C = 100$ μF gives:

$$R = \frac{t_{on}}{1.1C} = \frac{60\text{ s}}{1.1 \times 100\text{ μF}} = \frac{60}{110 \times 10^{-6}}$$

From which:

$$R = \frac{60}{110} \times 10^{6} = 0.545 \times 10^{6}\ \Omega = 545\text{ k}\Omega$$

In practice 560 kΩ (the nearest preferred value – see page 21) would be adequate.

The 'start' button needs to be connected between pin-2 and ground whilst the 'stop' button needs to be connected between pin-4 and ground. Each of the inputs requires a **pull-up resistor** to ensure that the input is taken high when the switch is not being operated. The precise value of the 'pull-up' resistor is unimportant and a value of 10 kΩ will be perfectly adequate in this application. The complete circuit of the 60 s timer is shown in Fig. 12.6.

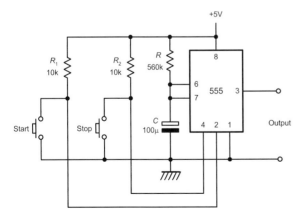

Figure 12.6   60 s timer (see Example 12.2)

Astable pulse generator

Figure 12.7 shows how the standard 555 can be used as an **astable pulse generator**. In order to understand how this circuit operates, assume that the output (pin-3) is initially high and that TR1 is in the non-conducting state. The capacitor, C, will begin to charge with current supplied by means of series resistors, $R1$ and $R2$.

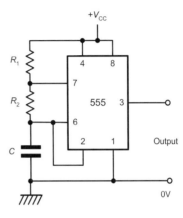

Figure 12.7   555 astable configuration

When the voltage at the **threshold** input (pin-6) exceeds two-thirds of the supply voltage the output of the upper comparator, IC1, will change state and the bistable will become **reset** due to voltage transition that appears at R. This, in turn, will make the inverted Q output go **high**, turning TR1 at the same time. Due to the inverting action of the buffer, IC4, the final output (pin-3) will go **low**.

The capacitor, C, will now discharge, with current flowing through $R2$ into the collector of TR1. At a certain point, the voltage appearing at the **trigger** input (pin-2) will have fallen back to one-third of the supply voltage at which point the lower comparator will change state and the voltage transition at S will return the bistable to its original **set** condition. The inverted Q output then goes low, TR1 switches **off** (no longer conducting), and the output (pin-3) goes **high**. Thereafter, the entire **charge/discharge cycle** is repeated indefinitely.

The output waveform produced by the circuit of Fig. 12.7 is shown in Fig. 12.8. The waveform has

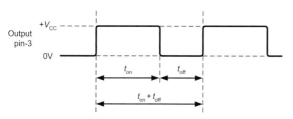

Figure 12.8   Waveforms for astable operation

the following properties:

Time for which output is high:

$$t_{on} = 0.693 \, C \, (R_1 + R_2)$$

Time for which output is low:

$$t_{off} = 0.693 \, C \, R_2$$

Period of output waveform:

$$t = t_{on} + t_{off} = 0.693 \, C \, (R_1 + 2R_2)$$

Pulse repetition frequency:

$$p.r.f. = \frac{1.44}{C(R_1 + 2R_2)}$$

Mark to space ratio:

$$\frac{t_{on}}{t_{off}} = \frac{R_1 + R_2}{R_2}$$

Duty cycle:

$$\frac{t_{on}}{t_{on} + t_{off}} = \frac{R_1 + R_2}{R_1 + 2R_2} \times 100\%$$

Where t is in seconds, C is in Farads, R_1 and R_2 are in ohms.

When $R_1 = R_2$ the duty cycle of the astable output from the timer can be found by letting $R = R_1 = R_2$. Hence:

$$\frac{t_{on}}{t_{off}} = \frac{R_1 + R_2}{R_2} = \frac{R + R}{R} = \frac{2}{1} = 2$$

In this case the duty cycle will be given by:

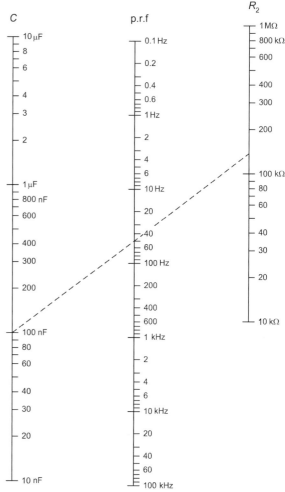

Figure 12.9   Chart for determining values of C, p.r.f. and R_2 for a 555 operating in astable mode where $R_2 \gg R_1$ (i.e. for a square wave output). The dotted line shows how a 50 Hz square wave will be produced when $C = 100$ nF and $R = 144$ kΩ (see Example 12.4)

$$\frac{t_{on}}{t_{on} + t_{off}} = \frac{R + R}{R + 2R} \times 100\% = \frac{2}{3} \times 100\% \simeq 67\%$$

The p.r.f. of the 555 astable output can be changed very easily by simply altering the values of R_1, R_2, and C. The values chosen for R_1, R_2 and C should normally be selected from within the following ranges in order to provide satisfactory performance:

$C = 10$ nF to 470 μF

$R_1 = 1$ kΩ to 1 MΩ

$R_2 = 1$ kΩ to 1 MΩ

The required values of C, R_1 and R_2 for any required p.r.f. and duty cycle can be determined from the formulae shown earlier. Alternatively, the graph shown in Fig. 12.9 can be used when R_1 and R_2 are equal in value (corresponding to a 67% duty cycle).

Square wave generators

Because the high time (t_{on}) is always greater than the low time (t_{off}), the mark to space ratio produced by a 555 timer can never be made equal to (or less than) unity. This could be a problem if we need to produce a precise square wave in which $t_{on} = t_{off}$. However, by making R_2 very much larger than R_1, the timer can be made to produce a reasonably symmetrical square wave output (note that the minimum recommended value for R_2 is 1 k —see earlier).

If $R_2 \gg R_1$ the expressions for p.r.f. and duty cycle simplify to:

$$\text{p.r.f.} \simeq \frac{0.72}{CR_2}$$

$$\frac{t_{on}}{t_{on} + t_{off}} \simeq \frac{R_2}{2R_2} \times 100\% = \frac{1}{2} \times 100\% = 50\%$$

Example 12.3

Design a pulse generator that will produce a p.r.f. of 10 Hz with a 67% duty cycle.

Solution

Using the circuit shown in Fig. 12.7, the value of p.r.f. can be calculated from:

$$\text{p.r.f.} = \frac{1.44}{C(R_1 + 2R_2)}$$

Since the specified duty cycle is 67% we can make R_1 equal to R_2. Hence if $R = R_1 = R_2$ we obtain the following relationship:

$$\text{p.r.f.} = \frac{1.44}{C(R + 2R)} = \frac{1.44}{3CR} = \frac{0.48}{CR}$$

We need to choose an appropriate value for C that is in the range stated earlier. Since we require a fairly low value of p.r.f. we will choose a value for C of 1μF. This should help to ensure that the value of R is neither too small nor too large. A value of 1 μF should also be easy to obtain. Making R the subject of the formula and substituting for $C = 1$ μF gives:

$$R = \frac{0.48}{\text{p.r.f.} \times C} = \frac{0.48}{\text{p.r.f.} \times 1 \times 10^{-6}}$$

$$R = \frac{480 \times 10^3}{100} = 4.8 \times 10^3 = 4.8 \text{ kΩ}$$

Example 12.4

Design a 5 V 50 Hz square wave generator using a 555 timer.

Solution

Using the circuit shown in Fig. 12.7, when $R_2 \gg R_1$, the value of p.r.f. can be calculated from:

$$\text{p.r.f.} \simeq \frac{0.72}{CR_2}$$

We shall use the minimum recommended value for R_1 (i.e. 10 kΩ) and ensure that the value of R_2 that we calculate from the formula is at least ten times larger in order to satisfy the criteria that R_2 should be very much larger than R_1. When selecting the value for C we need to choose a value that will keep the value of R_2 relatively large. A value of 100 nF should be about right and should also be easy to locate. Making R_2 the subject of the formula and substituting for $C = 100$ nF gives:

$$R_2 = \frac{0.72}{\text{p.r.f.} \times C} = \frac{0.72}{50 \times 100 \times 10^{-9}}$$

$$R_2 = \frac{0.72}{5 \times 10^{-6}} = 0.144 \times 10^6 = 144 \text{ k}\Omega$$

Alternatively, the chart shown in Fig. 12.9 can be used.

The value of R_2 is thus more than 100 times larger than the value that we are using for R_1. As a consequence the timer should produce a good square wave output.

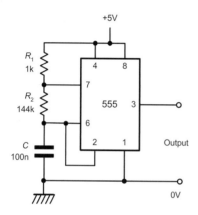

Figure 12.10 A 5 V 50 Hz square wave generator (see Example 12.4)

The complete circuit of the 5V 50 Hz square wave generator is shown in Fig. 12.10.

A variable pulse generator

Figure 12.11 shows how a variable pulse generator can be constructed using two 555 timer (or one 556 dual timer). The first timer, IC1, operates in astable mode whilst the second timer, IC2, operates in monostable mode. The p.r.f. generated by IC1 is adjustable by means of switch selected capacitors, C_1 to C_3, together with variable resistor, VR_1. The output from IC1 (pin-3) is fed via C_5 to the trigger input of IC2 (pin-2).

The monostable period of IC2 is adjustable by means of switch selected capacitors, C_6 to C_8, together with variable resistor, VR_2. The output from IC2 (pin-3) is fed to the output via VR_3.

The p.r.f. is adjustable over the range 10 Hz to 10 kHz whilst pulse widths can be varied from 50 μs to 50 ms. The output voltage is adjustable from 0 V to 10 V. Finally, $R5$, is included in order to limit the output current and provide a measure of protection in the event of a short-circuit present at the output.

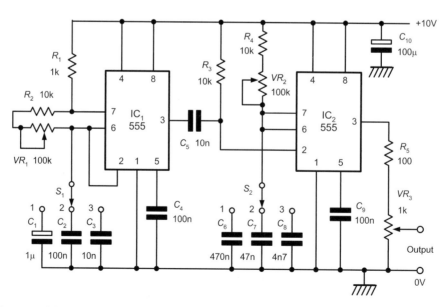

Figure 12.11 A variable pulse generator using two 555 timers

Practical investigation

Objective

To investigate the operation of a 555 monostable timer circuit.

Simulator

Breadboard, 5 V d.c. power supply, 555 timer, resistors of 10 kΩ (two required), 220 Ω, 100 kΩ, and 1 MΩ 5% 0.25 W, capacitors of 10 μF and 100 μF 16 V, LED, two normally open (NO) push-button switches, stopwatch or wristwatch with seconds display.

Procedure

Connect the circuit as shown in Fig. 12.12 with $C = 10$ μF and $R = 100$ kΩ. Connect the supply and press the 'stop' button. The LED should be off (indicating that the output is at 0 V).

Observe the time display and, at a convenient point, press the 'start' button. The LED should become illuminated after a period of about 1 s (this will probably be too short an interval to be measured accurately). Record the monostable time period (i.e. the time between pressing the 'start' button and the LED becoming illuminated) in Table 12.2.

Repeat the procedure for each of the remaining C–R values shown in Table 12.2. Note that you can interrupt the timing period at any time by pressing the 'stop' button.

Calculations and graph

Record your results in Table 12.2. For each pair of C–R values calculate the product of C (in μF) and R (in MΩ). Plot a graph showing corresponding values of monostable time plotted against corresponding values of $C \times R$ using the graph layout shown in Fig. 12.13.

Conclusions

Comment on the shape of the graph. Is this what you would expect? Measure the slope of the graph and use this to confirm the relationship for the monostable timing period quoted on page 219. If the graph is not linear can you suggest any reasons for this?

Further work

Connect a digital multimeter on the 20 V d.c. range so that you can accurately measure the d.c. voltage that appears between pin-6 (threshold input) and 0 V. With $C = 100$ μF and $R = 1$ MΩ press the 'start' button and then measure the voltage at pin-6 at intervals of 10 s over the range 0 to 120 s. Particularly note the voltage reached at the end of the monostable timing period (this should be exactly 2/3 of the supply voltage). Plot a graph of voltage against time and justify the shape of this graph.

Table 12.2   Table of results for the monostable timer circuit

C	R	Time
10 μF	100 kΩ	
10 μF	220 kΩ	
10 μF	470 kΩ	
10 μF	1 M kΩ	
100 μF	220 kΩ	
100 μF	470 kΩ	
100 μF	1 M kΩ	

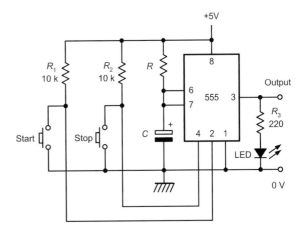

Figure 12.12   Monostable timer circuit used for the Practical investigation

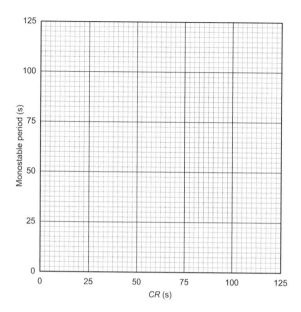

Figure 12.13   Graph layout for plotting the results

Important formulae introduced in this chapter

Monostable 555 timer:
(page 219)

$$t_{on} = 1.1\ CR$$

Astable 555 timer:
(page 222)

$$t_{on} = 0.693\ C(R_1 + R_2)$$

$$t_{off} = 0.693\ CR_2$$

$$t = 0.693\ C(R_1 + 2R_2)$$

$$\text{p.r.f.} = \frac{1.44}{C(R_1 + 2R_2)}$$

$$\frac{t_{on}}{t_{off}} = \frac{R_1 + R_2}{R_2}$$

$$\frac{t_{on}}{t_{on} + t_{off}} = \frac{R_1 + R_2}{R_1 + 2R_2} \times 100\%$$

When $R_2 \gg R_1$:
(page 223)

$$\text{p.r.f.} \simeq \frac{0.72}{CR_2}$$

$$\frac{t_{on}}{t_{on} + t_{off}} \simeq 50\%$$

Problems

12.1   Design a timer circuit that will produce a 10 V 2 ms pulse when a 10 V negative-going trigger pulse is applied to it.

12.2   Design a timer circuit that will produce time periods that can be varied from 1 s to 10 s. The timer circuit is to produce a +12V output.

12.3   Design a timer circuit that will produce a 67% duty cycle output at 400 Hz.

12.4   Design a timer circuit that will produce a square wave output at 1 kHz.

12.5   Refer to the variable pulse generator circuit shown in Fig. 12.11. Identify the component(s) that provides:
(a) variable adjustment of pulse width
(b) decade range selection of pulse width
(c) limits the range of variable adjustment of pulse width
(d) variable adjustment of p.r.f.
(e) decade range selection of p.r.f.
(f) limits the range of variable adjustment of p.r.f.
(g) variable adjustment of output amplitude
(h) protects IC_2 against a short-circuit connected at the output
(i) removes any unwanted signals appearing on the supply rail
(j) forms the trigger pulse required by the monostable stage

12.6   A 555 timer is rated for a maximum output current of 120 mA. What is the minimum value of load resistance that can be used if the device is to be operated from a 6 V d.c. supply?

Answers to these problem appear on page 376.

13

Radio

Maxwell first suggested the existence of electromagnetic waves in 1864. Later, Heinrich Rudolf Hertz used an arrangement of rudimentary resonators to demonstrate the existence of electromagnetic waves. Hertz's apparatus was extremely simple and comprised two resonant loops, one for transmitting and the other for receiving. Each loop acted both as a tuned circuit and as a resonant aerial. The transmitting loop was excited by means of an induction coil and battery. Some of the energy radiated by the transmitting loop was intercepted by the receiving loop and the received energy was conveyed to a spark gap where it could be released as an arc. The energy radiated by the transmitting loop was in the form of an **electromagnetic wave**—a wave that has both electric and magnetic field components and that travels at the speed of light.

In 1894, Marconi demonstrated the commercial potential of the phenomenon that Maxwell predicted and Hertz actually used in his apparatus. It was also Marconi that made radio a reality by pioneering the development of telegraphy without wires (i.e. wireless). Marconi was able to demonstrate very effectively that information could be exchanged between distant locations without the need for a 'land-line'.

Marconi's system of **wireless telegraphy** proved to be invaluable for maritime communications (ship to ship and ship to shore) and was to be instrumental in saving many lives. The military applications of radio were first exploited during the First World War (1914 to 1918) and, during that period, radio was first used in aircraft.

The radio frequency spectrum

Radio frequency signals are generally understood to occupy a frequency range that extends from a few tens of kilohertz (kHz) to several hundred Gigahertz (GHz). The lowest part of the radio frequency range that is of practical use (below 30 kHz) is only suitable for narrow-band communication. At this frequency, signals propagate as ground waves (following the curvature of the earth) over very long distances. At the other extreme, the highest frequency range that is of practical importance extends above 30GHz. At these microwave frequencies, considerable bandwidths are available (sufficient to transmit many television channels using point-to-point links or to permit very high definition radar systems) and signals tend to propagate strictly along line-of-sight paths.

At other frequencies signals may propagate by various means including reflection from ionized layers in the ionosphere. At frequencies between 3 MHz and 30 MHz ionospheric propagation regularly permits intercontinental broadcasting and communications.

For convenience, the radio frequency spectrum is divided into a number of bands, each spanning a decade of frequency. The use to which each frequency range is put depends upon a number of factors, paramount amongst which is the propagation characteristics within the band concerned. Other factors that need to be taken into account include the efficiency of practical aerial systems in the range concerned and the bandwidth available. It is also worth noting that, although it may appear from Fig. 13.1 that a great deal of the radio frequency spectrum is not used, it should be stressed that competition for frequency space is fierce. Frequency allocations are, therefore, ratified by international agreement and the various user services carefully safeguard their own areas of the spectrum.

Electromagnetic waves

As with light, radio waves propagate outwards from a source of energy (transmitter) and comprise electric (**E**) and magnetic (**H**) fields right angles to

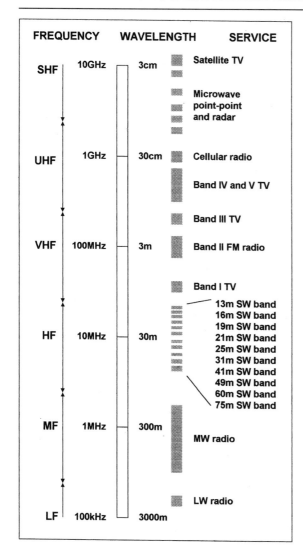

Figure 13.1 The radio frequency spectrum

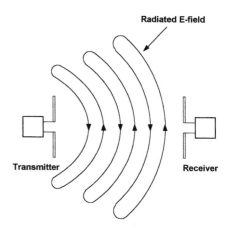

Figure 13.3   E-field lines between a transmitter and a receiver

one another. These two components, the **E-field** and the **H-field**, are inseparable. The resulting wave travels away from the source with the E and H lines mutually at right angles to the direction of **propagation**, as shown in Fig. 13.2.

Radio waves are said to be **polarized** in the plane of the electric (E) field. Thus, if the E-field is vertical, the signal is said to be vertically polarized whereas, if the E-field is horizontal, the signal is said to be horizontally polarized.

Figure 13.3 shows the electric E-field lines in the space between a transmitter and a receiver. The transmitter aerial (a simple dipole—see page 237) is supplied with a high frequency alternating current. This gives rise to an alternating electric field between the ends of the aerial and an alternating magnetic field around (and at right angles to) it.

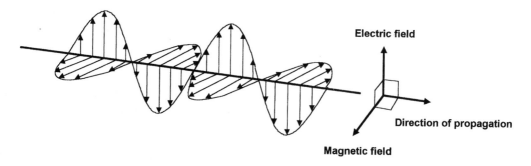

Figure 13.2 An electromagnetic wave

The direction of the E-field lines is reversed on each cycle of the signal as the **wavefront** moves outwards from the source. The receiving aerial intercepts the moving field and voltage and current is induced in it as a consequence. This voltage and current is similar (but of smaller amplitude) to that produced by the transmitter.

Frequency and wavelength

Radio waves propagate in air (or space) at the **speed of light** (300 million metres per second). The velocity of propagation, v, wavelength, λ, and frequency, f, of a radio wave are related by the equation:

$$v = f\lambda = 3 \times 10^8$$

This equation can be arranged to make f or λ the subject, as follows:

$$f = \frac{3 \times 10^8}{\lambda} \text{ Hz}$$

and

$$\lambda = \frac{3 \times 10^8}{f} \text{ m}$$

As an example, a signal at a frequency of 1 MHz will have a wavelength of 300 m whereas a signal at a frequency of 10 MHz will have a wavelength of 30 m.

When a radio wave travels in a cable (rather than in air or 'free space') it usually travels at a speed that is between 60% and 80% of that of the speed of light.

Example 13.1

Determine the frequency of a radio signal that has a wavelength of 15 m.

Solution

Using the formula:

$$f = \frac{3 \times 10^8}{\lambda} \text{ Hz}$$

where $\lambda = 15$ m gives:

$$f = \frac{3 \times 10^8}{15} = \frac{300 \times 10^6}{15} = 20 \times 10^6 \text{ Hz or 20 MHz}$$

Example 13.2

Determine the wavelength of a radio signal that has a frequency of 150 MHz.

Solution

Using the formula:

$$\lambda = \frac{3 \times 10^8}{f} \text{ m}$$

where $f = 150$ MHz gives:

$$\lambda = \frac{3 \times 10^8}{f} = \frac{3 \times 10^8}{150 \text{ MHz}} = \frac{300 \times 10^6}{150 \times 10^6} = \frac{300}{150} = 2 \text{ m}$$

Example 13.3

If the wavelength of a 30 MHz signal in a cable is 8 m, determine the velocity of propagation of the wave in the cable.

Solution

Using the formula:

$$v = f\lambda = 3 \times 10^8$$

where v is the velocity of propagation in the cable, gives:

$$v = f \lambda = 30 \text{ MHz} \times 8 \text{ m} = 2.4 \times 10^8 \text{m/s}$$

A simple CW transmitter and receiver

Figure 13.4 shows a simple radio communication system comprising a **transmitter** and **receiver** for use with **continuous wave** (CW) signals. Communication is achieved by simply switching (or 'keying') the radio frequency signal on and off. Keying can be achieved by interrupting the supply to the power amplifier stage or even the oscillator stage however it is normally applied within the driver stage that operates at a more modest power

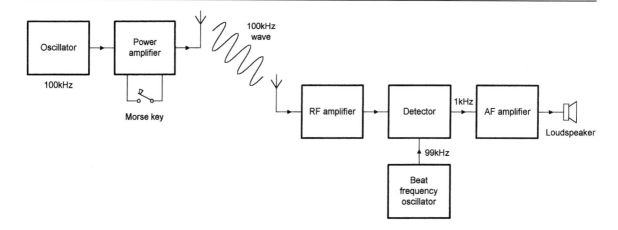

Figure 13.4 Simplified block schematic of a radio communication system comprising a continuous wave (CW) transmitter and a simple receiver with a beat frequency oscillator (BFO)

level. Keying the oscillator stage usually results in impaired frequency stability. On the other hand, attempting to interrupt the appreciable currents and/or voltages that appear in the power amplifier stage can also prove to be somewhat problematic.

The simplest form of CW receiver need consist of nothing more than a radio frequency amplifier (which provides gain and selectivity) followed by a detector and an audio amplifier. The **detector** stage mixes a locally generated radio frequency signal produced by the **beat frequency oscillator** (BFO) with the incoming signal to produce a signal within the audio frequency range.

As an example, assume that the incoming signal is at a frequency of 100 kHz and that the BFO is producing a signal at 99 kHz. A signal at the difference between these two frequencies (1 kHz) will appear at the output of the detector stage. This will then be amplified within the audio stage before being fed to the loudspeaker.

Example 13.4

A radio wave has a frequency of 162.5 kHz. If a beat frequency of 1.25 kHz is to be obtained, determine the two possible BFO frequencies.

Solution

The BFO can be above or below the incoming signal frequency by an amount that is equal to the beat frequency (i.e. the audible signal that results

from the 'beating' of the two frequencies and which appears at the output of the detector stage).

Hence, $f_{BFO} = f_{RF} \pm f_{AF}$

from which:

$f_{BFO} = 162.5 \text{ kHz} \pm 1.25 \text{ kHz}$
$= 160.75 \text{ kHz or } 163.25 \text{ kHz}$

Morse code

Transmitters and receivers for CW operation are extremely simple but nevertheless they can be extremely efficient. This makes them particularly useful for disaster and emergency communications or for any situation that requires optimum use of low power equipment. Signals are transmitted using the code invented by Samuel Morse (see Fig. 13.5).

The Morse code uses a combination of dots (short periods of transmission) and dashes (slightly longer periods of transmission) to represent characters. As an example, Fig. 13.6 shows how the radio frequency carrier is repeatedly switched on and off to transmit the character 'C'.

Modulation

In order to convey information using a radio frequency carrier, the signal information must be

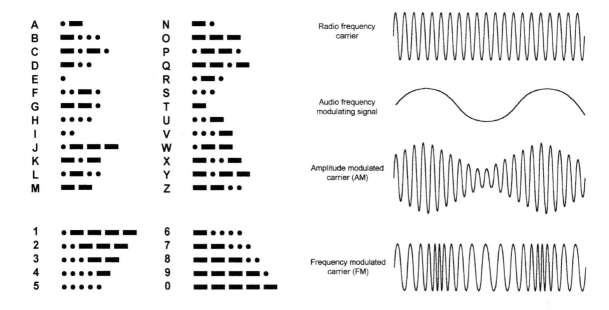

A	•▬	N	▬•
B	▬•••	O	▬▬▬
C	▬•▬•	P	•▬▬•
D	▬••	Q	▬▬•▬
E	•	R	•▬•
F	••▬•	S	•••
G	▬▬•	T	▬
H	••••	U	••▬
I	••	V	•••▬
J	•▬▬▬	W	•▬▬
K	▬•▬	X	▬••▬
L	•▬••	Y	▬•▬▬
M	▬▬	Z	▬▬••

1	•▬▬▬▬	6	▬••••
2	••▬▬▬	7	▬▬•••
3	•••▬▬	8	▬▬▬••
4	••••▬	9	▬▬▬▬•
5	•••••	0	▬▬▬▬▬

Figure 13.5 Morse code

Figure 13.7 Amplitude modulation (AM) and frequency modulation (FM)

Figure 13.6 RF signal for the Morse letter 'C'

superimposed or 'modulated' onto the carrier. **Modulation** is the name given to the process of changing a particular property of the carrier wave in sympathy with the instantaneous voltage (or current) a signal.

The most commonly used methods of modulation are **amplitude modulation** (AM) and **frequency modulation** (FM). In the former case, the carrier amplitude (its peak voltage) varies

according to the voltage, at any instant, of the modulating signal. In the latter case, the carrier frequency is varied in accordance with the voltage, at any instant, of the modulating signal.

Figure 13.7 shows the effect of amplitude and frequency modulating a sinusoidal carrier (note that the modulating signal is, in this case, also sinusoidal). In practice, many more cycles of the RF carrier would occur in the time-span of one cycle of the modulating signal.

Demodulation

Demodulation is the reverse of modulation and is the means by which the signal information is recovered from the modulated carrier. Demodulation is achieved by means of a **demodulator** (sometimes called a **detector**). The output of a demodulator consists of a reconstructed version of the original signal information present at the input of the modulator stage within the transmitter. We shall see how this works a little later.

An AM transmitter

Figure 13.8 shows the block schematic of a simple AM transmitter. An accurate and stable RF oscillator generates the radio frequency **carrier** signal. The output of this stage is then amplified and passed to a modulated RF power amplifier stage. The inclusion of an amplifier between the RF oscillator and the modulated stage also helps to improve frequency stability.

The low-level signal from the microphone is amplified using an AF amplifier before it is passed to an AF power amplifier. The output of the power amplifier is then fed as the supply to the modulated RF power amplifier stage. Increasing and reducing the supply to this stage is instrumental in increasing and reducing the amplitude of its RF output signal.

The modulated RF signal is then passed through an aerial tuning unit. This matches the aerial to the RF power amplifier and also reduces the level of unwanted harmonic components that may be present.

An FM transmitter

Figure 13.9 shows the block schematic of a simple FM transmitter. Once again, an accurate and stable RF oscillator generates the radio frequency carrier signal. As with the AM transmitter, the output of this stage is amplified and passed to an RF power amplifier stage. Here again, the inclusion of an amplifier between the RF oscillator and the RF power stage helps to improve frequency stability.

The low-level signal from the microphone is amplified using an AF amplifier before it is passed to a **variable reactance** element (e.g. a variable capacitance diode—see page 93) within the RF oscillator tuned circuit. The application of the AF signal to the variable reactance element causes the frequency of the RF oscillator to increase and decrease in sympathy with the AF signal.

The final RF signal from the power amplifier is passed through an aerial tuning unit that matches the aerial to the RF power amplifier and also helps to reduce the level of any unwanted harmonic components that may be present. As with the final stages of an AM transmitter, the RF power amplifier usually operates at an appreciable power level and this uses Class C to increase efficiency.

A tuned radio frequency (TRF) receiver

Tuned radio frequency (TRF) receivers provide a means of receiving local signals using fairly minimal circuitry. The simplified block schematic of a TRF receiver is shown in Fig. 13.10.

The signal from the aerial is applied to an RF amplifier stage. This stage provides a moderate amount of gain at the signal frequency. It also provides **selectivity** by incorporating one or more tuned circuits at the signal frequency. This helps the receiver to reject signals that may be present on adjacent channels.

The output of the RF amplifier stage is applied to the demodulator. This stage recovers the audio frequency signal from the modulated RF signal. The demodulator stage may also incorporate a tuned circuit to further improve the selectivity of the receiver.

The output of the demodulator stage is fed to the input of the AF amplifier stage. This stage increases the level of the audio signal from the demodulator so that it is sufficient to drive a loudspeaker.

TRF receivers have a number of limitations with regard to sensitivity and selectivity and this makes them generally unsuitable for use in commercial radio equipment.

A superhet receiver

Superhet receivers provide both improved **sensitivity** (the ability to receiver weak signals) and improved **selectivity** (the ability to discriminate signals on adjacent channels) when compared with TRF receivers. Superhet receivers are based on the **supersonic-heterodyne** principle where the wanted input signal is converted to a fixed **intermediate frequency** (IF) at which the majority of the gain and selectivity is applied. The intermediate frequency chosen is generally 455 kHz or 470 kHz for AM receivers and 10.7 MHz for communications and FM receivers. The simplified block schematic of a simple superhet receiver is shown in Fig. 13.11.

The signal from the aerial is applied to an **RF amplifier** stage. As with the TRF receiver, this stage provides a moderate amount of gain at the signal frequency. The stage also provides

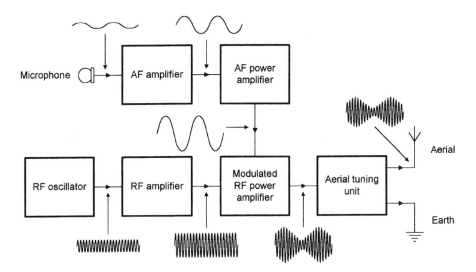

Figure 13.8 An AM transmitter

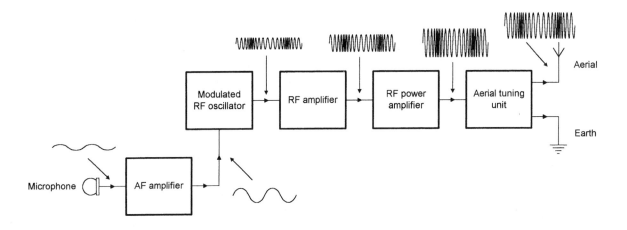

Figure 13.9 An FM transmitter

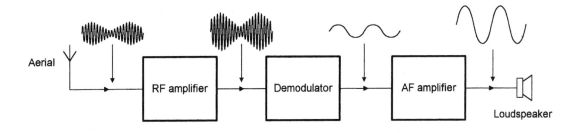

Figure 13.10 A TRF receiver

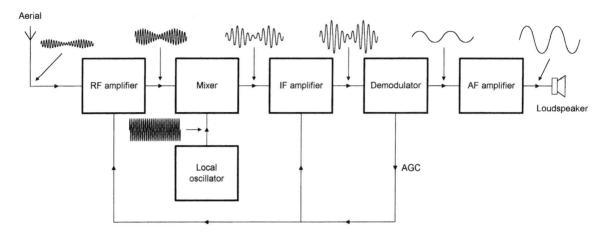

Figure 13.11   A superhet receiver

selectivity by incorporating one or more tuned circuits at the signal frequency.

The output of the RF amplifier stage is applied to the **mixer** stage. This stage combines the RF signal with the signal derived from the **local oscillator** stage in order to produce a signal at the **intermediate frequency** (IF). It is worth noting that the output signal produced by the mixer actually contains a number of signal components, including the sum and difference of the signal and local oscillator frequencies as well as the original signals plus harmonic components. The wanted signal (i.e. that which corresponds to the IF) is passed (usually by some form of filter) to the IF amplifier stage. This stage provides amplification as well as a high degree of selectivity.

The output of the IF amplifier stage is fed to the demodulator stage. As with the TRF receiver, this stage is used to recover the audio frequency signal from the modulated RF signal.

Finally, the AF signal from the demodulator stage is fed to the AF amplifier. As before, this stage increases the level of the audio signal from the demodulator so that it is sufficient to drive a loudspeaker.

In order to cope with a wide variation in signal amplitude, superhet receivers invariably incorporate some form of **automatic gain control** (AGC). In most circuits the d.c. level from the AM demodulator (see page 236) is used to control the gain of the IF and RF amplifier stages. As the signal level increases, the d.c. level from the demodulator stage increases and this is used to reduce the gain of both the RF and IF amplifiers.

The superhet receiver's intermediate frequency f_{IF}, is the difference between the signal frequency, f_{RF}, and the local oscillator frequency, f_{LO}. The desired local oscillator frequency can be calculated from the relationship:

$$f_{LO} = f_{RF} \pm f_{IF}$$

Note that in most cases (and in order to simplify tuning arrangements) the local oscillator operates above the signal frequency, i.e. $f_{LO} = f_{RF} + f_{IF}$
.

Example 13.5

A VHF Band II FM receiver with a 10.7 MHz IF covers the signal frequency range, 88 MHz to 108 MHz. Over what frequency range should the local oscillator be tuned?

Solution

Using $f_{LO} = f_{RF} + f_{IF}$ when $f_{RF} = 88$ MHz gives
$f_{LO} = 88$ MHz + 10.7 MHz = 98.7 MHz

Using $f_{LO} = f_{RF} + f_{IF}$ when $f_{RF} = 108$ MHz gives
$f_{LO} = 108$ MHz + 10.7 MHz = 118.7 MHz

The local oscillator tuning range should therefore be from 98.7 MHz to 118.7 MHz.

RF amplifiers

Figure 13.12 shows the circuit of a typical RF amplifier stage (this circuit can also be used as an IF amplifier in a superhet receiver). You might like to contrast this circuit with Fig. 7.34 shown on page 145. The amplifier operates in Class A and uses a small-signal NPN transistor connected in common-emitter mode. The essential difference between the circuit shown in Fig. 13.12 and that shown in Fig. 7.34 is that the RF amplifier uses a parallel tuned circuit as a collector load.

To improve matching and prevent damping of the tuned circuit (which results in a reduction in Q-factor and selectivity) the collector of TR_1 is **tapped** into the tuned circuit rather than connected straight across it. Since the tuned circuit has maximum impedance at resonance (see page 78), maximum gain will occur at the resonant frequency. By using a tuned circuit with high-Q factor it is possible to limit the response of the amplifier to a fairly narrow range of frequencies. The output (to the next stage) is taken from a secondary winding, L_2, on the main inductor, L_1.

In order to further improve **selectivity** (i.e. the ability to discriminate between signals on adjacent channels) several tuned circuits can be used together in order to form a more effective **bandpass filter**. Figure 13.13 shows one possible arrangement. When constructing an RF filter using several tuned circuits it is necessary to use the optimum coupling between the two tuned circuits. Figure 13.14 illustrates this point.

If the two tuned circuits are too 'loosely' coupled (they are said to be **under-coupled**) the frequency response characteristic becomes flat and insufficient output is obtained. On the other hand, if they are too 'tightly' coupled (they are said to be **over-coupled**) the response becomes broad and 'double-humped'. The optimum value of coupling (when the two tuned circuits are said to be **critically-coupled**) corresponds to a frequency response that has a relatively flat top and steep sides.

AM demodulators

Figure 13.15 shows the circuit of a typical AM demodulator stage. The RF input is applied to a

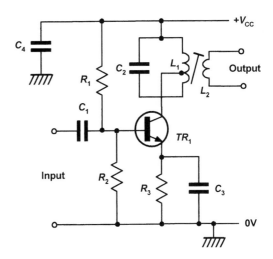

Figure 13.12   An RF amplifier

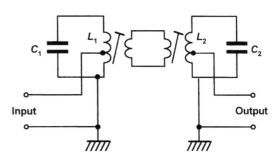

Figure 13.13   A tuned bandpass filter

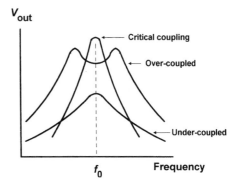

Figure 13.14   Frequency response for two coupled tuned circuits showing different amounts of coupling between the tuned circuits

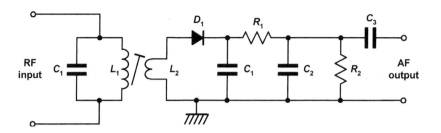

Figure 13.15 A diode AM demodulator

parallel tuned circuit (L_1 and C_1) which exhibits a very high impedance at the signal frequency. A secondary coupling winding, L_2, is used to match the relatively low impedance of the **diode demodulator** circuit to the high impedance of the input tuned circuit. Diode, D_1, acts as a half-wave rectifier conducting only on positive-going half-cycles of the radio frequency signal. Capacitor, C_1, charges to the peak value of each positive-going half-cycle that appears at the cathode of D_1.

The voltage that appears across C_1 roughly follows the peak of the half-cycles of rectified voltage. R_1 and C_2 form a simple filter circuit to remove unwanted RF signal components (this circuit works in just the same way as the smoothing filter that we met in Chapter 6—see pages 117 and 118). The final result is a voltage waveform appearing across C_2 that resembles the original modulating signal. As well as a providing a current path for D_1, R_2 forms a discharge path for C_1 and C_2. Coupling capacitor, C_3, is used to remove any d.c. component from the signal that appears at the output of the demodulator. Waveforms for the demodulator circuit are shown in Fig. 13.16. Figure 13.17 shows a complete IF amplifier together with an AM demodulator stage. Circuits of this type are used in simple superhet receivers.

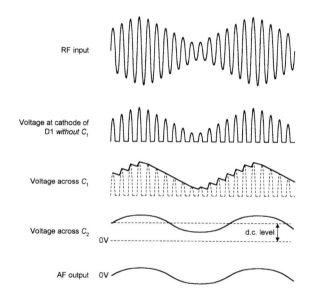

Figure 13.16 Waveforms for the AM demodulator shown in Fig. 13.15

Aerials

We shall start by describing one of the most fundamental types of aerial, the **half-wave dipole**. The basic half-wave dipole aerial (Fig. 13.18) consists of a single conductor having a length equal to one-half of the length of the wave being

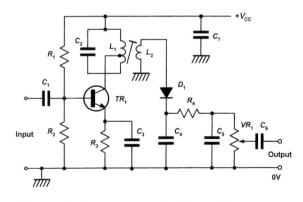

Figure 13.17 A complete RF/IF amplifier and AM demodulator

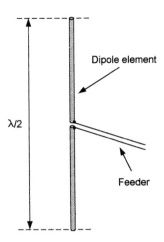

Figure 13.18 A half-wave dipole aerial

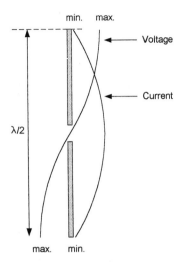

Figure 13.19 Voltage and current distribution in a half-wave dipole aerial

transmitted or received. The conductor is then split in the centre to enable connection to the feeder. In practice, because of the capacitance effects between the ends of the aerial and ground, the aerial is invariably cut a little shorter than a half wavelength.

The length of the aerial (from end to end) is equal to one half wavelength, hence:

$$l = \frac{\lambda}{2}$$

Now since $v = f \lambda$ we can conclude that, for a half-wave dipole,

$$l = \frac{v}{2f}$$

Note that l is the **electrical length** of the aerial rather than its actual **physical length**. End effects, or capacitance effects at the ends of the aerial, require that we reduce the actual length of the aerial and a 5% reduction in length is typically required for an aerial to be resonant at the centre of its designed tuning range.

Figure 13.19 shows the distribution of current and voltage along the length of a half-wave dipole aerial. The current is maximum at the centre and zero at the ends. The voltage is zero at the centre and maximum at the ends. This implies that the impedance is not constant along the length of aerial

but varies from a maximum at the ends (maximum voltage, minimum current) to a minimum at the centre.

The dipole aerial has directional properties illustrated in Fig. 13.20. Fig. 13.20(a) shows the radiation pattern of the aerial in the plane of the antenna's electric field whilst Fig. 13.20(b) shows the radiation pattern in the plane of the aerial's magnetic field. Things to note from these two diagrams are that:

(a) in the case of Fig. 13.20(a) minimum radiation occurs along the axis of the aerial whilst the two zones of maximum radiation are at 90° (i.e. are 'normal to') the dipole elements.

(b) in the case of Fig. 13.20(b) the aerial radiates uniformly in all directions.

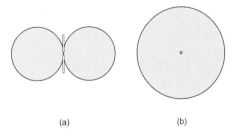

Figure 13.20 Radiation patterns for a half-wave dipole: (a) radiation in the electric field plane (b) radiation in the magnetic field plane

Hence, a vertical dipole, will have an **omni-directional** radiation pattern whilst a horizontal dipole will have a **bi-directional** radiation pattern. This is an important point as we shall see later. The combined effect of these two patterns in three-dimensional space will be a doughnut shape.

Example 13.6

Determine the length of a half-wave dipole aerial for use at a frequency of 150 MHz.

Solution

The length of a half-wave dipole for 150 MHz can be determined from:

$$l = \frac{v}{2f}$$

where $v = 3 \times 10^8$ m/s and $f = 150 \times 10^6$ Hz.

Hence:

$$l = \frac{v}{2f} = \frac{3 \times 10^8}{2 \times 150 \times 10^6} = \frac{3 \times 10^6}{3 \times 10^6} = 1 \text{ m}$$

Impedance and radiation resistance

Because voltage and current appear in an aerial (a minute voltage and current in the case of a receiving antenna and a much larger voltage and current in the case of a transmitting antenna) an aerial is said to have **impedance**. Here it's worth remembering here that impedance is a mixture of resistance, R, and reactance, X, both measured in ohms (Ω). Of these two quantities, X varies with frequency whilst R remains constant. This is an important concept because it explains why aerials are often designed for operation over a restricted range of frequencies.

The impedance, Z, of an aerial is the ratio of the voltage, E, across its terminals to the current, I, flowing in it. Hence:

$$Z = \frac{E}{I}$$

You might infer from Fig. 13.19 that the impedance at the centre of the half-wave dipole

should be zero. In practice the impedance is usually between 70 Ω and 75 Ω. Furthermore, at resonance the impedance is purely resistive and contains no reactive component (i.e. inductance and capacitance). In this case X is negligible compared with R. It is also worth noting that the d.c. resistance (or **ohmic resistance**) of an aerial is usually very small in comparison with its impedance and so it may be ignored. Ignoring the d.c. resistance of the aerial, the impedance of an antenna may be regarded as its **radiation resistance**, R_r (see Fig. 13.21).

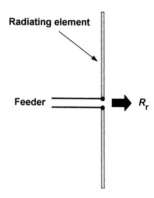

Figure 13.21 Radiation resistance

Radiation resistance is important because it is through this resistance that electrical power is transformed into radiated electromagnetic energy (in the case of a transmitting aerial) and incident electromagnetic energy is transformed into electrical power (in the case of a receiving aerial).

The equivalent circuit of an aerial is shown in Fig. 13.22. The three series-connected components that make up the aerial impedance are:

(a) the d.c. resistance, $R_{d.c.}$
(b) the radiation d.c. resistance, $R_{d.c.}$
(c) the 'off-tune' reactance, X

Note that when the antenna is operated at a frequency that lies in the centre of its pass-band (i.e. when it is *on-tune*) the off-tune reactance is zero. It is also worth bearing in mind that the radiation resistance of a half-wave dipole varies according to its height above ground. The 70 Ω to

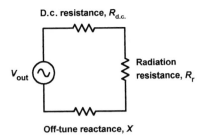

Figure 13.22 Equivalent circuit of an aerial

75 Ω impedance normally associated with a half-wave dipole is only realized when the aerial is mounted at an elevation of 0.2 wavelengths, or more.

Radiated power and efficiency

In the case of a transmitting aerial, the radiated power, P_r, produced by the antenna is given by:

$$P_r = I_a^2 \times R_r$$

where I_a is the aerial current, in amperes, and R_r is the radiation resistance in ohms. In most practical applications it is important to ensure that P_r is maximized and this is achieved by ensuring that R_r is much larger than the DC resistance of the antenna elements.

The efficiency of an antenna is given by the relationship:

$$\text{Radiation efficiency} = \frac{P_r}{P_r + P_{loss}} \times 100\%$$

where P_{loss} is the power dissipated in the DC resistance present. At this point it is worth stating that, whilst efficiency is vitally important in the case of a transmitting aerial it is generally unimportant in the case of a receiving aerial. This explains why a random length of wire can make a good receiving aerial but may not be very good as a transmitting antenna!

Example 13.7

An HF transmitting aerial has a radiation resistance of 12 Ω. If a current of 0.5 A is supplied to the aerial, determine the radiated power.

Solution

In this case, $I_a = 0.5$ A and $R_r = 12$ Ω

Now $P_r = I_a^2 \times R_r$ hence:

$$P_r = (0.5)^2 \times 12 = 0.25 \times 12 = 4 \text{ W}$$

Example 13.8

If the aerial in Example 13.7 has a d.c. resistance of 2 Ω, determine the power loss and the radiation efficiency of the aerial.

Solution

From the equivalent circuit shown in Fig. 13.22, the current that flows in the d.c. resistance of the aerial, $R_{d.c.}$, is the same as that which flows in its radiation resistance, R_r. Thus $I_a = 0.5$ A and $R_{d.c.} = 2$ Ω

Now $P_{loss} = I_a^2 \times R_{d.c.}$ hence:

$$P_{loss} = (0.5)^2 \times 2 = 0.25 \times 2 = 0.5 \text{ W}$$

The radiation efficiency of the aerial is given by:

$$\text{Radiation efficiency} = \frac{P_r}{P_r + P_{loss}} \times 100\%$$

$$= \frac{4}{4 + 0.5} \times 100\% = 89\%$$

In this example, more than 10% of the power output is actually wasted!

Aerial gain

The field strength produced by an aerial is proportional to the amount of current flowing in it. However, since different types of aerial produce different values of field strength for the same applied RF power level, we attribute a power gain to the aerial. This power gain is specified in relation to a **reference aerial** and it is usually specified in decibels (dB)—see Appendix 5.

Two types of reference aerial are used, an **isotropic radiator** and a **standard half-wave**

dipole. The former type of reference aerial is only a theoretical structure (it is assumed to produce a truly spherical radiation pattern and thus could only be realized in three-dimensional space well away from the earth). The latter type of aerial is a more practical reference since it is reasonably easy to produce a half-wave dipole for comparison purposes.

In order to distinguish between the two types of reference aerial we use subscripts **i** and **d** to denote isotropic and half-wave dipole reference aerials respectively. As an example, an aerial having a gain of 10 dB$_i$ produces ten times power gain when compared with a theoretical isotropic radiator. Similarly, an aerial having a gain of 13 dB$_d$ produces twenty times power gain when compared with a half-wave dipole. Putting this another way, to maintain the same field strength at a given point, you would have to apply 20 W to a half-wave dipole or just 1 W to the aerial in question! Some representative values of aerial gain are given in Table 13.1.

Figure 13.23 shows typical half-wave dipole aerials for domestic VHF Band II FM broadcast reception. The half-wave dipole in Fig. 13.23(a) is horizontally polarized (and therefore has a bi-directional characteristic) whilst the half-wave dipole in Fig. 13.23(b) is vertically polarized (and therefore has an omni-directional characteristic).

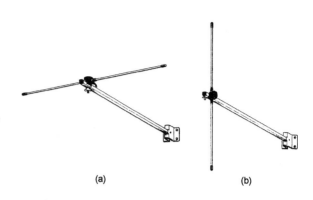

(a)   (b)

Figure 13.23 Typical dipole aerials for local VHF Band II FM reception: (a) horizontally polarized (b) vertically polarized. Note that (a) is bi-directional and (b) is omni-directional in the horizontal plane

Table 13.1 Some typical values of aerial gain

Application	Gain (dBd)
Half-wave wire dipole for VHF Band II FM broadcast reception	0
Dipole and reflector for Band III digital radio reception	3
Car roof mounted aerial for UHF private mobile radio (PMR)	4
Four-element Yagi for high-quality FM broadcast reception	6
Multi-element Yagi aerial for fringe area Band V terrestrial broadcast TV reception	12
Parabolic reflector antenna for satellite TV reception	24
3 m steerable parabolic dish reflector for tracking space vehciles at UHF	40

The Yagi beam aerial

Originally invented by two Japanese engineers, Yagi and Uda, the Yagi aerial has remained extremely popular in a wide variety of applications and, in particular, for fixed domestic FM radio and TV receiving aerials. In order to explain, in simple terms. how the Yagi aerial works we shall use a simple light analogy.

An ordinary filament lamp that radiates light in all directions. Just like an aerial, the lamp converts electrical energy into electromagnetic energy. The only real difference is that we can *see* the energy that it produces!

The action of the filament lamp is comparable with our fundamental dipole aerial. In the case of the dipole, electromagnetic radiation will occur all around the dipole elements (in three dimensions the radiation pattern will take on a doughnut shape). In the plane that we have shown in Fig. 13.20(a), the directional pattern will be a figure-of-eight that has two lobes of equal size. In order to concentrate the radiation into just one of the radiation lobes we could simply place a reflecting mirror on one side of the filament lamp. The radiation will be reflected (during which the reflected light will undergo a 180° phase change) and this will reinforce the light on one side of the filament lamp. In order to achieve the same effect in our aerial system we need to place a conducting element about one quarter of a wavelength behind the dipole element. This element is referred to as a **reflector** and it is said to be 'parasitic' (i.e. it is not actually connected to the feeder). The reflector needs to be cut slightly longer than the driven dipole element. The resulting directional pattern will now only have one **major lobe** because the energy radiated will be concentrated into just one half of the figure-of-eight pattern that we started with.

Continuing with our optical analogy, in order to further concentrate the light energy into a narrow beam we can add a lens in front of the lamp. This will have the effect of bending the light emerging from the lamp towards the normal line. In order to achieve the same effect in our aerial system we need to place a conducting element, known as a **director**, on the other side of the dipole and about one quarter of a wavelength from it. Once again, this element is parasitic but in this case it needs to be cut slightly shorter than the driven dipole element. The resulting directional pattern will now have a narrower major lobe as the energy becomes concentrated in the normal direction (i.e. at right angles to the dipole elements).

The resulting aerial is known as a three-element Yagi aerial, see Fig. 13.24. If desired, additional directors can be added to further increase the gain of the aerial and reduce the **beamwidth** of the major lobe. A typical three-element horizontally polarized Yagi suitable for VHF Band II broadcast reception is shown in Fig. 13.25.

Figure 13.25 A typical three-element Yagi aerial for VHF Band II FM reception

Further director elements can be added to increase the gain and reduce the **beamwidth** (i.e. the angle between the half-power or −3dB power points on the polar characteristic) of Yagi aerials. Some typical gain and beamwidth figures for Yagi aerials are given in Table 13.2. From this data it is worth noting that aerial gain increases by an approximate 3dB every time the antenna doubles in size. It is also worth noting the diminishing return as the Yagi becomes large (e.g. an increase of only 1 dB in gain as the aerial increases in size from 22 to 32 elements!).

Table 13.2 Typical gain and bandwidth figures for Yagi aerials

Number of elements	Gain (dBd)	Beamwidth (degrees)
3	4	70
4	6	60
8	9	40
16	11.5	20
22	12	13
32	13	10

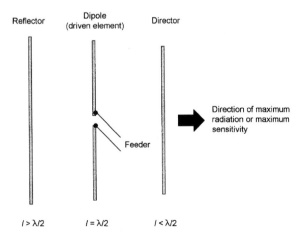

Figure 13.24 A three-element Yagi

Practical investigation

Objective

To investigate the operation of a 455 kHz RF/IF amplifier and demodulator stage.

Components and test equipment

Breadboard, 9 V d.c. power supply (or 9 V battery), BC548 (or similar NPN transistor), resistors of 1 kΩ, 4.7 kΩ, 10 kΩ, 22 kΩ 5% 0.25 W, capacitors of 100 pF, 470 pF, 10 nF (two required), 100 nF (two required), ferrite cored high-Q inductor of 220 µH (with series loss resistance of 2 Ω or less), oscilloscope, digital multimeter (for checking bias voltages), RF signal generator with amplitude modulated output, output attenuator and frequency adjustable over the range 200 kHz to 700 kHz, test leads and probes.

Procedure

Connect the circuit as shown in Fig. 13.26. With the signal generator disconnected, connect the d.c. supply and use the digital multimeter on the 20 V d.c. range to measure the collector, base and emitter voltages on TR_1. These should be approximately 9 V, 2.8 V, and 2.1 V respectively. If these voltages are substantially different you should carefully check the wiring and connections to TR_1.

Table 13.3   Table of results

Frequency (kHz)	200	300	400	455	500	600	700	800
Voltage (mV pk-pk)								

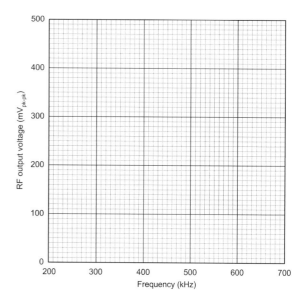

Figure 13.27   Graph layout for plotting the results

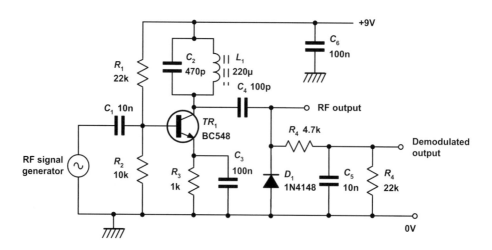

Figure 13.26   RF amplifier/demodulator circuit used in the Practical investigation

Switch the signal generator on and set the output to 2 mV peak-to-peak (707 µV RMS) at 455 kHz, unmodulated. Connect the oscilloscope (using macthed probes) to display the RF output on Channel 1 (or Y1) and the demodulated output on Channel 2 (or Y2). The oscilloscope timebase should be set to 1 µs/cm whilst the Channel 1 and 2 gain settings should both be set to 100 mV/cm. The Channel 1 input should be set to AC whilst the Channel 2 input should be set to DC.

Display the waveforms produced and sketch two or three cycles of both waveforms using the layout

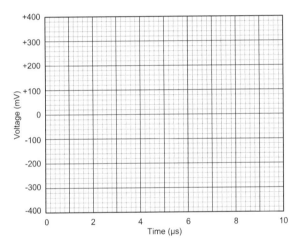

Figure 13.28 Layout for sketching the unmodulated RF output waveform

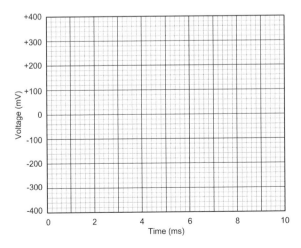

Figure 13.29 Layout for sketching the demodulated waveform

shown in Fig. 13.28. Now select modulated RF output on the signal generator and set to **modulation depth** to 30%. Change the oscilloscope timebase setting to 1 ms/cm and display the output waveforms produced. Sketch two or three cycles of the waveforms using the layout shown in Fig. 13.29.

Switch the modulation off and check that the output of the signal generator is still set to 455 kHz. Accurately measure the peak-peak RF output (Channel 1). This should be approximately 500 mV pk-pk. Record the output voltage in a table similar to that shown in Table 13.3.

Vary the signal generator output frequency over the range 200 kHz to 700 kHz in suitable steps and record the peak-peak output voltage at each step in your table.

Calculations and graphs

Record your results in Table 13.3. Plot a graph showing the frequency response of the RF/IF amplifier using the graph layout shown in Fig. 13.27. Determine the frequency at which maximum voltage gain is achieved and calculate the stage gain at this frequency. Also determine the bandwidth of the amplifier stage.

Conclusions

Comment on the shape of the frequency response and the waveform sketches. Were these what you would expect? Suggest a typical application for the circuit.

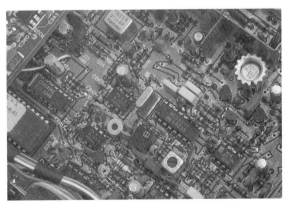

Figure 13.30 10.7 MHz and 455 kHz IF amplifier stages in a typical mobile transceiver

Formulae introduced in this chapter

Frequency and wavelength:
(page 229)

$$v = f\,\lambda \qquad f = \frac{v}{\lambda} \qquad \lambda = \frac{v}{f}$$

Velocity of propagation:
(page 229)

$v = 3 \times 10^8$ m/s for waves in air or space.

BFO frequency:
(page 230)

$$f_{BFO} = f_{RF} \pm f_{AF}$$

Local oscillator frequency:
(page 234)

$$f_{LO} = f_{RF} \pm f_{IF}$$

Half-wave dipole aerial:
(page 237)

$$l = \frac{\lambda}{2}$$

Aerial impedance:
(page 238)

$$Z = \frac{E}{I}$$

Radiated power:
(page 239)

$$P_r = I_a^2 \times R_r$$

Radiation efficiency:
(page 239)

$$\text{Efficiency} = \frac{P_r}{P_r + P_{loss}} \times 100\%$$

Problems

13.1   A broadcast transmitter produces a signal at 190 kHz. What will be the wavelength of the radiated signal?

Table 13.4 See Question 13.5

Frequency (MHz)	10.4	10.5	10.6	10.7	10.8	10.9	11.0
Voltage (V)	0.42	0.52	0.69	1.0	0.67	0.51	0.41

13.2   What frequency corresponds to the 13 m short wave band?

13.3   A signal in a cable propagates at two-thirds of the speed of light. If an RF signal at 50 MHz is fed to the cable, determine the wavelength in the cable.

13.4   An AM broadcast receiver has an IF of 470 kHz. If the receiver is to be tuned over the medium wave broadcast band from 550 kHz to 1.6 MHz, determine the required local oscillator tuning range.

13.5   The data shown in Table 13.4 was obtained during an experiment on an IF bandpass filter. Plot the frequency response characteristic and use it to determine the IF frequency, bandwidth and Q-factor of the filter.

13.6   Refer to the IF amplifier/AM demodulator circuit shown in Fig. 13.17. Identify the component(s) that provide:
(a) a tuned collector load for TR_1
(b) base bias for TR_1
(c) coupling of the signal from the IF amplifier stage to the demodulator stage
(d) a low-pass filter to remove RF signal components at the output of the demodulator
(e) a volume control
(f) removing the d.c. level on the signal at the output of the demodulator
(g) a bypass to the common rail for RF signals that may be present on the supply
(h) input coupling to the IF amplifier stage.

13.7   A half-wave dipole is to be constructed for a frequency of 50 MHz. Determine the approximate length of the aerial.

13.8   A power of 150 W is applied to a dipole aerial in order to produce a given signal strength at a remote location. What power, applied to a Yagi aerial with a gain of 8 dB$_d$, would be required to produce the same signal strength?

Answers to these problems appear on page 376.

14

Test equipment and measurements

This chapter is about making practical measurements on real electronic circuits. It describes and explains the use of the basic items of test equipment that you will find in any electronics laboratory or workshop. We begin the chapter by looking at how we use a moving coil meter to measure voltage, current and resistance and then quickly move on to more complex multi-range analogue and digital instruments and the oscilloscope. To help you make use of these test instruments we have included some DO's and DONT'S. If you intend to become an electronic engineer these will become your 'tools of the trade'. Being able to use them effectively is just the first step on the ladder!

Meters

Straightforward measurements of voltage, current and resistance can provide useful information on the state of almost any circuit. To get the best from a meter it is not only necessary to select an appropriate measurement function and range but also to be aware of the limitations of the instrument

and the effect that it might have on the circuit under investigation. When fault finding, it is interpretation that is put on the meter readings rather than the indications themselves.

Figures 14.2(a) and 14.2(b) respectively show the circuit of a simple **voltmeter** and a simple **ammeter**. Each instrument is based on the moving coil indicator shown in Fig. 14.1. The voltmeter consists of a **multiplier** resistor connected in series with the basic moving coil movement whilst the ammeter consists of a **shunt** resistor connected in parallel with the basic moving coil instrument. When determining the value of multiplier or shunt resistance (R_m and R_s respectively in Fig. 14.2) it is important to remember that the coil of the moving coil meter also has a resistance. We have shown this as a resistor, r, connected with the moving coil in Fig. 14.3. In both cases, the current required to produce full-scale deflection of the meter movement is I_m.

In the voltmeter circuit shown in Fig. 14.3(a):

$$V = I_m R_m + I_m r$$

from which:

$$I_m R_m = V - I_m r$$

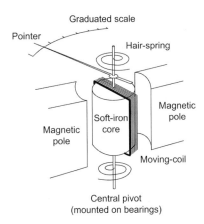

Figure 14.1 A moving coil meter movement

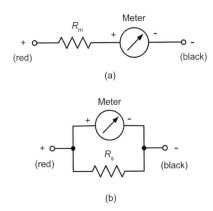

Figure 14.2 A moving coil meter connected (a) as a voltmeter and (b) an ammeter

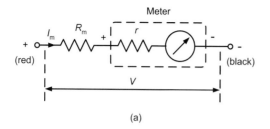

(a)

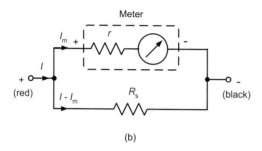

(b)

Figure 14.3 Circuit for determining the values of (a) a multiplier resistor and (b) a shunt resistor

Thus:

$$R_m = \frac{V - I_m r}{I_m}$$

In the ammeter circuit shown in Fig. 14.3(b):

$$(I - I_m) R_s = I_m r$$

from which:

$$R_s = \frac{I_m r}{I - I_m}$$

Example 14.1

A moving coil meter has a full-scale deflection current of 1 mA. If the meter coil has a resistance of 500 Ω, determine the value of multiplier resistor if the meter is to be used as a voltmeter reading 0 to 5 V.

Solution

Using $R_m = \dfrac{V - I_m r}{I_m}$ gives:

$$R_m = \frac{5 - \left(1 \times 10^{-3} \times 500\right)}{1 \times 10^{-3}} = \frac{5 - 0.5}{1 \times 10^{-3}} = 4.5 \times 10^3 = 4.5 \text{ k}\Omega$$

Example 14.2

A moving coil meter has a full-scale deflection current of 10 mA. If the meter coil has a resistance of 40 Ω, determine the value of shunt resistor if the meter is to be used as an ammeter reading 0 to 100 mA.

Solution

Using $R_s = \dfrac{I_m r}{I - I_m}$ gives:

$$R_s = \frac{10 \times 10^{-3} \times 40}{100 \times 10^{-3} - 10 \times 10^{-3}} = \frac{400 \times 10^{-3}}{90 \times 10^{-3}} = \frac{400}{90} = 4.44 \ \Omega$$

The circuit of a simple ohmmeter is shown in Fig. 14.4. The battery is used to supply a current that will flow in the unknown resistor, R_x, which is indicated on the moving coil meter. Before use, the variable resistor, RV, must be adjusted in order to produce full-scale deflection (corresponding to zero on the ohms scale). Zero resistance thus corresponds to maximum indication. Infinite resistance (i.e. when the two terminals are left open-circuit) corresponds to minimum indication. The ohms scale is thus reversed when compared with a voltage or current scale. The scale is also non-linear, as shown in Fig. 14.5.

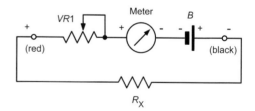

Figure 14.4 A simple ohmmeter

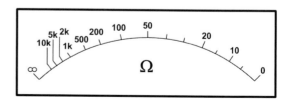

Figure 14.5 A typical ohmmeter scale

Multi-range meters

For practical measurements on electronic circuits it is often convenient to combine the functions of a voltmeter, ammeter and ohmmeter into a single instrument (known as a multi-range meter or simply a **multimeter**). In a conventional multimeter as many as eight or nine measuring functions may be provided with up to six or eight ranges for each measuring function. Besides the normal voltage, current and resistance functions, some meters also include facilities for checking transistors and measuring capacitance. Most multi-range meters normally operate from internal batteries and thus they are independent of the mains supply. This leads to a high degree of portability which can be all-important when measurements are to be made away from a laboratory or workshop.

Figure 14.6   Typical analogue (left) and digital (right) multimeters

Displays

Analogue instruments employ conventional moving coil meters and the display takes the form of a pointer moving across a calibrated scale. This arrangement is not so convenient to use as that employed in digital instruments because the position of the pointer is rarely exact and may require interpolation. Analogue instruments do, however, offer some advantages not the least of which lies in the fact that it is very easy to make adjustments to a circuit whilst observing the relative direction of the pointer; a movement in one direction representing an increase and in the other a decrease. Despite this, the principal disadvantage of many analogue meters is the rather cramped, and sometimes confusing, scale calibration. To determine the exact reading requires first an estimation of the pointer's position and then the application of some mental arithmetic based on the range switch setting.

Digital meters, on the other hand, are usually extremely easy to read and have displays that are clear, unambiguous, and capable of providing a very high resolution. It is thus possible to distinguish between readings that are very close. This is just not possible with an analogue instrument. Typical analogue and digital meters are shown in Fig. 14.6.

The type of display used in digital multi-range meters is either the liquid crystal display (LCD) or the light emitting diode (LED). The former type requires very little electrical power and thus is to be preferred on the grounds of low battery consumption. LCD displays are, however, somewhat difficult to read under certain light conditions and, furthermore, the display response can be rather slow. LED displays can be extremely bright but unfortunately consume considerable power and this makes them unsuitable for use in battery-powered portable instruments.

Loading

Another very significant difference between analogue and digital instruments is the input resistance that they present to the circuit under investigation when taking voltage measurements. The resistance of a reasonable quality analogue multi-range meter can be as low as 50 k on the 2.5 V. With a digital instrument, on the other hand, the input resistance is typically 10 M on the 2 V range. The digital instrument is thus to be preferred when accurate readings are to be taken. This is particularly important when measurements are to be made on high impedance circuits, as illustrated by the following:

Two multi-range meters are used to measure the voltage produced by the two potential dividers

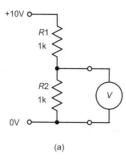

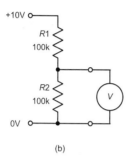

(a)

(b)

Figure 14.7   Examples of voltmeter loading: (a) a low resistance circuit (b) a high resistance circuit

shown in Fig. 14.7. One of the meters is an analogue type having an internal resistance of 10 kW on the 10 V range. The other is a digital type that has the much higher internal resistance of 10 MW. The two potential dividers each consist of resistors or identical value. However, the potential divider of Fig. 14.7(b) has a much lower resistance than that of Fig. 14.7(a). In both cases the 'true' voltage produced by the potential divider should be half the supply, i.e. exactly 5 V. The actual readings obtained from the instruments are shown in Table 14.1.

Table 14.1   Readings obtained from Fig. 14.7

Meter	High resistance circuit Fig. 14.7(a)	Low resistance circuit Fig. 14.7(b)
Analogue type (100 k)	4.9 V	3.3 V
Digital type (10 M)	4.99 V	4.97 V

The large difference in the case of Fig. 14.7(b) illustrates the effect of voltmeter **loading** on a high resistance circuit. An appreciable current is drawn away from the circuit into the measuring instrument. Clearly this is a very undesirable effect!

Sensitivity

The sensitivity of an analogue meter may be expressed in several ways. One is to specify the basic **full-scale deflection** (f.s.d.) current of the moving coil meter. This is typically 50 µA or less. An alternative method is that of quoting an **ohms-per-volt** rating. This is, in effect, the resistance presented by the meter when switched to the 1 V range.

The **ohms-per-volt** rating is inversely proportional to the basic full-scale sensitivity of the meter movement and, to determine the resistance of a meter on a particular voltage range, it is only necessary to multiply the range setting by the 'ohms-per-volt' rating. Table 14.2 shows how meter f.s.d. and ohms-per-volt are related. From this we can conclude that:

$$\text{Meter f.s.d.} = \frac{1}{\text{Ohms-per-volt}}$$

or

$$\text{Ohms-per-volt} = \frac{1}{\text{Meter f.s.d.}}$$

Table 14.2   Relationship between meter sensitivity and Ohms-per-volt

Meter f.s.d.	Ohms-per-volt
10 µA	100 k /V
20 µA	50 k /V
50 µA	20 k /V
100 µA	10 k /V
200 µA	5 k /V
500 µA	2 k /V
1 mA	1 k /V

Example 14.3

A meter has a full-scale deflection of 40 μA. What will its ohms-per-volt rating be?

Solution

$$\text{Ohms-per-volt} = \frac{1}{\text{Meter f.s.d.}} = \frac{1}{40 \times 10^{-6}} = 25 \text{ k}\Omega$$

Example 14.4

A 20 k /V meter is switched to the 10 V range. What current will flow in the meter when it is connected to a 6 V source?

Solution

The resistance of the meter will be given by:

$R_m = 10 \times 20 \text{ k} = 200 \text{ k}$

The current flowing in the meter will thus be given by:

$$I_m = \frac{6 \text{ V}}{R_m} = \frac{6}{200 \times 10^3} = 30 \times 10^{-6} = 30 \text{ μA}$$

Digital multi-range meters

Low-cost digital multi-range meters have been made possible by the advent of mass-produced LSI devices and liquid crystal displays. A 3½-digit display is the norm and this consists of three full digits that can display '0' to '9' and a fourth (most significant) digit which can only display '1'. Thus, the maximum display indication, ignoring the range switching and decimal point, is 1999; anything greater over-ranges the display.

The **resolution** of the instrument is the lowest increment that can be displayed and this would normally be an increase or decrease of one unit in the last (least significant) digit. The **sensitivity** of a digital instrument is generally defined as the smallest increment that can be displayed on the lowest (most sensitive) range. Sensitivity and resolution are thus not quite the same. To put this into context, consider the following example:

A digital multi-range meter has a 3½-digit display. When switched to the 2 V range, the maximum indication would be 1.999 V and any input of 2 V, or greater, would produce an over-

range indication. On the 2 V range, the instrument has a resolution of 0.001 V (or 1 mV). The lowest range of the instrument is 200 mV (corresponding to a maximum display of 199.9 mV) and thus the meter has a sensitivity of 0.1 mV (or 100 μV).

Nearly all digital meters have automatic zero and polarity indicating facilities and some also have **autoranging**. This feature, which is only found in the more expensive instruments, automatically changes the range setting so that maximum resolution is obtained without over-ranging. There is thus no need for manual operation of the range switch once the indicating mode has been selected. This is an extremely useful facility since it frees you from the need to make repeated adjustments to the range switch while measurements are being made.

Example 14.5

A digital multi-range meter has a 4½-digit display. When switched to the 200 V range, determine:

(a) the maximum indication that will appear on the display
(b) the resolution of the instrument.

Solution

(a) The maximum indication that will appear on the display is 199.99 V
(b) The resolution of the instrument will be 0.01 V or 10 mV.

Using an analogue multi-range meter

Figure 14.8 shows the controls and display provided by a simple analogue multi-range meter. The range selector allows you to select from a total of twenty ranges and six measurement functions. These functions are:

- DC voltage (DC, V)
- DC current (DC, mA)
- AC voltage (AC, V)
- Resistance (OHM)
- Continuity test (BUZZ)
- Battery check (BAT)

Note that more complex instruments may have several more ranges.

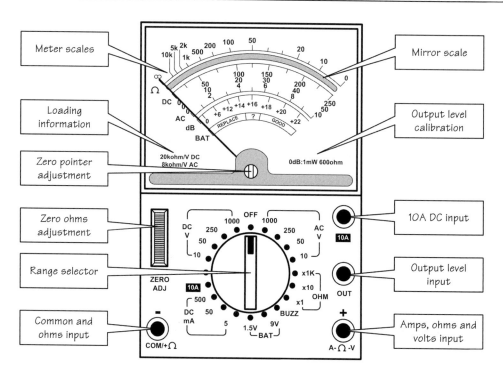

Figure 14.8 Analogue multimeter display and controls

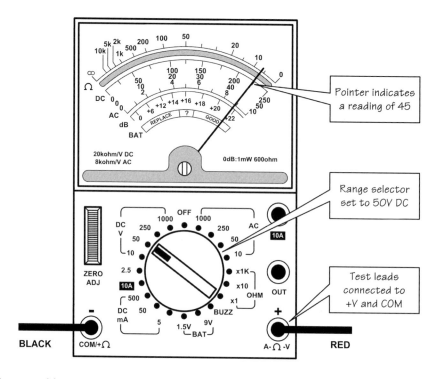

Figure 14.9 Analogue multimeter set to the DC, 50 V range

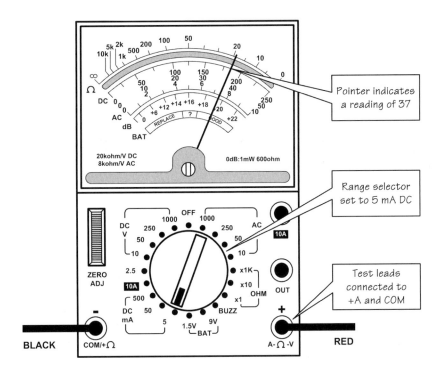

Figure 14.10 Analogue multimeter set to the DC, 5 mA range

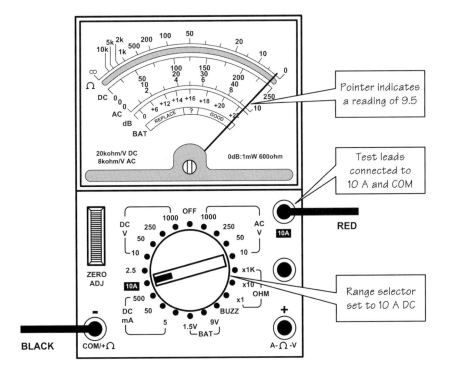

Figure 14.11 Analogue multimeter set to the DC, 10 A range

DC voltage measurement

Figure 14.9 shows how to make DC voltage measurements. In both cases, the red and black test leads are connected to the '+' and '−' sockets respectively. In Fig. 14.9, the range selector is set to DCV, 50 V. The pointer is reading just less than 45 on the range that has 50 as its full-scale indication (note that there are three calibrated voltage scales with maximum indications of 10 V, 50 V and 250 V respectively. The reading indicated is approximately 45 V.

DC current measurement

Figure 14.10 shows how to make a DC current measurement. Once again, the red and black test leads are connected to the '+' and '−' sockets respectively. The range selector is set to DC, 5 mA. In Fig. 14.10, the pointer is reading between 35 and 40 (and is actually a little closer to 35 than it is to 40) on the range that has 50 as its full-scale indication. The actual reading indicated is thus approximately 3.7 mA.

DC high-current measurement

In common with many simple multi-range meters, both analogue and digital, the high current range (e.g., 10 A) is not only selected using the range selector switch but a separate input connection must also be made. The reason for this is simply that the range switch and associated wiring is not designed to carry a high current. Instead, the high-current shunt is terminated separately at its own '10 A' socket.

Figure 14.11 shows the connections and range selector settings to permit high-current DC measurement. The range selector is set to DC 10 A and the red and black test leads are connected to '10 A' and '−' respectively. The pointer is reading mid-way between 9 and 10 on the range that has 10 as its full-scale indication. The actual reading indicated is thus 9.5 A.

AC voltage measurement

Figure 14.12 shows how to make AC voltage measurements. Once again, the red and black test

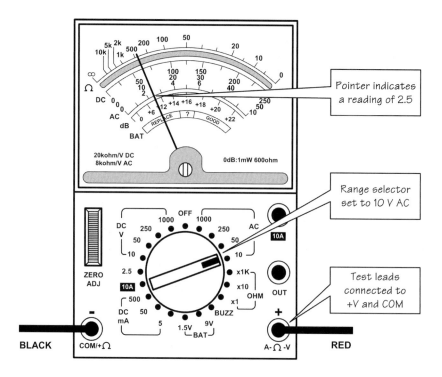

Figure 14.12 Analogue multimeter set to the AC, 10 V range

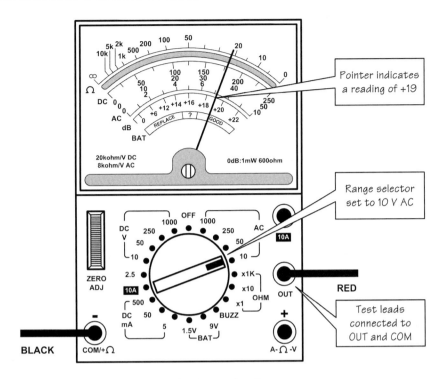

Figure 14.13   Analogue multimeter set to the dB (output level) range

leads are connected to the '+' and '−' sockets respectively. In Fig. 14.12, the range selector is set to AC, 10 V. The pointer is reading mid-way between 2 and 3 and the indicated reading is approximately 2.5 V.

Output level measurement

Figure 14.13 shows how to make output level measurements. The red and black test leads are respectively connected to 'OUT' and '−' respectively. The range selector is set to AC, 10 V (note that the output level facility is actually based on AC voltage measurement).

Output level indications are indicated in decibels (dB) where 0 dB (the **reference level**) corresponds to a power level of 1mW in a resistance of 600 . The pointer is reading mid-way between +18 and +20 on the dB scale and the indicated reading is +19 dB.

Resistance

Figure 14.14 shows how to make resistance measurements. In all three cases, the red and black

test leads are connected to the '+' and '−' sockets respectively. Before making any measurements it is absolutely essential to zero the meter. This is achieved by shorting the test leads together and adjusting the ZERO ADJ control until the meter reads full-scale (i.e., zero on the ohms scale). In Fig. 14.14, the range selector is set to OHM, ×1. The pointer is reading mid-way between 0 and 10 and the resistance indicated is approximately 5 .

DO's and DON'Ts of using an analogue multimeter

DO ensure that you have selected the correct range and measuring function before attempting to connect the meter into a circuit.

DO select a higher range than expected and then progressively increase the sensitivity as necessary to obtain a meaningful indication.

DO remember to zero on the ohms range before measuring resistance.

DO switch the meter to the 'off' position (if one is available) before attempting to transport the meter.

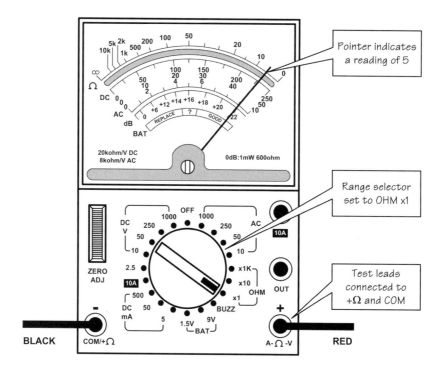

Figure 14.14 Analogue multimeter set to the OHM, ×1 range

DO check and, if necessary, replace the internal batteries regularly.

DO use properly insulated test leads and prods.

DON'T attempt to measure resistance in a circuit that has the power applied to it.

DON'T rely on voltage readings made on high-impedance circuits (the meter's own internal resistance may have a significant effect on the voltages that you measure).

DON'T rely on voltage and current readings made on circuits where high-frequency signals may be present (in such cases an analogue meter may produce readings that are wildly inaccurate or misleading).

DON'T subject the instrument to excessive mechanical shock or vibration (this can damage the sensitive meter movement).

DON'T leave the instrument for very long periods without removing the batteries (these may leak and cause damage).

Using a digital multi-range meter

Digital multi-range meters offer a number of significant advantages when compared with their more humble analogue counterparts. The display fitted to a digital multi-range meter usually consists of a 3½-digit seven-segment display – the ½ simply indicates that the first digit is either blank (zero) or 1. Consequently, the maximum indication on the 2 V range will be 1.999 V and this shows that the instrument is capable of offering a resolution of 1 mV on the 2 V range (i.e. the smallest increment in voltage increment that can be measured is 1 mV). The resolution obtained from a comparable analogue meter would be of the order of 50 mV, or so, and thus the digital instrument provides a resolution that is many times greater than its analogue counterpart.

Figure 14.15 shows the controls and display provided by a simple digital multi-range meter. The mode switch and range selector allow you to select from a total of twenty ranges and eight measurement functions. These functions are:

- DC voltage (DC, V)
- DC current (DC, A)
- AC voltage (AC, V)
- AC current (AC, A)
- Resistance (OHM)
- Capacitance (CAP)
- Continuity test (buzzer)
- Transistor current gain (h_{FE})

DC voltage measurement

Figure 14.16 shows how to make DC voltage measurements using a digital multi-range meter. The red and black test leads are connected to the 'V-OHM' and 'COM' sockets respectively. In Fig. 14.6, the mode switch and range selector is set to DCV, 200 V, and the display indicates a reading of 124.5 V.

DC current measurements

Figure 14.17 shows how to make a DC current measurement. Here, the red and black test leads are connected to the 'mA' and 'COM' sockets respectively. The mode switch and range selectors are set to DC, 200 mA, and the display indicates a reading of 85.9 mA.

DC high-current measurement

In common with simple analogue multi-range meters, the meter uses a shunt which is directly connected to a separate '10 A' terminal. Figure 14.18 shows the connections, mode switch and range selector settings to permit high-current DC measurement. The mode switch and range selectors are set to DC, 2,000 mA (2 A) and the red and black test leads are connected to '10 A' and 'COM' respectively. The display indicates a reading of 2.99 A.

AC voltage measurement

Figure 14.19 shows how to make AC voltage measurements. Once again, the red and black test leads are connected to the 'V-OHM' and 'COM' sockets respectively. In Fig. 14.19, the mode switch and range selectors are set to AC, 10 V, and the display indicates a reading of 1.736 V.

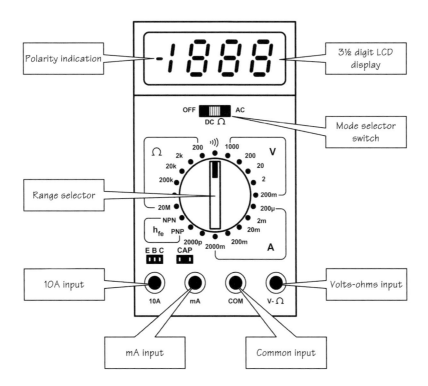

Figure 14.15 Digital multimeter display and controls

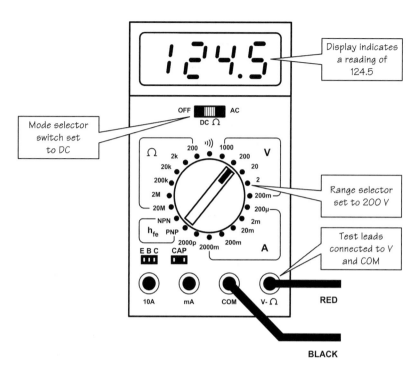

Figure 14.16   Digital multimeter set to the DC, 200 V range

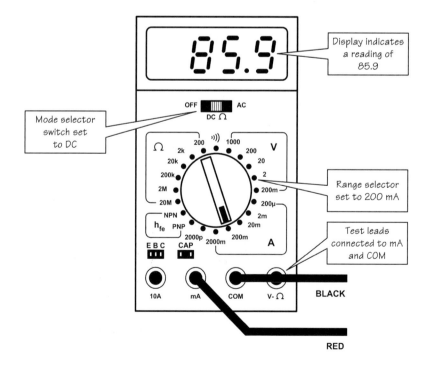

Figure 14.17   Digital multimeter set to the DC, 200 mA range

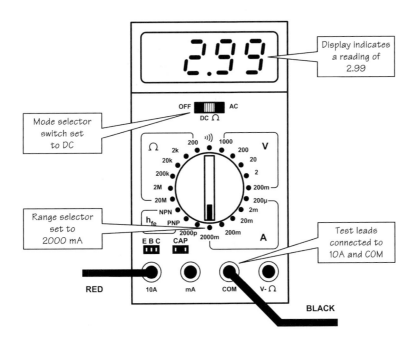

Figure 14.18 Digital multimeter set to the DC, 10 A range

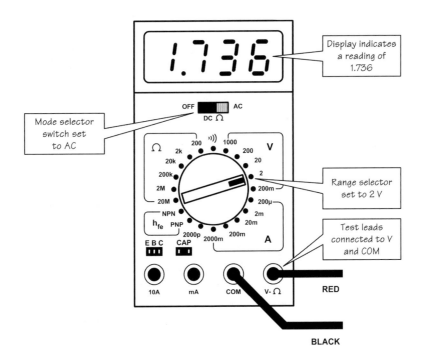

Figure 14.19 Digital multimeter set to the AC, 2 V range

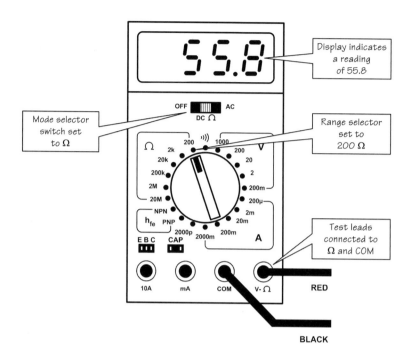

Figure 14.20   Digital multimeter set to the OHM, 200   range

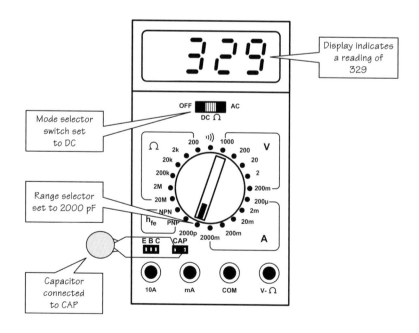

Figure 14.21   Digital multimeter set to the capacitance, 2,000 pF range

Resistance measurement

Figure 14.20 shows how to make resistance measurements. As before, the red and black test leads are respectively connected to 'V-OHM' and 'COM' respectively. In Fig. 14.20, the mode switch and range selectors are set to OHM, 200ohm, and the meter indicates a reading of 55.8 . Note that it is not necessary to 'zero' the meter by shorting the test probes together before taking any measurements (as would be the case with an analogue meter).

Capacitance measurement

Many modern digital multi-range meters incorporate a capacitance measuring although this may be limited to just one or two ranges. Figure 14.21 shows how to carry out a capacitance measurement. The capacitor on test is inserted into the two-way connector marked 'CAP' whilst the mode switch and range selector controls are set to DC, 2,000 pF. The display indication shown in Fig. 14.21 corresponds to a capacitance of 329 pF.

Transistor current gain (h_{FE}) measurement

Many modern digital multi-range meters also provide some (fairly basic) facilities for checking transistors. Figure 14.22 shows how to measure the current gain (h_{FE}) of an NPN transistor (see page 142). The transistor is inserted into the three-way connector marked 'EBC', taking care to ensure that the emitter lead is connected to 'E', the base lead to 'B', and the collector lead to 'C'. The mode switch and range selector controls are set to DC, NPN respectively. The display indication in Fig. 14.22 shows that the device has a current gain of 93. This means that, for the device in question, the ratio of collector current (I_C) to base current (I_B) is 93.

Connecting a meter into a circuit

Figure 14.23 shows how a meter is connected to read the voltages and currents present at various points in a simple transistor audio amplifier.

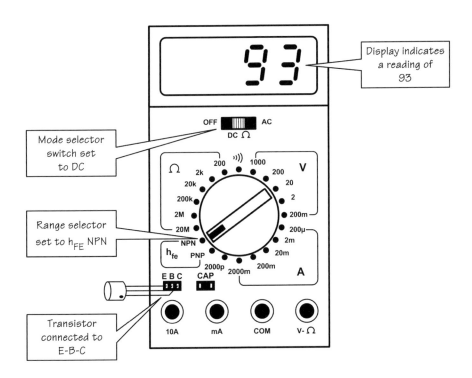

Figure 14.22 Digital multimeter set to the transistor current gain, NPN range

(a) At A, the supply current is measured by removing the **supply rail fuse** and inserting the meter in its place. A suitable meter range is 2 A, DC.

(b) At B, the collector current for TR4 is measured by removing the **service link** and inserting the meter in its place. A suitable meter range is 2 A, DC.

(c) At C, the emitter current for TR4 can be measured by connecting the meter across the emitter resistor, R11, and measuring the **voltage drop** across it. The emitter current can then be calculated using Ohm's Law). This can be much quicker than disconnecting the emitter lead and inserting the meter switched to the current range. A suitable range for the meter is 2 V, DC (note that 2 V is dropped across 1 when a current of 2 A is flowing).

(d) At D, the voltage at the junction of R11 and R12 is measured by connecting the meter between the junction of R11 and R12 and common/ground. A suitable meter range is 20 V, DC.

(e) At E, the base-emitter voltage for TR2 is measured by connecting the meter with its positive lead to the base of TR2 and its negative lead to the emitter of TR2. A suitable meter range is 2 V, DC.

DO's and DON'Ts of using a digital multimeter

DO ensure that you have selected the correct range and measuring function before attempting to connect the meter into a circuit.

DO select a higher range than expected and then progressively increase the sensitivity as necessary to obtain a meaningful indication.

DO switch the meter to the 'off' position in order to conserve battery life when the instrument is not being used.

DO check and, if necessary, replace the internal battery regularly.

DO use properly insulated test leads and prods.

DO check that a suitably rated fuse is used in conjunction with the current ranges (if the current ranges aren't working it's probably the fuse that's blown!).

DON'T attempt to measure resistance in a circuit that has the power applied to it.

DON'T rely on voltage and current readings made on circuits where high-frequency signals may be present (as with analogue instruments, digital meters may produce readings that are wildly inaccurate or misleading in such circumstances).

DON'T rely on measurements made when voltage/current is changing or when a significant amount of AC may be present superimposed on a DC level.

The oscilloscope

An oscilloscope is an extremely comprehensive and versatile item of test equipment which can be used in a variety of measuring applications, the most important of which is the display of time related voltage waveforms.

The oscilloscope display is provided by a **cathode ray tube** (CRT) that has a typical screen area of 8 cm × 10 cm. The CRT is fitted with a **graticule** that may either be integral with the tube face or a separate translucent sheet. The graticule is usually ruled with a 1 cm grid to which further bold lines may be added to mark the major axes on the central viewing area. Accurate voltage and time measurements may be made with reference to the graticule, applying a scale factor derived from the appropriate range switch.

A word of caution is appropriate at this stage, however. Before taking meaningful measurements from the CRT screen it is absolutely essential to ensure that the front panel variable controls are set in the **calibrate** (CAL) position. Results will almost certainly be inaccurate if this is not the case!

The use of the graticule is illustrated by the following example:

An oscilloscope screen is depicted in Fig. 14.24. This diagram is reproduced actual size and the fine graticule markings are shown every 2 mm along the central vertical and horizontal axes.

The oscilloscope is operated with all relevant controls in the 'CAL' position. The timebase (horizontal deflection) is switched to the 1 ms/cm range and the vertical attenuator (vertical

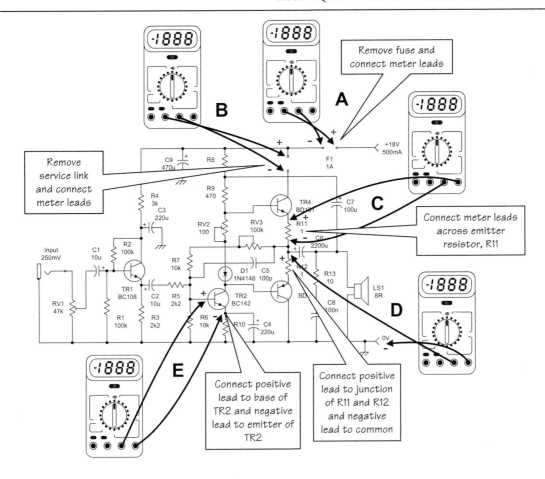

Figure 14.23 Connecting a meter to display currents and voltages in a simple audio amplifier

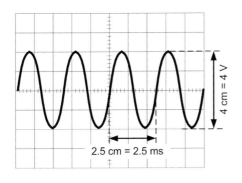

Timebase: 1 ms/cm

Vertical attenuator: 1 V/cm

Figure 14.24 Using an oscilloscope graticule

deflection) is switched to the 1 V/cm range. The overall height of the trace is 4 cm and thus the peak-peak voltage is 4 × 1 V = 4 V. Similarly, the time for one complete cycle (period) is 2.5 × 1 ms = 2.5 ms. One further important piece of information is the shape of the waveform that, in this case, is sinusoidal.

The front panel layout for a typical general-purpose two-channel oscilloscope is shown in Figs 14.25 and 14.26. The controls and adjustments are summarized in Table 14.3.

Using an oscilloscope

An oscilloscope can provide a great deal of information about what is going on in a circuit. In

Table 14.3   Oscilloscope controls and adjustments

Control	Adjustment
Cathode ray tube display	
Focus	Provides a correctly focused display on the CRT screen.
Intensity	Adjusts the brightness of the display.
Astigmatism	Provides a uniformly defined display over the entire screen area and in both x and y-directions. The control is normally used in conjunction with the focus and intensity controls.
Trace rotation	Permits accurate alignment of the display with respect to the graticule.
Scale illumination	Controls the brightness of the graticule lines.
Horizontal deflection system	
Timebase (time/cm)	Adjusts the timebase range and sets the horizontal time scale. Usually this control takes the form of a multi-position rotary switch and an additional continuously variable control is often provided. The 'CAL' position is usually at one, or other, extreme setting of this control.
Stability	Adjusts the timebase so that a stable displayed waveform is obtained.
Trigger level	Selects the particular level on the triggering signal at which the timebase sweep commences.
Trigger slope	This usually takes the form of a switch that determines whether triggering occurs on the positive or negative going edge of the triggering signal.
Trigger source	This switch allows selection of one of several waveforms for use as the timebase trigger. The options usually include an internal signal derived from the vertical amplifier, a 50 Hz signal derived from the supply mains, and a signal which may be applied to an External Trigger input.
Horizontal position	Positions the display along the horizontal axis of the CRT.
Vertical deflection system	
Vertical attenuator (V/cm)	Adjusts the magnitude of the signal attenuator (V/cm) displayed (V/cm) and sets the vertical voltage scale. This control is invariably a multi-position rotary switch; however, an additional variable gain control is sometimes also provided. Often this control is concentric with the main control and the 'CAL' position is usually at one, or other, extreme setting of the control.
Vertical position	Positions the display along the vertical axis of the CRT.
a.c.-d.c.-ground	Normally an oscilloscope employs d.c. coupling throughout the vertical amplifier; hence a shift along the vertical axis will occur whenever a direct voltage is present at the input. When investigating waveforms in a circuit one often encounters a.c. superimposed on d.c. levels; the latter may be removed by inserting a capacitor in series with the signal. With the a.c.-d.c.-ground switch in the d.c. position a capacitor is inserted in the input lead, whereas in the DC position the capacitor is shorted. If ground is selected, the vertical input is taken to common (0 V) and the oscilloscope input is left floating. This last facility is useful in allowing the accurate positioning of the vertical position control along the central axis. The switch may then be set to d.c. and the magnitude of any d.c. level present at the input may be easily measured by examining the shift along the vertical axis.
Chopped-alternate	This control, which is only used in dual beam oscilloscopes, provides selection of the beam splitting mode. In the chopped position, the trace displays a small portion of one vertical channel waveform followed by an equally small portion of the other. The traces are, in effect, sampled at a relatively fast rate, the result being two apparently continuous displays. In the alternate position, a complete horizontal sweep is devoted to each channel alternately.

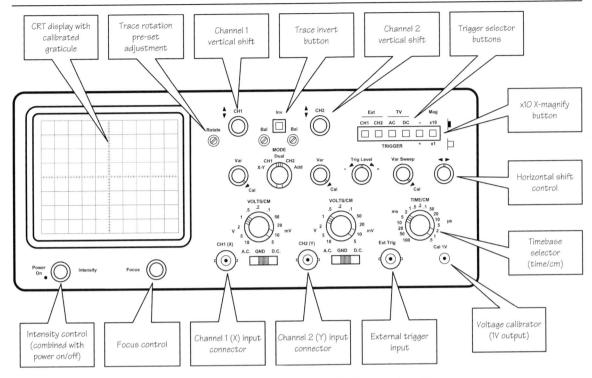

Figure 14.25 Front panel controls and displays on a typical dual-channel oscilloscope

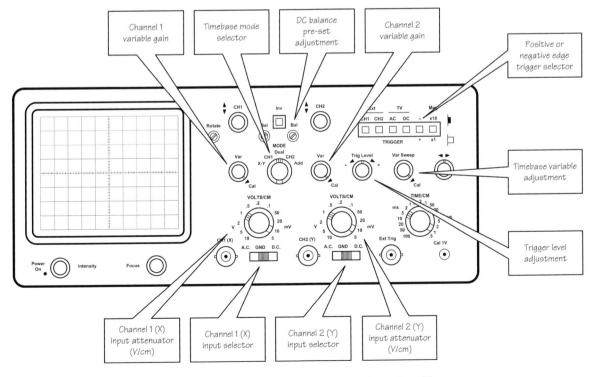

Figure 14.26 Front panel controls and displays on a typical dual-channel oscilloscope

effect, it allows you to 'see' into the circuit, displaying waveforms that correspond to the signals that are present. The procedure and adjustments differ according to the type of waveform being investigated and whether the oscilloscope is being used to display a single waveform (i.e. single-channel operation) or whether it is being used to display two waveforms simultaneously (i.e. dual-channel operation).

Sinusoidal waveforms (single-channel operation)

The procedure for displaying a repetitive sine wave waveform is shown in Fig. 14.27. The signal is connected to the Channel 1 input (with 'AC' input selected) and the mode switch in the Channel 1 position. 'Channel 1' must be selected as the trigger source and the trigger level control adjusted for a stable display. Where accurate measurements are required it is essential to ensure that the 'Cal' position is selected for both the variable gain and time controls.

Square waveforms (single-channel operation)

The procedure for displaying a repetitive square waveform is shown in Fig. 14.28. Once again, the signal is connected to the Channel 1 input (but this time with 'DC' input selected) and the mode switch in the Channel 1 position. 'Channel 1' must be selected as the trigger source and the trigger level control adjusted for a stable display (which can be triggered on the positive or negative going edge of the waveform according to the setting of the trigger polarity button). Any DC level present on the input can be measured from the offset produced on the Y-axis. To do this, you must first select 'GND' on the input selector then centre the trace along the Y-axis before switching to 'DC' and noting how far up or down the trace moves (above or below 0 V). This may sound a little difficult but it is actually quite easy to do! The same technique can be used for measuring any DC offset present on a sinusoidal signal.

Sine/square waveforms (dual–channel operation)

The procedure for displaying two waveforms (either sine or square or any other repetitive signal) simultaneously is shown in Fig. 14.29. The two signals are connected to their respective inputs (Channel 1 and Channel 2) and the mode switch set to the 'Dual' position. The oscilloscope can be triggered by either of the signals (Channel 1 or Channel 2) as desired. Once again, the display can be triggered on the positive or negative-going edge of the waveform depending upon the setting of the trigger polarity button. Dual-channel operation can be invaluable when it is necessary to compare two waveforms, e.g. the input and output waveforms of an amplifier.

Connecting an oscilloscope into a circuit

Figure 14.30 shows how an oscilloscope can be connected to display the waveforms present at various points in a simple transistor audio amplifier. To reduce the likelihood of picking up of hum and noise, the input to the oscilloscope is via a screened lead fitted with a probe. The ground (outer screen) is connected to the common 0V rail whilst the probe is simply moved around the circuit from point to point. Note that, because of the ground connection to the oscilloscope it is NOT usually possible to display a waveform that appear 'across' a component (e.g. between the base and emitter of a transistor). For this reason, waveforms are nearly always displayed relative to ground (or common).

Checking distortion

Oscilloscopes are frequently used to investigate **distortion** in amplifiers and other electronic systems. Different forms of distortion have a different effect on a waveform and thus it is possible to determine which type of distortion is present. A 'pure' sine wave is used as an input signal and the output is then displayed on the oscilloscope. Figure 14.31 shows waveforms that correspond to the most common forms of distortion.

Checking frequency response

An oscilloscope can also be used to provide a rapid assessment of the frequency response of an amplifier or other electronic system. Instead of using a sine wave as an input signal a square wave input is used. Different frequency response

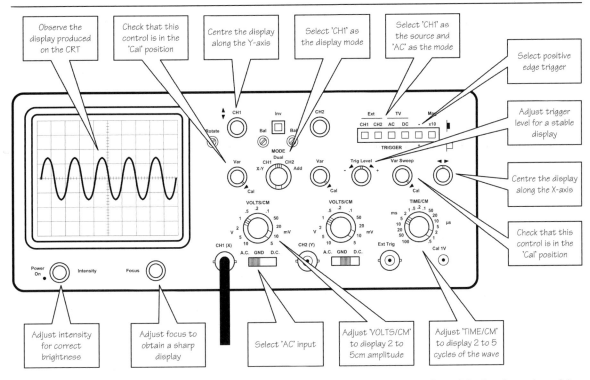

Figure 14.27 Procedure for adjusting the controls to display a sinusoidal waveform (single-channel mode)

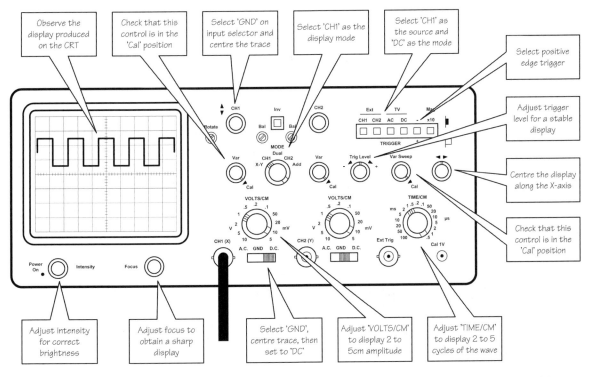

Figure 14.28 Procedure for adjusting the controls to display a square waveform (single-channel mode)

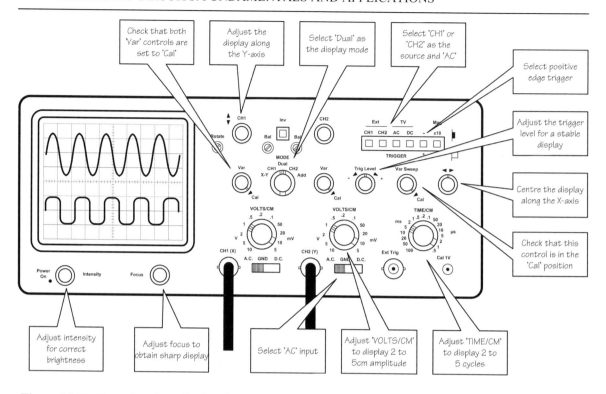

Figure 14.29 Procedure for adjusting the controls to display two waveforms (dual-channel mode)

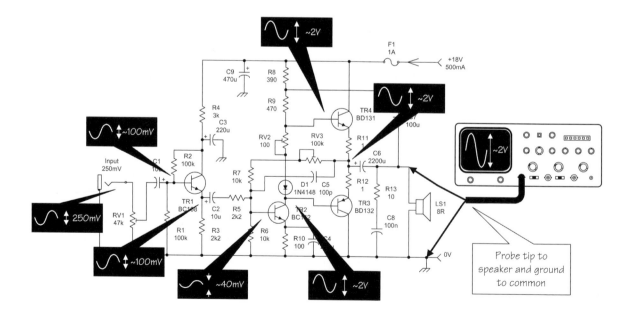

Figure 14.30 Using an oscilloscope to display the waveforms in a simple audio amplifier

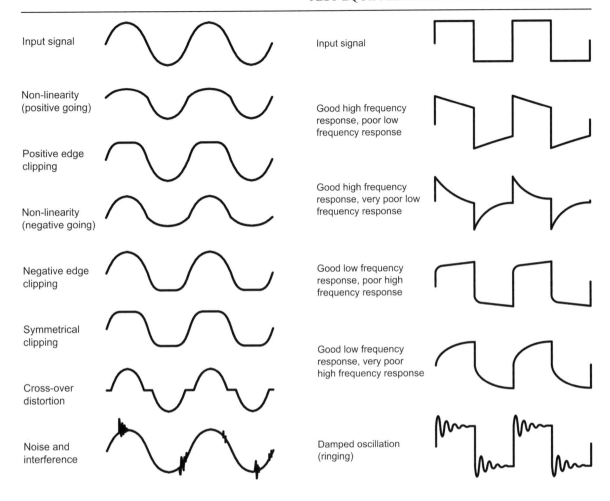

Input signal	Input signal
Non-linearity (positive going)	Good high frequency response, poor low frequency response
Positive edge clipping	Good high frequency response, very poor low frequency response
Non-linearity (negative going)	Good low frequency response, poor high frequency response
Negative edge clipping	Good low frequency response, very poor high frequency response
Symmetrical clipping	
Cross-over distortion	Damped oscillation (ringing)
Noise and interference	

Figure 14.31 Typical waveforms produced by different types of distortion

Figure 14.32 Using a square waveform to rapidly assess frequency response

produces a different effect on a waveform and thus it is possible to assess whether the frequency response is good or poor (a perfect square wave output corresponds to a perfect frequency response). Figure 14.32 shows waveforms that correspond to different frequency response characteristics.

Measuring pulse parameters

When dealing with rectangular waveforms and pulses it is often necessary to be able to use an oscilloscope to measure parameters such as:

Periodic time, t

This is the time (measured at the 50% amplitude points) for one a complete cycle of a repetitive pulse waveform. The periodic time is sometimes referred to as the **period** (see page 70). Note also that the **pulse repetition frequency (p.r.f.)** is the reciprocal of the periodic time.

On time, t_{on}

This is the time for which the pulse amplitude exceeds 50% of its amplitude. The on-time is sometimes referred to as the **high time**.

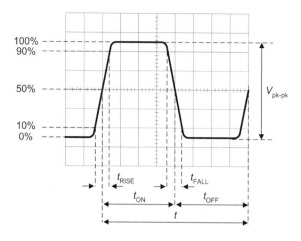

Figure 14.33   Pulse parameters

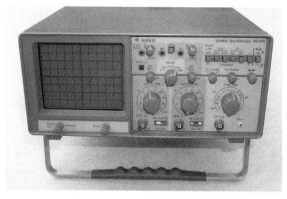

Figure 14.34   A typical bench oscilloscope

Off time, t_{off}

This is the time for which the pulse amplitude falls below 50% of its amplitude. The off time is sometimes referred to as the **low time**.

Rise time, t_{rise}

This is the time measured between the 10% and 90% points on the rising (or **positive-going**) edge of the pulse.

Fall time, t_{fall}

This is the time measured between the 90% and 10% points on the falling (or **negative-going**) edge of the pulse.

These **pulse parameters** are shown in Fig. 14.33.

DO's and DON'Ts of using an oscilloscope

DO ensure that the vertical gain and variable timebase controls are set to the calibrate (CAL) positions before attempting to make accurate measurements based on the attenuator/timebase settings and graticule.

DO check that you have the correct trigger source selected for the type of waveform under investigation.

DO remember to align the trace with the X-axis of the graticule with the input selector set to 'GND' prior to taking DC offset measurements.

DO make use of the built-in calibrator facility to check the accuracy of the attenuator and the 'CAL' setting of the variable gain control.

DO use properly screened leads and a suitable probe for connecting to the circuit under investigation.

DO check that you have made a proper connection to ground or 0 V prior to taking measurements.

DON'T leave the intensity control set at a high level for any length of time.

DON'T leave a bright spot on the display for even the shortest time (this may burn the phosphor coating of the screen).

DON'T rely on voltage measurements on circuits where high-frequency signals may be outside the bandwidth of the oscilloscope.

DON'T forget to set the a.c.-d.c.-gnd switch to the d.c. position when making measurements of the d.c. offset present on a.c. signals.

DON'T place the oscilloscope where there is a strong local magnetic field. This may cause unwanted deflection of electron beam!

Practical investigation

Objective

To investigate the voltages, currents and waveforms in a simple audio amplifier.

Components and test equipment

Breadboard, 17 V d.c. power supply (with a current limited output of up to at least 500 mA), BC141, BD131 and BD132 transistors (the latter to be fitted with small heatsinks), 1N4148 diode, fixed resistors of 1   (two required), 100   , 1.5 k , 10 k , 22 k , and 33 k , 5% 0.25 W, capacitors of 10 μF, 220 μF, 1,000 μF and 4,700 μF at 25 V d.c., oscilloscope, digital multimeter, AF signal generator with sine wave output at 1 kHz, test leads and probes.

Procedure

Connect the circuit as shown in Fig. 14.35. With the signal generator disconnected, connect the d.c. supply and use the digital multimeter on the 20 V d.c. range to measure the d.c. voltage at the junction of R6 and R7. Adjust RV1 so that this voltage is exactly half that of the supply (i.e. 8.5 V).

Now remove the link and use the digital multimeter set to the 200 mA d.c. range to measure the collector current for TR3. Adjust RV2 for a current of exactly 24 mA. Repeat these two adjustments until no further change is noted. Replace the link and set the digital multimeter to the 20 V d.c. range. Measure the d.c. voltages present at the collector, base and emitter of TR1, TR2 and TR3. Also measure the d.c. voltage at the anode and cathode of D1. Record your measurements in a table.

Connect the AF signal generator to the input of the amplifier and set the output to 200 mV pk-pk at 1 kHz sine wave. Set the Channel 1 and Channel 2 inputs of the oscilloscope to 100 mV/cm with the a.c.-d.c.-ground switches in the a.c. position. Set the trigger to internal and the timebase to 1 ms/cm. Display the waveforms at the input and output of the amplifier and sketch these using the layout shown in Fig. 14.36.

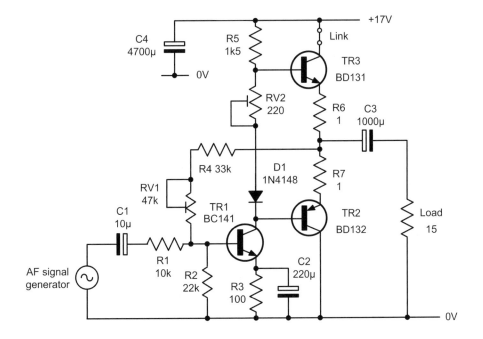

Figure 14.35 Circuit used in the Practical investigation

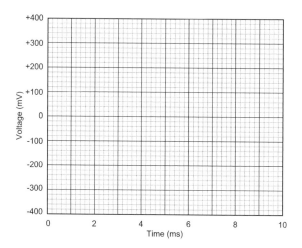

Figure 14.36 Layout for the waveform sketches

Conclusions

Comment on the measured voltages. Are these what you would have expected? Justify them by calculation and state any assumptions made. Comment on the waveforms obtained. What was the voltage gain of the amplifier at 1 kHz? What output power was produced in the load with 200 mV pk-pk input? Suggest a typical application for the circuit.

Formulae introduced in this chapter

Voltmeter multiplier resistor
(page 246)

$$R_m = \frac{V - I_m r}{I_m}$$

Ammeter shunt resistor
(page 246)

$$R_s = \frac{I_m r}{I - I_m}$$

Ohms-per-volt
(page 248)

$$\text{Ohms-per-volt} = \frac{1}{\text{Meter f.s.d.}}$$

Problems

14.1 A moving coil meter has a full-scale deflection current of 500 µA. If the meter coil has a resistance of 400 , determine the value of multiplier resistor if the meter is to be used as a voltmeter reading 0 to 10 V.

14.2 A moving coil meter has a full-scale deflection current of 5 mA. If the meter coil has a resistance of 100 , determine the value of shunt resistor if the meter is to be used as an ammeter reading 0 to 50 mA.

14.3 A meter has a full-scale deflection of 60 µA. What will its ohms-per-volt rating be?

14.4 A 50 k /V meter is switched to the 3 V range. What current will flow in the meter when it is connected to a 2 V source?

14.5 A digital multi-range meter has a 3½-digit display. When switched to the 20 V range, determine:
(a) the maximum indication that will appear on the display
(b) the resolution of the instrument.

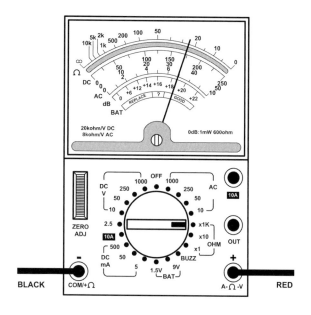

Figure 14.37 See Question 14.6

14.6 Determine the reading on the multi-range meter shown in Fig. 14.37.

14.7 Determine the reading on the multi-range meter shown in Fig. 14.38.

14.8 Determine the reading on the multi-range meter shown in Fig. 14.39

14.9 Figure 14.40 shows the display of an oscilloscope in which the graticule is

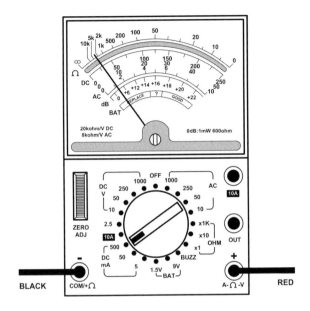

Figure 14.38 See Question 14.7

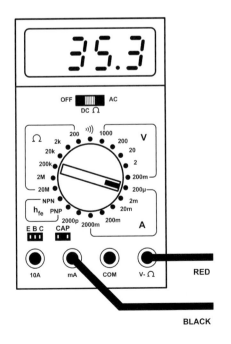

Figure 14.39 See Question 14.8

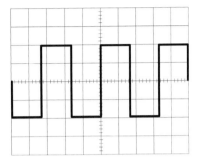

(a) Timebase: 1 ms/cm
Y-attenuator: 1 V/cm

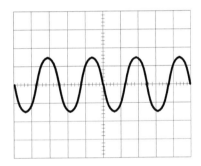

(b) Timebase: 50 ns/cm
Y-attenuator: 50 mV/cm

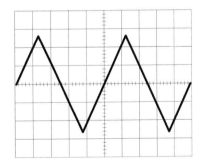

(c) Timebase: 100 ms/cm
Y-attenuator: 50 V/cm

Figure 14.40 See Question 14.9

divided into squares of 1 cm. Determine the amplitude and period of each waveform.

14.10 Figure 14.41 shows the display of an oscilloscope in which the graticule is divided into squares of 1 cm. For the square wave pulse shown determine:
(a) the periodic time
(b) the high time
(c) the low time
(d) the rise time
(e) the fall time
(f) the pulse amplitude.

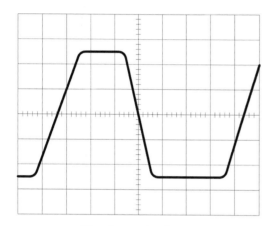

Timebase: 1μs/cm

Y-attenuator: 1 V/cm

Figure 14.41 See Question 14.10

Figure 14.42 See Question 14.11

Figure 14.43 See Question 14.12

14.11 Figure 14.42 shows the display provided by an RF power meter. Explain why the scale is non-linear and illustrate your answer with reference to a formula.

14.12 Figure 14.43 shows a test set-up for radio servicing. Identify the test instruments and state a typical application for each.

14.13 Using a sine wave reference, sketch waveforms to show the typical appearance of:
(a) cross-over distortion
(b) noise
(c) positive edge clipping
(d) negative-going non-linearity.

14.14 With the aid of sketches, show how a reference square wave input can be used to quickly assess the frequency response of an amplifier.

14.15 Explain the purpose of each of the following facilities/adjustments provided on a bench oscilloscope:
(a) timebase selector
(b) trigger source
(c) trigger slope
(d) a.c.-d.c.-ground selector
(e) scale illumination
(f) astigmatism
(g) calibrator.

14.16 Explain, with the aide of a sketch, how you would use an oscilloscope to measure the rise time and period of a rectangular pulse.

Answers to these problems appear on page 376.

15

Fault finding

Fault finding is a disciplined and logical process in which 'trial fixing' should never be contemplated. The generalized process of fault finding is illustrated in the flowchart of Fig. 15.1. First you need to verify that the equipment really is faulty and that you haven't overlooked something obvious (such as a defective battery or disconnected signal cable). This may sound rather obvious but in some cases a fault may simply be attributable to maladjustment or misconnection. Furthermore, where several items of equipment are connected together, it may not be easy to pinpoint the single item of faulty equipment.

The second stage is that of gathering all relevant information. This process involves asking questions such as:

- In what circumstances did the circuit fail?
- Has the circuit operated correctly before and exactly what has changed?
- Has the deterioration in performance been sudden or progressive?
- What fault symptoms do you notice?

The answers to these questions are crucial and, once the information has been analysed, the next stage involves separating the 'effects' from the 'cause'. Here you should list each of the *possible* causes. Once this has been done, you should be able to identify and focus upon the *most probable* cause. *Corrective action* (such as component removal and replacement, adjustment or alignment) can then be applied before further functional checks are carried out. It should then be possible to determine whether or not the fault has been correctly identified. Note, however, that the failure of one component can often result in the malfunction or complete failure of another. As an example, a short-circuit capacitor will often cause a fuse to blow. When the fuse is replaced and the supply is reconnected the fuse will once again blow because the capacitor is still faulty. It is therefore important to consider what other problems may be present when a fault is located.

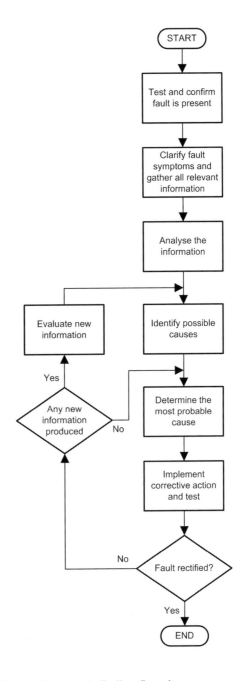

Figure 15.1 Fault finding flowchart

Safety considerations

Before we outline the basic steps for fault finding on some simple electronic circuits, it is vitally important that you are aware of the potential hazards associated with equipment which uses high voltages or is operated from the a.c. mains supply.

Whereas many electronic circuits operate from low voltage supplies and can thus be handled quite safely, the high a.c. voltages present in mains operated equipment represent a potentially lethal shock hazard. The following general rules should *always* be followed when handling such equipment:

1. Switch off the mains supply *and* remove the mains power connector whenever *any* of the following tasks are being performed:
    (a) Dismantling the equipment.
    (b) Inspecting fuses.
    (c) Disconnecting or connecting internal modules.
    (d) Desoldering or soldering components.
    (e) Carrying out continuity tests on switches, transformer windings, bridge rectifiers, etc.
2. When measuring a.c. and d.c. voltages present within the power unit take the following precautions:
    (a) Avoid direct contact with incoming mains wiring.
    (b) Check that the equipment is connected to an effective earth.
    (c) Use insulated test prods.
    (d) Select appropriate meter ranges *before* attempting to take any measurements.
    (e) If in any doubt about what you are doing, switch off at the mains, disconnect the mains connector and *think*.

Approach to fault finding

Fault finding on most circuits is a relatively straightforward process. Furthermore, and assuming that the circuit has been correctly assembled and wired, there is usually a limited number of 'stock faults' (such as transistor or integrated circuit failure) that can occur. To assist in this process, it is a good idea to identify a number of **test points** at which the voltages present (both a.c. and d.c.) can be used as indicators of the circuit's functioning. Such test points should he identified prior to circuit construction and marked with appropriate terminal pins.

Readers should note that the most rapid method of fault diagnosis is not necessarily that of following voltages or signals from one end to the other. Most textbooks on fault finding discuss the relative methods of the so-called **end-to-end** and **half split** methods. The former method involves tracing the signal and measuring voltages from the input to the output (or vice versa) whilst the latter method involves dividing the circuit into two halves and then eliminating the half that is operational before making a further sub-division of the half in which the fault resides.

Transistor faults

As long as a few basic rules can be remembered, recognizing the correct voltages present at the terminals of a transistor is a fairly simple process. The potentials applied to the transistor terminals are instrumental in determining the correct bias conditions for operation as an amplifier, oscillator, or switch. If the transistor is defective, the usual bias voltages will be substantially changed. The functional state of the transistor may thus be quickly determined by measuring the d.c. potentials present at the transistor's electrodes while it is still in circuit.

The potential developed across the forward biased base-emitter junction of a silicon transistor is approximately 0.6 V. In the case of an NPN transistor, the base will be positive with respect to the emitter whilst, for a PNP transistor, the base will be negative with respect to the emitter, as shown in Fig. 15.2. The base-emitter voltage drop tends to be larger when the transistor is operated as a saturated switch (and is in the 'on' state) or when it is a power type carrying an appreciable collector current. In these applications, base-emitter voltages of between 0.65 V and 0.7 V are not unusual. Small-signal amplifiers, on the other hand, operate with significantly lower values of collector current and values of base-emitter voltage in the range 0.55 V to 0.65 V are typical.

A measured base-emitter voltage greatly in excess of 0.6 V is an indication of a defective transistor with an open-circuit base-emitter junction. A measured base-emitter potential very much less than 0.6 V indicates that the transistor is not being supplied with a base bias and, while this may be normal for a switching transistor in the 'off' state, it is indicative of a circuit fault in the case of a linear small-signal amplifier stage. In the case of oscillators and medium/high power r.f. amplifiers operating in Class C little or no bias will usually be present and, furthermore, the presence of r.f. drive during measurement can produce some very misleading results. In such cases it is probably worth removing a transistor suspected of being defective and testing it out of circuit.

Unfortunately, it is not so easy to predict the voltage present at the collector-base junction of a transistor. The junction is invariably reverse biased and the potential present will vary considerably depending upon the magnitude of collector current, supply voltage, and circuit conditions. As a rule-of-thumb, small-signal amplifiers using resistive collector loads usually operate with a collector-emitter voltage which is approximately half that of the collector supply. The collector-base voltage will be slightly less than this. Hence, for a stage that is operated from a decoupled supply rail of, say, 8.5 V a reasonable collector-base voltage would be somewhere in the range 3 V to 4 V. Tuned amplifiers having inductive collector loads generally operate with a somewhat higher value of collector-emitter voltage since there is no appreciable direct voltage drop across the load. As a result, the collector-base voltage drop is greater (a typical value being 6 V). Figure 15.3 shows the electrode potentials normally associated with transistors operating as linear small-signal amplifiers.

Where a transistor is operated as a saturated switch, the junction potentials depend upon whether the transistor is in the 'on' or 'off' state. In the 'off' condition, the base-emitter voltage will be very low (typically 0 V) whereas the collector-base voltage will be high and, in many cases, almost equal to the collector supply voltage. In the 'on' state, the base-emitter voltage will be relatively large (typically 0.7 V) and the collector-base voltage will fall to a very low value (typically 0.5 V). It should be noted that, in normal saturated switching, the collector-emitter voltage may fall to as low as 0.2 V and thus the base-emitter voltage will become reversed in polarity (i.e. base positive with respect to collector in the case of an NPN transistor). The junction potentials associated with

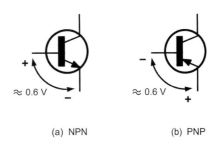

(a) NPN   (b) PNP

Figure 15.2   Base-emitter voltages for NPN and PNP silicon transistors

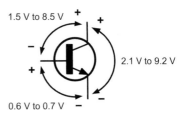

Figure 15.3   Typical voltages found in a small-signal transistor amplifier (polarities will be reversed in the case of a PNP transistor)

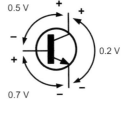

(a) 'On' (conducting)

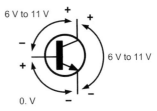

(b) 'Off' (non-conducting)

Figure 15.4   Typical voltages found in a transistor switching circuit (polarities will be reversed in the case of a PNP transistor)

transistors operating as saturated switches are shown in Fig. 15.4.

Transistors may fail in a number of ways. Individual junctions may become open-circuit or short-circuit. In some cases the entire device may become short-circuit or exhibit a very low value of internal resistance. The precise nature of the fault will usually depend upon the electrical conditions prevalent at the time of failure. Excessive reverse voltage, for example, is likely to cause a collector-base or base-emitter junction to become open-circuit. Momentary excessive collector current is likely to rupture the base-emitter junction while long-term over-dissipation is likely to cause an internal short-circuit or very low resistance condition which may, in some cases, affect all three terminals of the device.

In order to illustrate the effects of various transistor fault conditions consider the circuit of a typical tuned amplifier stage shown in Fig. 15.5. The transistor is operated in Class A with conventional base bias potential divider and a tuned transformer collector load. Normal working voltages are shown in the circuit diagram. Note that the junction potentials (0.6 V and 4.7 V for the base-emitter and collector-base junctions respectively) are in agreement with the voltages given in Fig. 15.4. The circuit voltages corresponding to six different transistor fault conditions are shown in Table 15.1.

Each fault will now be discussed individually:

Fault 1

The collector-base short-circuit gives rise to identical base and collector voltage. The base-

emitter junction is still intact and thus the normal forward voltage drop of 0.6 V is present. In this condition a relatively high value of current is drawn by the stage and thus the supply voltage falls slightly.

Fault 2

The base-emitter short-circuit produces identical base and emitter voltages. No collector current flows and the collector voltage rises to almost the full supply voltage. The base and emitter voltages are relatively low since the emitter resistor effectively appears in parallel with the lower section of the base bias potential divider, thus pulling the base voltage down.

Fault 3

Identical voltages at the collector and emitter result from a collector-emitter short-circuit. The emitter voltage rises above the base voltage and the base-emitter junction is well and truly turned off. The short-circuit causes a higher value of current to be drawn from the supply and hence the supply voltage falls slightly.

Fault 4

Perhaps the most obvious fault condition is when a short-circuit condition affects all three terminals. The voltages are, naturally, identical and, as with fault 3, more current than usual is taken from the supply and, as a consequence, the supply voltage falls slightly.

Fault 5

With the base-emitter junction open-circuit, the base-emitter voltage rises well above the 0.6 V which would normally be expected. No collector or emitter current is flowing and thus the collector voltage rises, while the emitter voltage falls.

Fault 6

No collector current flows when the collector-base junction is open-circuit and, as with fault 2, the collector voltage rises towards the supply. Note that, since the base-emitter junction is intact, the normal forward bias of 0.6 V is present and this condition distinguishes this fault from that described in Fault 2.

Table 15.1 Voltage for various fault conditions

Fault number	Transistor voltages			Fault condition
	e	b	c	
1	2.7	3.3	3.3	b-c short
2	0.2	0.2	8.3	b-e short
3	3.0	0.8	3.0	c-e short
4	2.9	2.9	2.9	c-b-e short
5	0.0	2.2	8.3	b-e open
6	0.2	0.8	8.2	c-b open

Integrated circuit faults

Integrated circuits may fail in various ways. Occasionally, the manifestation of the fault is simply a chip which is chronically overheated— the judicious application of a finger tip to the centre of the plastic package will usually help you to identify such a failure. Any chip that is noticeably hotter than others of a similar type should be considered suspect. Where integrated circuits are fitted in sockets, it will be eminently possible to remove and replace them with known functional devices (but, do remember to switch 'off' and disconnect the supply during the process).

In the case of digital circuitry, the task of identifying a logic gate which is failing to perform its logical function can be accomplished by various means but the simplest and most expedient is with the aid of a logic probe (see page 282). This invaluable tool comprises a hand-held probe fitted with LED to indicate the logical state of its probe tip.

In use, the logic probe is moved from point to point within a logic circuit and the state of each node is noted and compared with the expected logic level. In order to carry out checks on more complex logic arrangements a logic pulser may be used in conjunction with the logic probe. The pulser provides a means of momentarily changing the state of a node (regardless of its actual state) and thus permits, for example, the clocking of a bistable element (see page 284).

Operational amplifiers can usually be checked using simple d.c. voltage measurements. The d.c. voltages appearing at the inverting and non-inverting inputs should be accurately measured and compared. Where the voltage at the inverting input is more positive with respect to that at the non-inverting input, the output voltage will be high (positive if the operational amplifier is operated from a dual supply rail).

Conversely, if the voltage at the inverting input is negative, with respect to that at the non-inverting input, the output voltage will be high (positive if the operational amplifier is operating from a dual supply).

Finally, if both inputs are at 0 V and there is virtually no difference in the input voltages, the output should also be close to 0 V. If it is high or low (or sitting at one or other of the supply voltages) the device should be considered suspect.

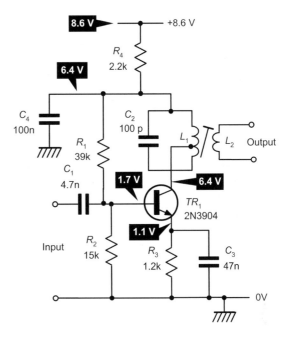

Figure 15.5   Circuit to illustrate the effects of various transistor faults on the d.c. voltages present in a small-signal amplifier stage (normal d.c. working voltages are shown)

The detection of faults within other linear integrated circuits can be rather more difficult. However, a good starting point is that of disconnecting the supply and inserting a meter to determine the supply current under quiescent conditions. The value should be compared with that given by the manufacturer as 'typical'. Where there is a substantial deviation from this figure the device (or its immediate circuitry) should be considered suspect.

Diode testing

Go/no-go checks provide you with a very rapid method of assessing whether a junction semiconductor device, such as a diode or bipolar transistor, is functional. The basic principle is simply that of checking the forward and reverse resistance of each P–N junction that may be present. The source of current for the resistance checks is simply the internal battery in your analogue multimeter.

Figure 15.6 shows the simplified internal circuit of an analogue multimeter on the resistance (ohms) range. This circuit consists of a battery (usually 1.5 V, 3 V, 4.5 V or 9 V) connected in series with a preset 'set-zero ohms' control and a fixed, current-limiting resistor, and a sensitive moving coil meter movement. There is one important thing to note from this arrangement and it relates to the polarity of the meter leads—the *red meter test lead will have a negative polarity* whilst the *black meter test lead will exhibit a positive polarity* when the instrument is switched to the 'ohms' range. (Note that the same does *not* apply to digital multimeters).

Unfortunately, simple go/no-go checks are unreliable when using a digital multimeter. Indeed, some instruments sometimes fail to provide any meaningful indication when carrying out resistance checks on diodes and transistors! However, if you have a more sophisticated digital instrument, you may find that it already has a built-in diode or transistor checker—in which case you should use this facility to make measurements rather than attempt to use simple go/no-go checking based on resistance measurements.

Because the output of the analogue multimeter is polarity conscious, it can be used to check the ability of a component to conduct in both directions. Thus it can, for example, be used to distinguish between a resistor and a diode. The former will exhibit the same resistance whichever way it is connected to the multimeter. The diode, on the other hand, will have a resistance which will vary significantly depending upon which way round it is connected.

Diode 'go/no-go' checks

When the red (positive) side of the ohmmeter is connected to the anode and the black (negative) side is connected to the cathode, the diode under test is forward-biased, and the ohmmeter will indicate a low-resistance reading. When the leads are reversed, the diode will be reverse-biased, and the ohmmeter indicates a very high, or infinite resistance. The ohmmeter should be switched to one of the middle resistance ranges (×10, ×100 or ×1k) for testing diodes.

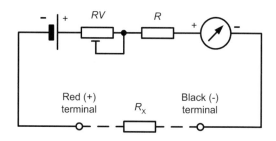

(a) Ohmmeter circuit

(b) Polarity appearing at the terminals

Figure 15.6   Polarity of a multimeter terminals when used as an ohmmeter (note how the red terminal is negative and the black terminal is positive)

A zener diode may usually be tested in the same manner as is a general-purpose diode. If it tests the same as does a general-purpose diode, it is probably good. However, the most rapid method of checking a zener diode is to apply power to the circuit it is used in and measure the voltage across it. If this voltage reading is the correct zener voltage (as rated by the manufacturer), the zener is good. If the voltage across the zener is incorrect, either the supply to the zener is low, or not present, or the zener itself has failed.

Typical meter indications for silicon diodes are shown in Fig. 15.7 (forward bias is applied in Fig. 15.7(a) and reverse bias is applied in Fig. 15.7(b)) . Corresponding indications for germanium diodes will be somewhat lower than those shown for silicon diodes.

Encapsulated bridge rectifiers are supplied with all four diode connections accessible. It is thus possible to check each individual diode for both forward and reverse resistance. It is also possible to check that the bridge as a whole is neither short-circuited nor open-circuited.

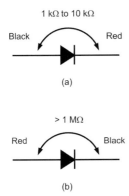

(a)

(b)

Figure 15.7 Diode 'go/no-go' checks with typical resistance values in the 'go' case

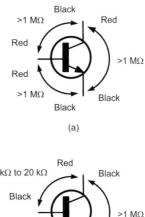

(a)

(b)

Figure 15.8 NPN transistor 'go/no-go' checks with typical resistance values in the 'go' case

Bipolar transistor 'go/no-go' checks

Typical meter indications for silicon NPN transistors are shown in Fig. 15.8 (reverse bias is applied in Fig. 15.8(a) and forward bias is applied in Fig. 15.8(b)). Corresponding indications for germanium NPN transistors will be somewhat lower. Note also that the ×100 resistance range should be used for measurements below 20 kΩ, and the ×1k resistance range should be used for measurements above 20 kΩ.

Typical meter indications for silicon PNP transistors are shown in Fig. 15.9 (forward bias is applied in Fig. 15.9(a) and reverse bias is applied in Fig. 15.9(b)). As for NPN transistors, corresponding indications for germanium PNP will be somewhat lower. Once again, the ×100 resistance range should be used for measurements below 20 kΩ, and the ×1k resistance range should be used for measurements above 20 kΩ.

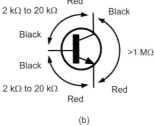

(a)

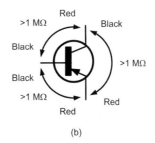

(b)

Figure 15.9 PNP transistor 'go/no-go' checks with typical resistance values in the 'go' case

Bipolar transistor current gain checks

Simple go/no-go checks can also be used to determine whether a transistor is providing current gain. The principle is illustrated in Fig. 15.10. For an NPN transistor this is achieved by measuring the resistance between collector and emitter, as shown in Fig. 15.10(a), before applying a small amount of base current via a 100 kΩ resistor, as shown in Fig. 15.10(b). The application of base current should

have the effect of producing a significant increase in collector current and this, in turn, will result in a much reduced resistance measured between the collector and emitter. It is, of course, not possible to provide an accurate indication of current gain by this method however it will tell you whether or not a transistor is producing current gain. The method will also let you compare the current gain of one (known) transistor with another. The equivalent arrangement for testing a PNP transistor is shown in Fig. 15.11.

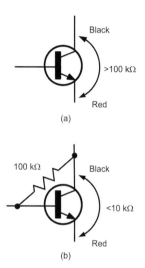

(a)

(b)

Figure 15.10   Checking NPN transistor current gain using an ohmmeter and a 100 kΩ resistor

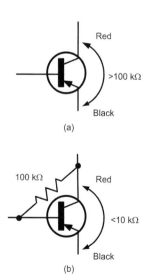

(a)

(b)

Figure 15.11   Checking PNP transistor current gain using an ohmmeter and a 100 kΩ resistor

A transistor/diode checker

A simple transistor/diode checker can make an excellent project for the beginner to electronic circuit construction. The instrument shown in Fig. 15.14 measures current gain for NPN and PNP transistors up to 1,000 and leakage current up to 100 µA. It also has a diode check facility which can be useful for rapid testing of signal, rectifier, and zener diodes. The values of R1, R2 and R3 can be made from series combinations of preferred value resistors (for example, R2 can be made from a 150 kΩ resistor connected in series with an 18 kΩ resistor) whilst the meter shunt resistor, R5, should be calculated in order to provide a full-scale deflection on the meter of 10 mA whenever any of the current gain ranges are selected (see page 246). Typical front panel layouts and meter scales are shown in Figs 15.12 and 15.13.

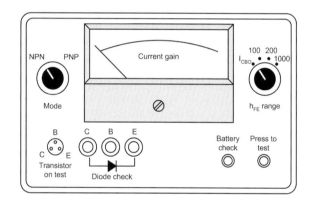

Figure 15.12   Typical front panel layout for the transistor and diode checker

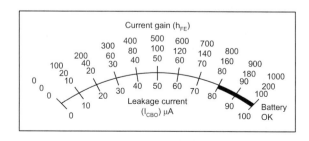

Figure 15.13   A suitable meter scale for the transistor and diode checker

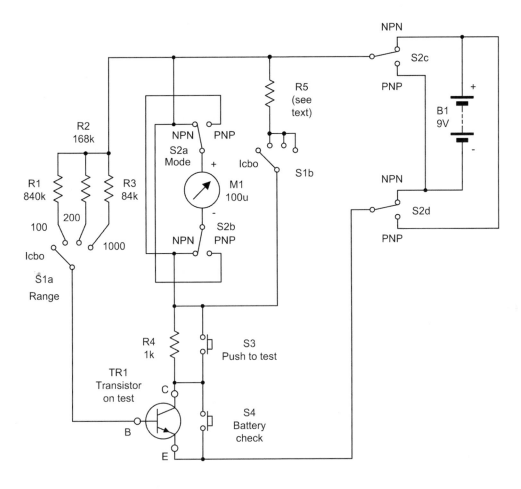

Figure 15.14 Complete circuit of the transistor and diode checker

Logic circuit faults

Whilst multimeters can be used for checking the voltages present on the supply rails within digital logic circuits they are generally unsuitable for measuring the logic states of the points or **nodes** within a circuit. There are several reasons for this, including the fact that logic states are represented by voltage levels that vary according to the integrated circuit technology employed. However the principal reason for misleading indications is that the logic levels can be rapidly changing or may take the form of short duration pulses that simply cannot be detected by a conventional voltage measuring instrument. That said, where logic levels

do remain static for periods of several seconds, a multimeter may be used on the DC voltage ranges to detect the presence of either a logic 0 or logic 1 state. In the case of standard TTL devices (which operate from a +5V supply) the following voltage levels apply:

Voltage measured	*Logic level indicated*
< 0.8 V	0
0.8 V to 2.0 V	indeterminate
> 2.0 V	1

It should be noted that an 'indeterminate' logic level may result from a **tri-state** condition where several logic device place their outputs on a shared signal (or bus) line. To prevent a conflict occurring

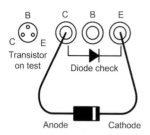

(a) Conventional signal, switching and rectifier diodes

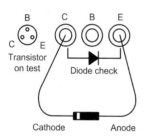

(b) Miniature wire ended diodes

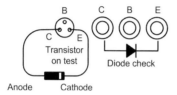

(c) Zener diodes

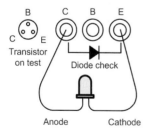

(d) Light emitting diodes

Figure 15.15 Transistor and diode checker connections for various different types of diode (note particularly that the connections to a zener diode must be reversed)

between the devices, the outputs of any inactive devices are placed in a high-impedance state in which the voltage at the output can safely take any value within the logic range without causing damage to the inactive device. Modern high-impedance instruments will usually produce a misleading fluctuating indication in such circumstances and this can sometimes be confused with nodes that are actually pulsing.

Logic probes

The simplest, and by far the most convenient, method of tracing logic states involves the use of a logic probe. This invaluable tool comprises a compact hand-held probe fitted with LEDs that indicate the logical state of its probe tip.

Unlike multi-range meters, logic probes can distinguish between lines which are actively pulsing and those which are in a permanently tri-state condition. In the case of a line which is being pulsed, the logic 0 and logic 1 indicators will both be illuminated (though not necessarily with the same brightness) whereas, in the case of a tri-state line neither indicator should be illuminated.

Logic probes usually also provide a means of displaying pulses having a very short duration which may otherwise go undetected. This is accomplished by the inclusion of a pulse stretching circuit (i.e. a monostable). This elongates short duration pulses so that a visible indication is produced on a separate 'pulse' LED.

Logic probes invariably derive their power supply from the circuit under test and are connected by means of a short length of twin flex fitted with insulated crocodile clips. While almost any convenient connecting point may be used, the positive supply and ground terminals make ideal connecting points which can be easily identified.

A typical logic probe circuit is shown in Fig. 15.16. This circuit uses a dual comparator to sense the logic 0 and logic 1 levels and a timer that acts as a **monostable pulse stretcher** to indicate the presence of a pulse input rather than a continuous logic 0 or logic 1 condition. Typical logic probe indications and waveforms are shown in Fig. 15.18.

Figure 15.19 shows how a logic probe can be used to check a typical combinational logic arrangement. In use, the probe is simply moved from node to node and the logic level is displayed and compared with the expected level.

Logic pulsers

We sometimes need to simulate the logic levels generated by a peripheral device or a sensor. A permanent logic level can easily be generated by pulling a line up to + 5 V via a 1 kΩ resistor or by temporarily tying a line to 0 V. However, on other occasions, you may need to simulate a pulse rather than a permanent logic state and this can be achieved by means of a logic pulser.

A logic pulser provides a means of momentarily forcing a logic level transition into a circuit regardless of its current state and thus it overcomes the need to disconnect or de-soldering any of the devices. The polarity of the pulse (produced at the touch of a button) is adjusted so that the node under investigation is momentarily forced into the opposite logical state. During the period before the button is depressed and for the period after the

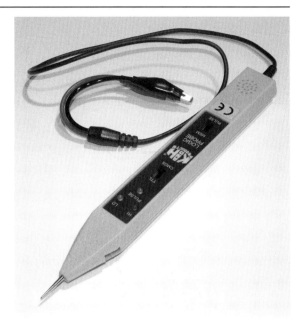

Figure 15.17   A versatile logic probe with logic level and pulse indicating facilities. Note the crocodile leads for connecting to the TTL or CMOS supply and ground connections

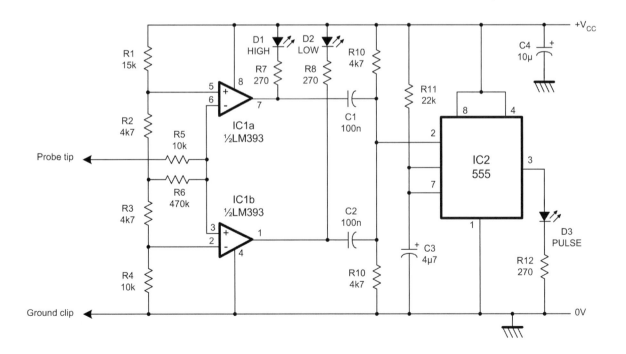

Figure 15.16   A simple logic probe for TTL and CMOS circuits with pulse stretching facilities

LED INDICATOR			STATE INDICATED	WAVEFORM
LOW	PULSE	HIGH		
OFF	OFF	ON	Steady logic 1	1 ———— 0
ON	OFF	OFF	Steady logic 0	1 0 ————
OFF	OFF	OFF	Open circuit or undefined level	1 ———— 0
OFF	BLINK	OFF	Pulse train of near 50% duty cycle at >1MHz	1 ⊓⊓⊓⊓⊓⊓ 0
ON	BLINK	ON	Pulse train of near 50% duty cycle at <1MHz	1 ⊓ ⊓ ⊓ 0
OFF	BLINK	ON	Pulse train of high mark:space ratio	1 ⊓ ⊓ ⊓ 0
ON	BLINK	OFF	Pulse train of low mark:space ratio	1 ⊔ ⊔ ⊔ 0

Figure 15.18   Typical logic probe indications

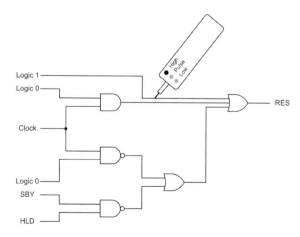

Figure 15.19   Using a logic probe to trace logic levels in a simple logic circuit (note that the supply and ground connections are not shown)

pulse has been completed, the probe tip adopts a tri-state (high impedance) condition. Hence the probe does not permanently affect the logical state of the point in question.

Logic pulsers derive their power supply from the circuit under test in the same manner as logic probes. Here again, the supply and ground connections usually make suitable connecting points.

A typical logic pulser circuit is shown in Fig. 15.20. The circuit comprises a 555 monostable pulse generator triggered from a push-button. The output of the pulse generator is fed to a complementary transistor arrangement in order to make it fully TTL-compatible. As with the logic pulser, this circuit derives its power from the circuit under test (usually +5V).

Figure 15.21 shows an example of the combined use of a logic probe and a logic pulser for testing a simple J-K bistable. The logic probe is used to check the initial state of the bistable outputs (see Fig. 15.21(a) and (b)). Note that the outputs should be complementary. Next, the logic pulser is applied to the clock (CK) input of the bistable (Fig. 15.21(c)) and the Q output is checked using the logic probe. The application of a pulse (using the trigger button) should cause the Q output of the bistable to change state (see Fig. 15.21(d)).

Practical investigation

Objective

To investigate faults in a typical linear regulated power supply.

Components and test equipment

Breadboard, mains transformer with 220 V (or 110 V where appropriate) primary and 15 V 2 A secondary winding, four 1N5401 diodes, TIP41C transistor mounted on a heatsink of less than 2.5 °C/W, TL081 operational amplifier, fixed resistors of 1 kΩ (two required) and 2.2 kΩ (two required), 500 mA quick-blow fuse, 2 A slow-blow fuse, BZX79 7.5 V zener diode, capacitors of 6,800 μF, 470 μF, and 100 μF (all rated at 50 V d.c. working), digital multimeter, test leads and probes.

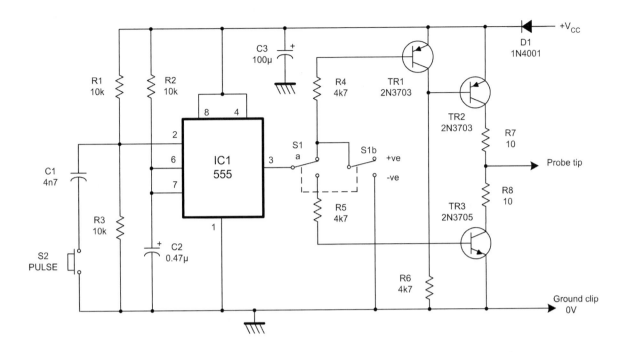

Figure 15.20 A logic pulser suitable for fault tracing on TTL and CMOS logic circuits

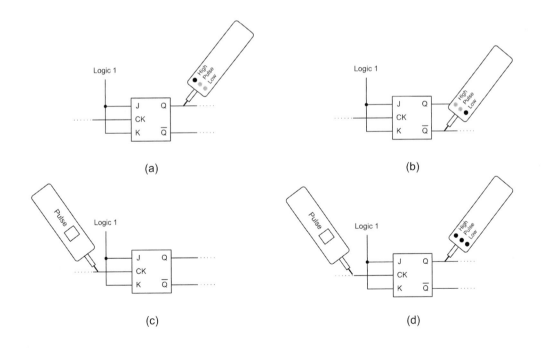

Figure 15.21 Using a logic probe and a logic pulser to check a bistable

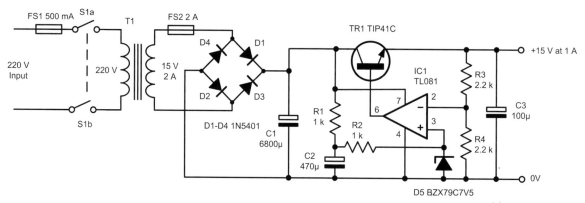

Figure 15.22 Regulated power supply used in the Practical investigation

Procedure

Connect the circuit and measure the voltages at the following test points with no fault present; TR1 collector, base and emitter; IC1 inverting input (pin-2) and non-inverting input (pin-3). Record your results in a suitable table.

Repeat the voltage measurements with the following fault conditions present:
(a) R1 open circuit
(b) C2 short circuit
(c) D5 open circuit
(d) R3 open circuit.

In each case, use your knowledge of how the circuit operates to justify the readings obtained.

Problems

15.1   Figure 15.23 shows a two-stage high-gain amplifier. Identify the faulty component and the nature of the fault (e.g. open-circuit or short-circuit) for each of the faults shown in Table 15.2.

15.2   With the aid of a simple diagram, explain how each of the following instruments can be used to aid fault finding in digital circuits:
(a) logic probe
(b) logic pulser.

15.3   Explain how an ohmmeter can be used to perform simple 'go/no-go' checks on:
(a) a diode
(b) an NPN transistor.

Answers to these problems appear on page 376.

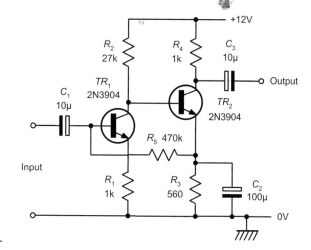

Figure 15.23   See Question 15.1

Table 15.2   See Question 15.1

Condition	TR1			TR2		
	c	b	e	c	b	e
Normal	3.3	0.9	0.3	7.4	3.3	2.6
Fault 1	0.4	0.0	0.0	12.0	0.4	0.0
Fault 2	4.2	1.3	0.7	4.2	4.2	4.2
Fault 3	3.2	1.0	0.3	12.0	3.2	2.7
Fault 4	4.2	4.2	0.3	5.9	4.2	3.5
Fault 5	0.8	0.0	0.0	0.1	0.8	0.0
Fault 6	5.1	0.3	0.0	4.5	5.1	4.4

16

Sensors and interfacing

Sensors provide us with a means of generating signals that can be used as inputs to electronic circuits. The things that we might want to sense include physical parameters such as temperature, light level, and pressure. Being able to generate an electrical signal that accurately represents these quantities allows us not only to measure and record these values but also to control them.

Sensors are, in fact, a subset of a larger family of devices known as **transducers** so we will consider these before we look at sensors and how we condition the signals that they produce in greater detail. We begin, however, with a brief introduction to the instrumentation and control systems in which sensors, transducers, and signal conditioning circuits are used.

Instrumentation and control systems

Figure 16.1 shows the arrangement of an instrumentation system. The physical quantity to be measured (e.g. temperature) acts upon a sensor that produces an electrical output signal. This signal is an electrical analogue of the physical input but note that there may not be a linear relationship between the physical quantity and its electrical equivalent. Because of this and since the output produced by the sensor may be small or may suffer from the presence of noise (i.e. unwanted signals) further signal conditioning will be required before the signal will be at an acceptable level and in an acceptable form for signal processing, display and

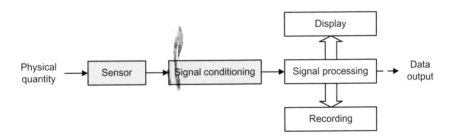

(a) An instrumentation system

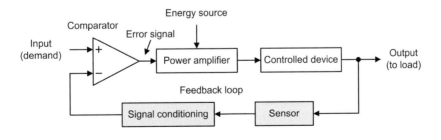

(b) A control system

Figure 16.1 Instrumentation and control systems

recording. Furthermore, because the signal processing may use digital rather than analogue signals an additional stage of analogue-to-analogue conversion may be required.

Figure 16.1(b) shows the arrangement of a control system. This uses **negative feedback** in order to regulate and stabilize the output. It thus becomes possible to set the input or **demand** (i.e. what we desire the output to be) and leave the system to regulate itself by comparing it with a signal derived from the output (via a sensor and appropriate signal conditioning).

A **comparator** is used to sense the difference in these two signals and where any discrepancy is detected the input to the power amplifier is adjusted accordingly. This signal is referred to as an **error signal** (it should be zero when the output exactly matches the demand). The input (demand) is often derived from a simple potentiometer connected across a stable d.c. voltage source whilst the controlled device can take many forms (e.g. a d.c. motor, linear actuator, heater, etc.).

Transducers

Transducers are devices that convert energy in the form of sound, light, heat, etc., into an equivalent electrical signal, or vice versa.

Before we go further, let's consider a couple of examples that you will already be familiar with. A loudspeaker is a transducer that converts low-frequency electric current into audible sounds. A microphone, on the other hand, is a transducer that performs the reverse function, i.e. that of converting sound pressure variations into voltage or current. Loudspeakers and microphones can thus be considered as complementary transducers.

Transducers may be used as both inputs to electronic circuits and outputs from them. From the two previous examples, it should be obvious that a loudspeaker is an **output transducer** designed for use in conjunction with an audio system. Whereas, a microphone is an **input transducer** designed for use with a recording or sound reinforcing system.

There are many different types of transducer and Tables 16.1 and 16.2 provide some examples of transducers that can be used to input and output three important physical quantities; sound, temperature, and angular position.

Sensors

A *sensor* is a special kind of transducer that is used to generate an input signal to a measurement, instrumentation or control system. The signal produced by a sensor is an **electrical analogy** of a physical quantity, such as distance, velocity, acceleration, temperature, pressure, light level, etc. The signals returned from a sensor, together with control inputs from the user or controller (as appropriate) will subsequently be used to determine the output from the system. The choice of sensor is governed by a number of factors including accuracy, resolution, cost, and physical size.

Sensors can be categorized as either **active** or **passive**. An active sensor *generates* a current or voltage output. A passive transducer *requires a source of current or voltage* and it modifies this in some way (e.g. by virtue of a change in the sensor's resistance). The result may still be a voltage or current *but it is not generated by the sensor on its own*.

Sensors can also be classed as either **digital** or **analogue**. The output of a digital sensor can exist in only two discrete states, either 'on' or 'off', 'low' or 'high', 'logic 1' or 'logic 0', etc. The output of an analogue sensor can take any one of an infinite number of voltage or current levels. It is thus said to be *continuously variable*. Table 16.3 provides details of some common types of sensor.

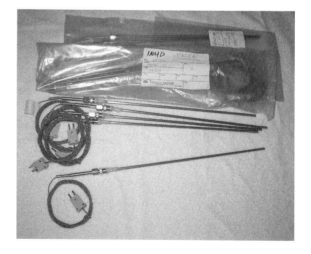

Figure 16.2 A selection of thermocouple probes

Table 16.1   Some examples of input transducers

Physical quantity	Input transducer	Notes
Sound (pressure change)	Dynamic microphone (see Fig. 16.3)	Diaphragm attached to a coil is suspended in a magnetic field. Movement of the diaphragm causes current to be induced in the coil.
Temperature	Thermocouple (see Fig. 16.2)	Small e.m.f. generated at the junction between two dissimilar metals (e.g. copper and constantan). Requires reference junction and compensated cables for accurate measurement.
Angular position	Rotary potentiometer	Fine wire resistive element is wound around a circular former. Slider attached to the control shaft makes contact with the resistive element. A stable d.c. voltage source is connected across the ends of the potentiometer. Voltage appearing at the slider will then be proportional to angular position.

Table 16.2   Some examples of output transducers

Physical quantity	Output transducer	Notes
Sound (pressure change)	Loudspeaker (see Fig. 16.3)	Diaphragm attached to a coil is suspended in a magnetic field. Current in the coil causes movement of the diaphragm which alternately compresses and rarefies the air mass in front of it.
Temperature	Heating element (resistor)	Metallic conductor is wound onto a ceramic or mica former. Current flowing in the conductor produces heat.
Angular position	Rotary potentiometer	Multi-phase motor provides precise rotation in discrete steps of 15° (24 steps per revolution), 7.5° (48 steps per revolution) and 1.8° (200 steps per revolution).

Figure 16.3   A selection of audible transducers

Figure 16.4   Various switch sensors

Table 16.3   Some examples of output transducers

Physical quantity	Output transducer	Notes
Angular position	Resistive rotary position sensor (see Fig. 16.5)	Rotary track potentiometer with linear law produces analogue voltage proportional to angular position.
	Optical shaft encoder	Encoded disk interposed between optical transmitter and receiver (infra-red LED and photodiode or photo-transistor).
Angular velocity	Tachogenerator	Small d.c. generator with linear output characteristic. Analogue output voltage proportional to shaft speed.
	Toothed rotor tachometer	Magnetic pick-up responds to the movement of a toothed ferrous disk. The pulse repetition frequency of the output is proportional to the angular velocity.
Flow	Rotating vane flow sensor (see Fig. 16.9)	Turbine rotor driven by fluid. Turbine interrupts infra-red beam. Pulse repetition frequency of output is proportional to flow rate.
Linear position	Resistive linear position sensor	Linear track potentiometer with linear law produces analogue voltage proportional to linear position. Limited linear range.
	Linear variable differential transformer (LVDT)	Miniature transformer with split secondary windings and moving core attached to a plunger. Requires a.c. excitation and phase-sensitive detector.
	Magnetic linear position sensor	Magnetic pick-up responds to movement of a toothed ferrous track. Pulses are counted as the sensor moves along the track.
Light level	Photocell	Voltage-generating device. The analogue output voltage produced is proportional to light level.
	Light dependent resistor (LDR) (see Fig. 16.8)	An analogue output voltage results from a change of resistance within a cadmium sulphide (CdS) sensing element. Usually connected as part of a potential divider or bridge.
	Photodiode (see Fig, 16.8)	Two-terminal device connected as a current source. An analogue output voltage is developed across a series resistor of appropriate value.
	Phototransistor (see Fig. 16.8)	Three-terminal device connected as a current source. An analogue output voltage is developed across a series resistor of appropriate value.
Liquid level	Float switch (see Fig. 16.7)	Simple switch element which operates when a particular level is detected.
	Capacitive proximity switch	Switching device which operates when a particular level is detected. Ineffective with some liquids.
	Diffuse scan proximity switch	Switching device which operates when a particular level is detected. Ineffective with some liquids.

Table 16.3   (continued)

Physical quantity	Output transducer	Notes
Pressure	Microswitch pressure sensor (see Fig. 16.4)	Microswitch fitted with actuator mechanism and range setting springs. Suitable for high-pressure applications.
	Differential pressure vacuum switch	Microswitch with actuator driven by a diaphragm. May be used to sense differential pressure. Alternatively, one chamber may be evacuated and the sensed pressure applied to a second input.
	Piezo-resistive pressure sensor	Pressure exerted on diaphragm causes changes of resistance in attached piezo-resistive transducers. Transducers are usually arranged in the form of a four active element bridge which produces an analogue output voltage.
Proximity	Reed switch (see Fig. 16.4)	Reed switch and permanent magnet actuator. Only effective over short distances.
	Inductive proximity switch	Target object modifies magnetic field generated by the sensor. Only suitable for metals (non-ferrous metals with reduced sensitivity).
	Capacitive proximity switch	Target object modifies electric field generated by the sensor. Suitable for metals, plastics, wood, and some liquids and powders.
	Optical proximity switch (see Fig. 16.4)	Available in diffuse and through scan types. Diffuse scan types require reflective targets. Both types employ optical transmitters and receivers (usually infra-red emitting LEDs and photo-diodes or photo-transistors). Digital input port required.
Strain	Resistive strain gauge	Foil type resistive element with polyester backing for attachment to body under stress. Normally connected in full bridge configuration with temperature-compensating gauges to provide an analogue output voltage.
	Semiconductor strain gauge	Piezo-resistive elements provide greater outputs than comparable resistive foil types. More prone to temperature changes and also inherently non-linear.
Temperature	Thermocouple (see Fig. 16.2)	Small e.m.f. generated by a junction between two dissimilar metals. For accurate measurement, requires compensated connecting cables and specialized interface.
	Thermistor (see Fig. 16.6)	Usually connected as part of a potential divider or bridge. An analogue output voltage results from resistance changes within the sensing element.
	Semiconductor temperature sensor (see Fig. 16.6)	Two-terminal device connected as a current source. An analogue output voltage is developed across a series resistor of appropriate value.
Weight	Load cell	Usually comprises four strain gauges attached to a metal frame. This assembly is then loaded and the analogue output voltage produced is proportional to the weight of the load.
Vibration	Electromagnetic vibration sensor	Permanent magnet seismic mass suspended by springs within a cylindrical coil. The frequency and amplitude of the analogue output voltage are respectively proportional to the frequency and amplitude of vibration.

The contact 'bounce' that occurs when a switch is operated results in rapid making and breaking of the switch until it settles into its new state. Figure 16.13 shows the waveform generated by the simple switch input circuit of Fig. 16.11 as the contacts close. The spurious states can cause problems if the switch is sensed during the period in which the switch contacts are in motion, and hence steps must be taken to minimize the effects of bounce. This may be achieved by using some extra circuitry in the form of a **debounce circuit** or by including appropriate software delays (of typically 4 to 20 ms) so that spurious switching states are ignored. We shall discuss these two techniques separately.

Immunity to transient switching states is generally enhanced by the use of active-low inputs (i.e. a logic 0 state at the input is used to assert the condition required). The debounce circuit shown in Fig. 16.14 is adequate for most toggle, slide and push-button type switches. The value of the 100 Ω resistor takes into account the low-state sink current required by IC1 (normally 1.6 mA for standard TTL and 400 µA for LS-TTL). This resistor should not be allowed to exceed approximately 470 Ω in order to maintain a valid logic 0 input state. The values quoted generate an approximate 1 ms delay (during which the switch contacts will have settled into their final state). It should be noted that, on power-up, this circuit generates a logic 1 level for approximately 1 ms before the output reverts to a logic 0 in the inactive state. The circuit obeys the following state table:

Switch condition	Logic output
closed	1
open	0

An alternative, but somewhat more complex, switch de-bouncing arrangement is shown in Fig. 16.15. Here a single-pole double-throw (SPDT) changeover switch is employed. This arrangement has the advantage of providing complementary outputs and it obeys the following state table:

Switch condition	Q
$Q \rightarrow 1$	1
$Q \rightarrow 0$	0

Rather than use an integrated circuit R-S bistable in the configuration of Fig. 16.15 it is often expedient to make use of 'spare' two-input NAND or NOR

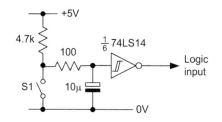

Figure 16.14 Simple debounce circuit

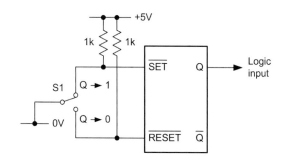

Figure 16.15 Debounce circuit based on an R-S bistable

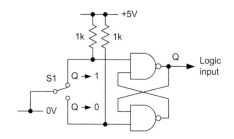

(a) Based on NAND gates

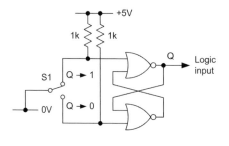

(b) Based on NOR gates

Figure 16.16 Alternative switch debounce circuits: (a) based on NAND gates; (b) based on NOR gates

gates arranged to form bistables using the circuits shown in Figs 16.16(a) and 16.16(b), respectively. Figure 16.17 shows a rather neat extension of this theme in the form of a touch-operated switch. This arrangement is based on a 4011 CMOS quad two-input NAND gate (though only two gates of the package are actually used in this particular configuration).

Finally, it is some times necessary to generate a latching action from a normally-open push-button switch. Figure 16.18 shows an arrangement in which a 74LS73 JK bistable is clocked from the output of a debounced switch.

Pressing the switch causes the bistable to change state. The bistable then remains in that state until the switch is depressed a second time. If desired, the complementary outputs provided by the bistable may be used to good effect by allowing the unused output to drive an LED. This will become illuminated whenever the Q output is high.

Semiconductor temperature sensors

Semiconductor temperature sensors are ideal for a wide range of temperature-sensing applications. The popular AD590 semiconductor temperature sensor, for example, produces an output current which is proportional to absolute temperature and which increases at the rate of 1 μA/K. The characteristic of the device is illustrated in Fig. 16.19. The AD590 is laser trimmed to produce a current of 298.2 μA ($\pm$2.5 μA) at a temperature of 298.2°C (i.e. 25°C). A typical interface between the AD590 and an analogue input is shown in Fig. 16.20. The Practical investigation (page 301) makes further use of this device.

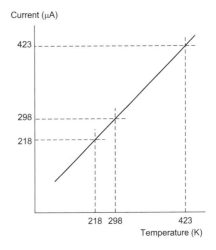

Figure 16.19 AD590 semiconductor temperature sensor characteristic

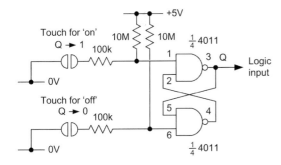

Figure 16.17 Touch-operated switch

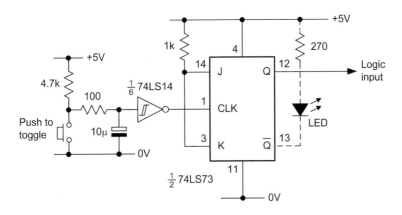

Figure 16.18 Latching action switch

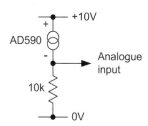

Figure 16.20   Typical input interface for the AD590 temperature sensor (the output voltage will increase at the rate of 10 mV per °C)

Thermocouples

Thermocouples comprise a junction of dissimilar metals which generate an e.m.f. proportional to the temperature differential which exists between the measuring junction and a reference junction. Since the measuring junction is usually at a greater temperature than that of the reference junction, it is sometimes referred to as the **hot junction**. Furthermore, the reference junction (i.e. the **cold junction**) is often omitted in which case the sensing junction is simply terminated at the signal conditioning board. This board is usually maintained at, or near, normal room temperatures.

Thermocouples are suitable for use over a very wide range of temperatures (from -100°C to +1100°C). Industry standard 'type K' thermocouples comprise a positive arm (conventionally coloured brown) manufactured from nickel/chromium alloy whilst the negative arm (conventionally coloured blue) is manufactured from nickel/aluminium.

The characteristic of a **type K thermocouple** is defined in BS 4937 Part 4 of 1973 (International Thermocouple Reference Tables) and this standard gives tables of e.m.f. versus temperature over the range 0°C to +1100°C. In order to minimize errors, it is usually necessary to connect thermocouples to appropriate signal conditioning using compensated cables and matching connectors. Such cables and connectors are available from a variety of suppliers and are usually specified for use with type K thermocouples. A selection of typical thermocouple probes for high temperature measurement was shown earlier in Fig. 16.2.

Threshold detection

Analogue sensors are sometimes used in situations where it is only necessary to respond to a pre-determined threshold value. In effect, a two-state digital output is required. In such cases a simple one-bit analogue-to-digital converter based on a comparator can be used. Such an arrangement is, of course, very much simpler and more cost-effective than making use of a conventional analogue input port!

Simple threshold detectors for light level and temperature are shown in Figs 16.21, 16.22 and 16.24. These circuits produce TTL-compatible outputs suitable for direct connection to a logic circuit or digital input port.

Figure 16.21 shows a light level threshold detector based on a comparator and light-dependent resistor (LDR). This arrangement generates a logic 0 input whenever the light level exceeds the threshold setting, and vice versa. Figure 16.22 shows how light level can be sensed using a photodiode. This circuit behaves in the same manner as the LDR equivalent but it is important to be aware that circuit achieves peak sensitivity in the near infra-red region. Figure 16.23 shows how the spectral response of a typical light-dependent resistor (NORP12) compares with that of a conventional photodiode (BPX48). Note that the BPX48 can also be supplied with an integral daylight filter (BPX48F).

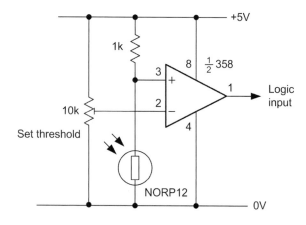

Figure 16.21   Light-level threshold detector based on a light-dependent resistor (LDR)

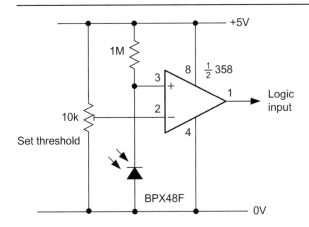

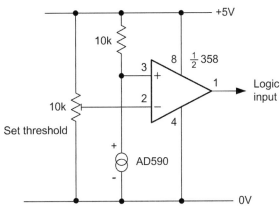

Figure 16.22 Light level threshold detector based on a photodiode

Figure 16.24 Temperature threshold detector based on an AD590 semiconductor temperature sensor

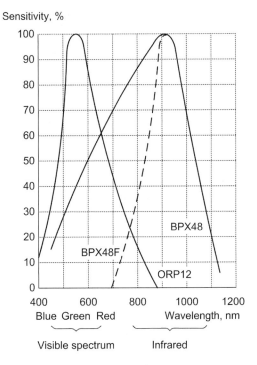

Figure 16.23 Comparison of the spectral response of an LDR and some common photodiodes

Figure 16.24 shows how temperature thresholds can be sensed using the AD590 sensor described earlier. This arrangement generates a logic 0 input whenever the temperature level exceeds the threshold setting, and vice versa.

Outputs

Having dealt at some length with input sensors, we shall now focus our attention on output devices (such as relays, loudspeakers and LED indicators) and the methods used for interfacing them. Integrated circuit output drivers are available for more complex devices, such as LCD displays and stepper motor. Many simple applications will, however, only require a handful of components in order to provide an effective interface.

LED indicators

Indicators based on light emitting diodes (LEDs) are inherently more reliable than small filament lamps and also consume considerably less power. They are thus ideal for providing visual status and warning displays. LEDs are available in a variety of styles and colours and 'high brightness' types can be employed where high-intensity displays are required.

A typical red LED requires a current of around 10 mA to provide a reasonably bright display and such a device may be directly driven from a buffered digital output port. Different connections are used depending upon whether the LED is to be illuminated for a logic 0 or logic 1 state. Several possibilities are shown in Fig. 16.25.

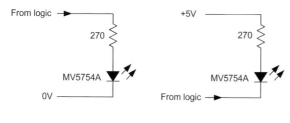

(a) Logic 1 to illuminate the LED   (b) Logic 0 to illuminate the LED

Figure 16.25 Driving an LED from a buffered logic gate or digital I/O port

Table 16.4 Typical waveform produced by a switch closure

Voltage	Series resistance (all 0.25 W)
3V to 4V	100 Ω
4V to 5V	150 Ω
5V to 8V	220 Ω
8V to 12V	470 Ω
12V to 15V	820 Ω
15V to 20V	1.2 kΩ
20V to 28V	1.5 kΩ

Where drive current is insufficient to operate an LED, an auxiliary transistor can be used as shown in Fig. 16.26. The LED will operate when the output from a logic circuit card is taken to logic 1 and the operating current should be approximately 15 mA (thereby providing a brighter display than the arrangements previously described). The value of LED series resistance will be dependent upon the supply voltage and can be determined from the equation shown on page 95 or it can be selected from the data shown in Table 16.4.

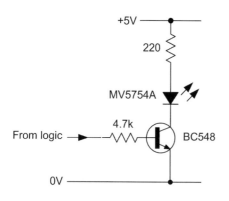

Figure 16.26 Using an auxiliary transistor to drive an LED where current drive is limited

Driving high-current loads

Due to the limited output current and voltage capability of most standard logic devices and I/O

ports, external circuitry will normally be required to drive anything other than the most modest of loads. Figure 16.27 shows some typical arrangements for operating various types of medium- and high-current load. Fig. 16.27(a) shows how an NPN transistor can be used to operate a low-power relay. Where the relay requires an appreciable operating current (say, 150 mA, or more) a plastic encapsulated Darlington power transistor should be used as shown in Figure 16.27(b). Alternatively, a power MOSFET may be preferred, as shown in Fig. 16.27(c). Such devices offer very low values of 'on' resistance coupled with a very high 'off' resistance. Furthermore, unlike conventional bipolar transistors, a power FET will impose a negligible load on an I/O port. Figure 16.27(d) shows a filament lamp driver based on a plastic Darlington power transistor. This circuit will drive lamps rated at up to 24 V, 500 mA. Finally, where visual indication of the state of a relay is desirable it is a simple matter to add an LED indicator to the driver stage, as shown in Fig. 16.28.

Audible outputs

Where simple audible warnings are required, miniature piezoelectric transducers may be used. Such devices operate at low voltages (typically in the range 3V to 15V) and can be interfaced with the aid of a buffer, open-collector logic gate, or

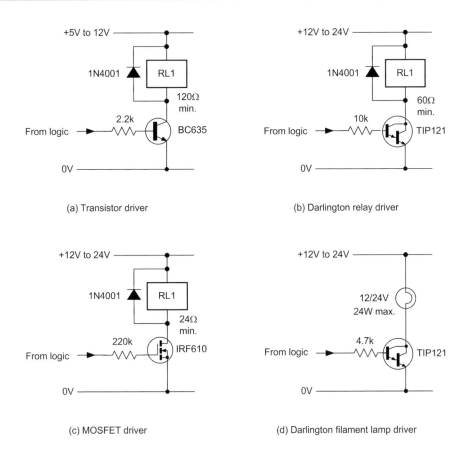

(a) Transistor driver

(b) Darlington relay driver

(c) MOSFET driver

(d) Darlington filament lamp driver

Figure 16.27 Typical medium- and high-current driver circuits: (a) transistor low-current relay driver; (b) Darlington medium/high-current relay driver; (c) MOSFET relay driver; (d) Darlington low-voltage filament lamp driver

transistor. Figures 16.29(a)-(c) show typical interface circuits which produce an audible output when the port output line is at logic 1.

Where a pulsed rather than continuous audible alarm is required, a circuit of the type shown in Fig. 16.31 can be employed. This circuit is based on a standard 555 timer operating in astable mode and operates at approximately 1 Hz. A logic 1 from the port output enables the 555 and activates the pulsed audio output.

Finally, the circuit shown in Fig. 16.32 can be used where a conventional moving-coil loudspeaker is to be used in preference to a piezo-electric transducer. This circuit is again based on the 555 timer and provides a continuous output at approximately 1 kHz whenever the port output is at logic 1.

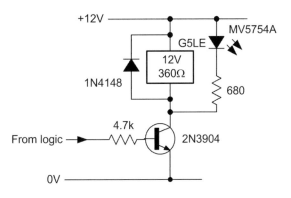

Figure 16.28 Showing how an LED indicator can easily be added to a relay driver

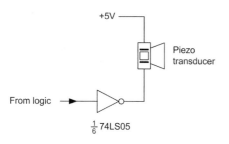

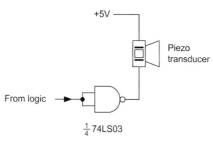

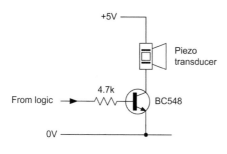

Figure 16.29 Audible output driver circuits

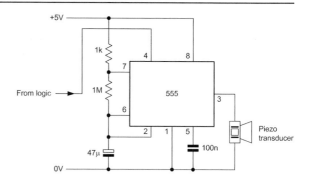

Figure 16.31 Audible alarm circuit based on a 555 astable oscillator and a piezoelectric transducer

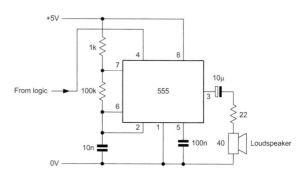

Figure 16.32 Audible alarm circuit based on a 555 astable oscillator and a 40 Ω loudspeaker

Motors

Circuit arrangements used for driving DC motors generally follow the same lines as those described earlier for use with relays. As an example, the circuits shown in Fig. 16.30 show how a Darlington

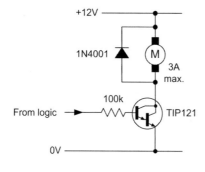

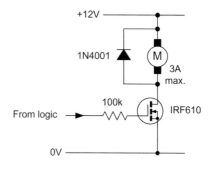

Figure 16.30   Motor driver circuits

driver and a power MOSFET can be used to drive a low-voltage DC motor. These circuits are suitable for use with DC motors rated at up to 12 V with stalled currents of up to 3 A. In both cases, a logic 1 from the output port will operate the motor.

Driving mains connected loads

Control systems are often used in conjunction with mains connected loads. Modern **solid-state relays** (SSRs) offer superior performance and reliability when compared with conventional relays in such applications. SSRs are available in a variety of encapsulations (including DIL, SIL, flat-pack, and plug-in octal) and may be rated for RMS currents between 1 A and 40 A.

In order to provide a high degree of isolation between input and output, SSRs are optically coupled. Such devices require minimal input

currents (typically 5 mA, or so, when driven from 5 V) and they can thus be readily interfaced with an I/O port that offers sufficient drive current. In other cases, it may be necessary to drive the SSR from an unbuffered I/O port using an open-collector logic gate. Typical arrangements are shown in Fig. 16.33. Finally, it is important to note that, when an inductive load is to be controlled, a **snubber network** should be fitted, as shown in Fig. 16.34.

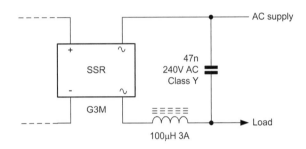

Figure 16.34 Using a snubber circuit with an inductive load

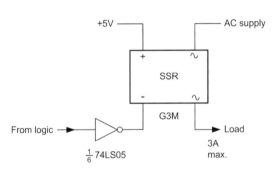

(a) Using an open-collector buffer

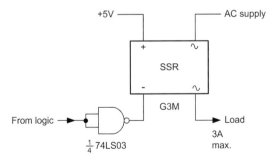

(b) Using an open-collector logic gate

Figure 16.33 Interface circuits for driving solid state relays

Practical investigation

Objective

To investigate the operation of an AD590 semiconductor temperature sensor and its interface/signal conditioning circuit.

Components and test equipment

Breadboard, d.c. power supply with ± 5 V output at 100 mA (or more), 10 kΩ linear variable resistor (preferably multi-turn type), fixed resistors of 10 kΩ (two required) and 100 kΩ 0.25 W 5%, AD590 temperature sensor, TL081 operational amplifier, digital multimeter, thermometer, test leads and probes, small hand-held hair drier.

Procedure

Connect the circuit, set VR1 to mid-position, and measure the output voltage produced by the circuit. Adjust VR1 for an output voltage reading of exactly zero at a temperature of 20°C. Place the

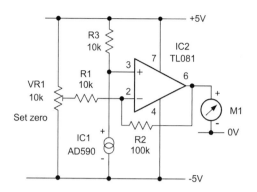

Figure 16.35   Temperature sensor and interface circuit used in the Practical investigation

temperature sensor in close proximity to the thermometer in the air flow from the hair drier. Slowly increase the temperature and record the output voltage for each increase of 10°C indicated on the thermometer using the table shown in Table 16.5.

Graph and calculations

Plot a graph showing output voltage against temperature (see Fig. 16.36) and measure the slope of the graph.

Conclusions

Comment on the shape of the graph. Is this what you would expect? Suggest an application for the circuit.

Table 16.5 Table of results

Temperature (°C)	Output voltage (V)
20	
30	
40	
50	
60	
70	

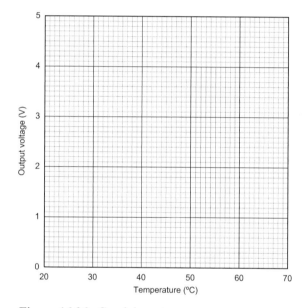

Figure 16.36 Graph layout

Problems

16.1   For each of the automotive applications listed below, state:
(a) the quantity sensed or measured
(b) the name of the transducer used for sensing or measurement
(c) the output quantity produced by the transducer

Applications:

1. Sensing the amount of fuel in a tank
2. Sensing the flow rate of fuel in a fuel feed pipe
3. Warning a driver that a road surface is liable to be icy
4. Warning a driver that a door is open.

16.2   Sketch circuit diagrams to show how the following devices can be interfaced to a logic circuit based on conventional LS-TTL logic:
(a) an LED
(b) a normally-open push button
(c) a relay
(d) a piezoelectric audible transducer.

16.3   With the aid of a circuit diagram, explain the operation of a threshold light level sensor. Specify the type of sensor used and explain how the response might vary with light of different wavelengths.

Answers to these problems appear on page 376.

17

Circuit simulation

Computer simulation provides you with a powerful and cost-effective tool for designing, simulating, and analysing a wide variety of electronic circuits. In recent years, the computer software packages designed for this task have not only become increasingly sophisticated but also have become increasingly easy to use. Furthermore, several of the most powerful and popular packages are now available at low cost either in evaluation, 'lite' or student versions. In addition, there are several excellent freeware and shareware packages (see Appendix 7).

Whereas early electronic simulation software required that circuits were entered using a complex **netlist** that described all of the components and connections present in a circuit, most modern packages use an on-screen graphical representation of the circuit on test. This, in turn, generates a netlist (or its equivalent) for submission to the computational engine that actually performs the circuit analysis using mathematical models and **algorithms**. In order to describe the characteristics and behaviour of components such as diodes and transistors, manufacturers often provide models in the form of a standard list of parameters.

Most programs that simulate electronic circuits use a set of algorithms that describe the behaviour of electronic components. The most commonly used algorithm was developed at the Berkeley Institute in the United States and it is known as

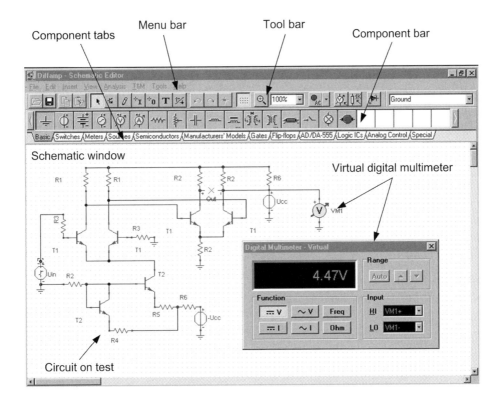

Figure 17.1 Using Tina Pro to construct and test a circuit prior to detailed analysis

SPICE (Simulation Program with Integrated Circuit Emphasis).

Results of circuit analysis can be displayed in various ways, including displays that simulate those of real test instruments (these are sometimes referred to as **virtual instruments**). A further benefit of using electronic circuit simulation software is that, when a circuit design has been finalized, it is usually possible to export a file from the design/simulation software to a PCB layout package. It may also be possible to export files for use in screen printing or CNC drilling. This greatly reduces the time that it takes to produce a finished electronic circuit.

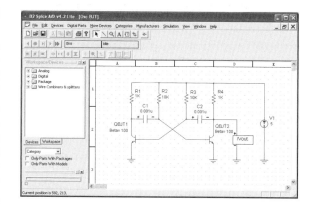

Figure 17.2 An astable multivibrator circuit being simulated using B2 Spice

Types of analysis

Various types of analysis are available within modern SPICE-based circuit simulation packages (see Appendix 7). These usually include:

DC analysis

DC analysis determines the DC operating point of the circuit under investigation. In this mode any wound components (e.g. inductors and transformers) are short-circuited and any capacitors that may be present are left open-circuit. In order to determine the initial conditions, a DC analysis is usually automatically performed prior to a transient analysis. It is also usually performed prior to an AC small-signal analysis in order to obtain the linearized, small-signal models for non-linear devices. Furthermore, if specified, the DC small-signal value of a transfer function (ratio of output variable to input source), input resistance, and output resistance is also computed as a part of the DC solution. The DC analysis can also be used to generate DC transfer curves in which a specified independent voltage or current source is stepped over a user-specified range and the DC output variables are stored for each sequential source value.

AC small-signal analysis

The AC small-signal analysis feature of SPICE software computes the AC output variables as a function of frequency. The program first computes the DC operating point of the circuit and

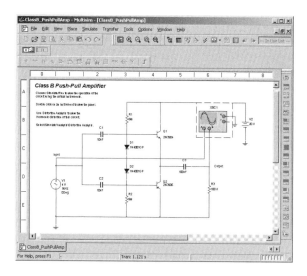

Figure 17.3 A Class B push-pull amplifier circuit being simulated by Multisim

determines linearized, small-signal models for all of the non-linear devices in the circuit (e.g. diodes and transistors). The resultant linear circuit is then analysed over a user-specified range of frequencies. The desired output of an AC small-signal analysis is usually a transfer function (voltage gain, transimpedance, etc.). If the circuit has only one AC input, it is convenient to set that input to unity and zero phase, so that output variables have the same value as the transfer function of the output variable with respect to the input.

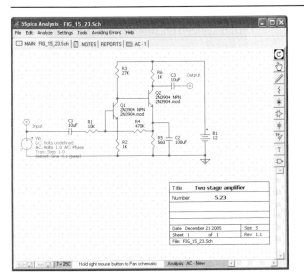

Figure 17.4 High-gain amplifier (Fig. 15.23) being analysed using the 5Spice Analysis package

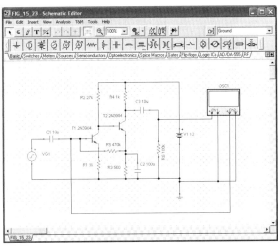

Figure 17.6 High-gain amplifier (Fig. 15.23) being analysed using the Tina Pro package

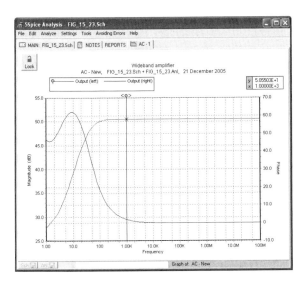

Figure 17.5 Gain and phase plotted as a result of small-signal AC analysis of the circuit in Fig. 17.4

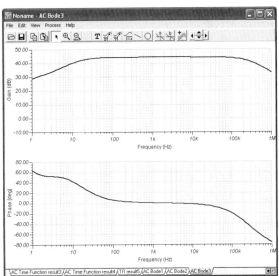

Figure 17.7 Gain and phase plotted as a result of small-signal AC analysis of the circuit in Fig. 17.6

Transient analysis

The transient analysis feature of a SPICE package computes the transient output variables as a function of time over a user-specified time interval. The initial conditions are automatically determined by a DC analysis. All sources that are not time dependent (for example, power supplies) are set to their DC value.

Pole-zero analysis

The pole-zero analysis facility computes the poles and/or zeros in the small-signal AC transfer function. The program first computes the DC operating point and then determines the linearized, small-signal models for all the non-linear devices in the circuit. This circuit is then used to find the poles and zeros of the transfer function.

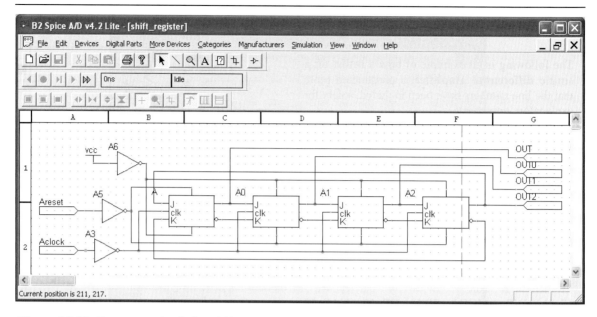

Figure 17.17   Four-stage circulating shift register simulated using B2 Spice

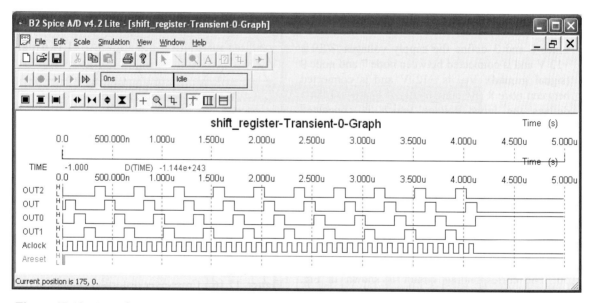

Figure 17.18   Waveforms for the our stage circulating shift register in Fig. 17.17

analyse logic and also 'mixed-mode' (i.e. analogue and digital) circuits. Several examples of digital logic analysis are shown in Figs 17.17, 17.18 and 17.19.

Figure 17.17 shows a four-stage shift register based on J-K bistables. The result of carrying out an analysis of this circuit is shown in Fig. 17.18.

Finally, Fig. 17.19 shows how a simple combinational logic circuit can be rapidly 'assembled' and tested and its logical function checked. This circuit arrangement provides a solution to Question 10.20 on page 198 and it shows how the exclusive-OR function can be realized using only two-input NAND gates.

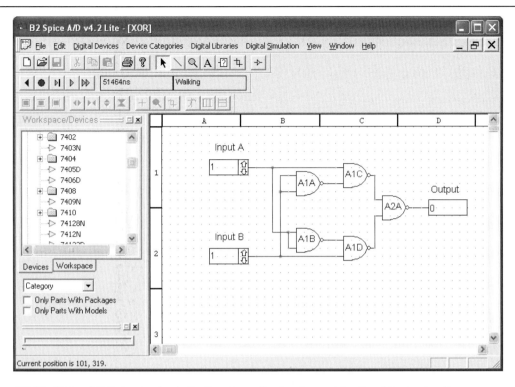

Figure 17.19 Using B2 Spice to check the function of a simple combinational logic circuit

Practical investigation

Objective

To investigate the use of SPICE software to analyse a DC coupled power amplifier circuit.

Components and test equipment

PC with SPICE software (e.g. Tina Pro, Multisim, or B2 Spice, see Appendix 9).

Procedure

Construct the circuit shown in Fig. 17.20 using the screen drawing facilities provided by the SPICE program. Carry out a DC analysis of the circuit and record the voltages in Table 17.1. Check that these are what you would expect.

Use the data from the DC analysis to determine:
  (a) the quiescent (no-signal) collector current in T7 and T8
  (b) the output offset voltage (i.e. the output voltage when the input voltage is zero).

Carry out an AC analysis of the amplifier and determine the voltage gain of the circuit (this should be the same as the ratio of R12 to R2). Check that the voltage gain is what you would have expected. Also determine the maximum input voltage before the output voltage becomes clipped.

Table 17.1 Table of results

Transistor	Collector	Base	Emitter
T1			
T2			
T3			
T4			
T5			
T6			
T7			
T8			

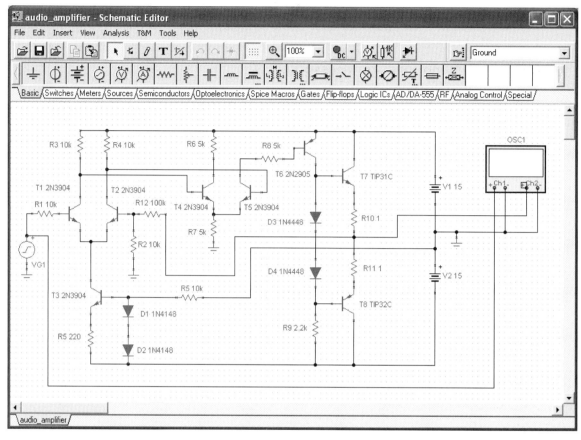

Figure 17.20 DC coupled power amplifier circuit used in the Practical Investigation (Tina Pro)

Problems

17.1   In relation to SPICE software explain the meaning of the following terms:
(a) node
(b) netlist
(c) algorithm
(d) model
(e) mixed-mode.

17.2   Explain why it is usually necessary to carry out a DC analysis *before* an AC analysis is performed.

17.3   Explain the purpose of transient analysis using SPICE software. How are the results of this analysis usually displayed?

17.4   Define the terms transimpedance and transconductance in relation to SPICE software.

17.5   The netlist used in a SPICE analysis program is as follows:

```
CE AMPLIFIER
VCC 5 0 -5
VIN 1 0 AC 1
R1 1 2 10K
R2 3 5 1K
R3 2 3 1M
Q1 3 2 0 MOD1
MODEL MOD1 PNP BF=75 VAF=50
IS=1.E-12 RB=100 CJC=.5PF TF=.6NS
.TF V(5) VIN
.AC DEC 10 1 10MEG
.END
```

Sketch the corresponding circuit diagram and label the nodes.

18

The PIC microcontroller

Earlier in Chapter 11 we introduced the PIC microcontroller as a device that is normally used in stand-alone (or **embedded**) applications to perform simple logic, timing and input/output control. In recent years such devices have become increasingly sophisticated in terms of both the variety of I/O peripheral facilities provided and the ability to connect to other devices in bus and networked configurations. This latter feature makes it possible to have several microcontrollers working together, exchanging data and control information, in **distributed systems** that can be of virtually any size.

PIC architecture

PIC microcontrollers first became popular in the mid-1990s. Since then the range of devices available has increased dramatically. Nowadays a PIC device exists for almost any embedded application, from small 6-pin devices ideal for simple control applications, through to powerful high-speed devices packed with I/O features and large amounts of memory.

PIC are based on **RISC** (Reduced Instruction Set Computer) architecture and, as a consequence, they use a relatively small number of instructions. In fact, some PIC chips have as few as 33 instructions compared with some general-purpose microprocessors (such as the Z80) that may have several hundred. Because it is only necessary to remember a relatively small number of commands, it is relatively easy to learn to program a PIC using its own assembly language (note, however, that it may take several instructions to achieve on a PIC what can be done in a single instruction using a fully-fledged microprocessor). And, if you don't like the idea of having to learn assembly code programming you can can make use of C, BASIC or Flowcode (a language based almost entirely on flowcharts).

Another feature of PIC devices is that they use **Harvard architecture**. This means that they contain completely separate memory and buses for program instructions and data memory. Program memory can be ROM, EPROM or Flash, whereas data memory must be read/write memory (commonly known as RAM).

Mainstream microprocessors (like the Z80 that we met in Chapter 11) are categorized in terms of the number of data bits that they can manipulate (i.e. the 'width' of the data bus). They can be 8-bit, 16-bit, 32-bit and 64-bit and 128-bit devices. PIC microcontrollers, by contrast, are all based on an 8-bit data bus and, because of this, they can *only* operate on 8-bits of data at a time (despite this, you will sometimes encounter instructions that reference more data bits, i.e. 12-bits, 14-bits and 16-bits). Furthermore, although the data memory is 8-bits wide for all PIC microcontrollers, the program memory varies from 12-bits to 16-bits.

PIC families

In the early 1990s, PIC microcontrollers were grouped in three families, often referred to as 'Base-Line', 'Mid-Range', and 'High-End'. Many of these devices are incompatible with the latest development platforms and software tools. Despite this, it is usually possible to locate a modern device that is compatible with (and will generally out-perform) one of these early devices.

The base-line family comprised just four devices, which all featured 12-bit instructions. The PIC16C54 and PIC16C56 were both 18-pin devices having 12 digital I/O lines. The PIC16C54 had 512 × 12-bit words of EPROM program memory, whilst the PIC16C56 had 1K × 12-bit words. Both of these devices had just 32 bytes of RAM data memory (ample for most simple control applications). The larger PIC16C55 and PIC16C57 28-pin devices had 20 digital I/O lines. The PIC16C55 had 512 × 12-bit words of EPROM program memory and 32 bytes of RAM data memory, whilst the PIC16C57 had 2K × 12-bit words of EPROM program memory and 80 bytes of RAM data memory. All of these devices also featured an 8-bit real-time clock/counter and a watchdog timer.

The mid-range family of PIC devices contained a number of devices that expanded on the capabilities of the base-line devices. Of particular note are the PIC16C71, which added A/D inputs capable of reading analogue signals, and the popular PIC16C84, which featured EEPROM program and data memory. Because of the fact that the PIC16C84 could be erased electronically (without the need for the ultraviolet eraser required for EPROM erasure), it soon became a favourite with hobbyists and must have been the most popular PIC for many years.

The high-end device was the PIC17C42. This device came in a 40-pin package and featured 16-bit wide instructions. It also allowed for external memory to be interfaced and supported up to 64K × 6-bit of addressable program memory space. The device provided 33 digital I/O pins, EEPROM data memory, a serial port (USART), three 16-bit timer/counters and two high-speed pulse width modulation (PWM) outputs.

There are now more than 200 different PIC chips available and the 'Base-Line', 'Mid-Range' and 'High-End' classification has now been replaced by several different ranges of PIC families such as the PIC-10, PIC-12, PIC-16 and PIC-18. Generally speaking, the higher the family number, the more powerful the device in terms of processing capabilities, memory sizes and I/O features. Another feature of many of the devices is **Flash program memory**. These devices can be erased and reprogrammed electrically, as with the previously popular PIC16C84. These Flash memory devices are signified by a letter 'F' in their part number (e.g. PIC16F877A). Note that Flash devices are much easier to work with for one-off prototyping because erasure and reprogramming is greatly simplified.

Choosing a PIC device

When choosing a PIC device for a particular project it is important to select a device that is well supported, both in terms of being a member of one of the current PIC families but also in terms of the

Table 18.1   Some popular PIC families

PIC family	Characteristics
PIC10F	The PIC10F range features just a few 6/8-pin microcontrollers. These chips are supplied in DIP packages.
PIC12F	The PIC12F devices include the PIC12F629, PIC12F675 and PIC12F683. They are all 8-pin devices featuring six digital I/O lines and differ in the amount of program and data memory that they provide. The PIC12F683 also provides analogue inputs.
PIC16F	The PIC16F family is very large and includes devices in 14-, 18-, 28- and 40-pin packages. The chips vary in terms of the amount of memory they provide and also in their I/O features. Some devices worth noting include the PIC16F630, PIC16F676, PIC16F684 and PIC16F688, which are 14-pin equivalents to the PIC12F devices mentioned above, the PIC16F84A, which is the most current equivalent to the popular PIC16C84, and the PIC16F877A, which is a powerful 40-pin device that has become extremely popular. The PIC16F877A features 8192 × 14-bit Flash program memory, 368 bytes of RAM data memory, 256 bytes of EEPROM data memory, 33 digital I/O lines, eight 10-bit A/D channels, three counter/timers and a serial port (USART).
PIC18F	The PIC18F devices form the new high-end range. One feature that they all share is that their instruction set is optimized for C compilers. Many feature large amounts of memory and special I/O features including Universal Serial Bus (USB) and Controller Area Network (CAN) interfaces. One interesting device is the 40-pin PIC18F452, which is similar in many respects and pin-compatible with the PIC16F877A. It is also worth noting that some PIC18F devices are supplied in 80-pin packages and provide more than 70 digital I/O lines. However, these are in surface-mount packages and so are difficult to experiment with!

programming and environment that you intend to use for software development. It is also important to ensure that the device incorporates all of the peripheral I/O facilities that you will need. Such facilities might include:

- Bus and communication interfaces (such as RS232/RS485, I²C, CAN, LIN, USB, TCP/IP)
- Display peripheral interfaces (such as LED or LCD drivers)
- Capture/compare facilities
- Pulse Width Modulators (PWMs)

- Counters/timers
- Watchdog facilities
- Analogue-to-digital (A/D) converters
- Analogue comparators and operational amplifiers
- Brown-out detectors
- Low-voltage detectors
- Temperature sensors
- Oscillators
- Voltage references
- Digital-to-analogue (D/A) converters

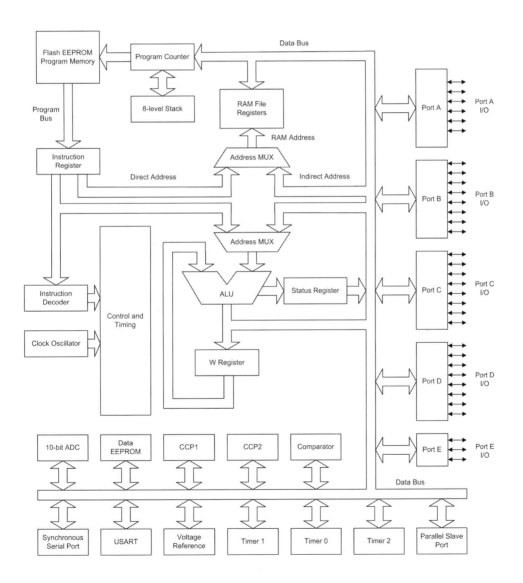

Figure 18.1 Simplified architecture of a 16F877A PIC microcontroller

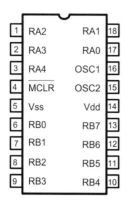

Figure 18.2 Pin connections for the PIC16C84 and PIC16F84 devices

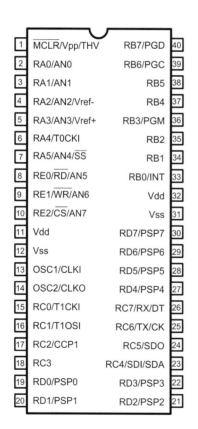

Figure 18.3 Pin connections for the PIC16F877 device

Features of the PIC16F84

The PIC16F84 is an excellent choice for most simple microcontroller projects. This device is used in many current PIC designs and it has the following features:

- 35 single-word instructions (see Table 18.3)
- 1k × 14-bit EEPROM (Flash) program memory
- 68 × 8-bit general purpose SRAM registers
- 15 × 8-bit special function hardware registers
- 64 × 8-bit EEPROM data memory
- 1,000,000 data memory erase/write cycles (typical)
- Data retention > 40 years
- 5 data input/output pins, Port A
- 8 data input/output pins, Port B
- 25 mA current sink max. per pin
- 20 mA current source per pin
- 8-bit timer/counter with prescaler
- Power-on reset (POR)
- Power-up timer (PWRT)
- Oscillator start-up timer (OST)
- Watchdog timer (WDT) with own on-chip RC oscillator
- Power saving 'sleep' function
- Serial in-system programming
- Selectable oscillator options:
  RC: low cost RC oscillator
  XT: standard crystal resonator (100 kHz to 4 MHz)
  HS: high-speed crystal/resonator (4 MHz to 10 MHz - PIC16F84-10 only)
  LP: power-saving low-frequency crystal (32 kHz to 200 kHz)
- Interrupts:
  External, RB0/INT pin
  TMR0 timer overflow
  Port B RB4 to RB7 interrupt on change
  Data EEPROM write complete
- Operating voltage range: 2.0 V to 6.0 V
- Power consumption: < 2 mA at 5 V, 4 MHz
- 60 µA typical at 2 V, 32 kHz
- < 1 µA typical standby at 2 V (PIC16F84)

Figure 18.2 shows the pin connections for the PIC16F84 whilst Fig. 18.4 shows how the device can be used in a simple control application that uses four switched inputs (S1 to S4) and three relay outputs (RL1 to RL3).

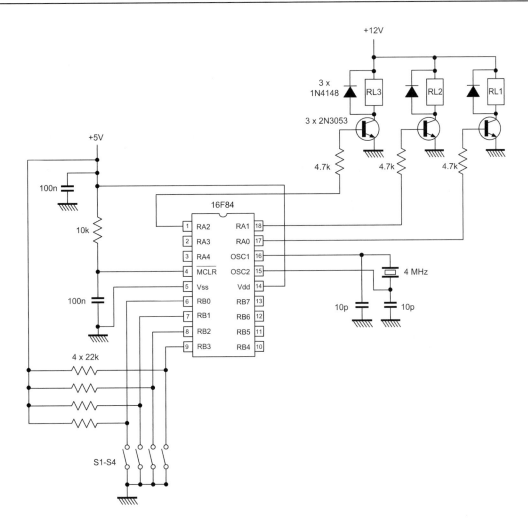

Figure 18.4 A simple microcontroller application based on the PIC16F84 device which four switch inputs (S1-S4) to control three relays (RL1-RL3). The program is automatically reset at power-up.

Features of the PIC16F877

The PIC16F877 is used for many current PIC designs. This chip has the following features:

- 35 single-word instructions (see Table 18.3)
- 8k x 14-bit EEPROM (Flash)
- 368 bytes of general purpose SRAM registers
- 256 x 8-bit EEPROM data memory
- 1,000,000 data memory erase/write cycles (typical)
- Data retention >40 years
- 6 data input/output pins, Port A

- 8 data input/output pins, Port B
- 8 data input/output pins, Port C
- 8 data input/output pins, Port D
- 3 data input/output pins, Port E
- Two 8-bit and one 16-bit timers
- Two capture and compare PWM modules
- Universal synchronous/asynchronous receiver transmitter (USART)
- Synchronous serial port (SSP)
- Parallel slave port (PSP)
- Brown-out detection and reset facility
- 10-bit, 8-channel ADC

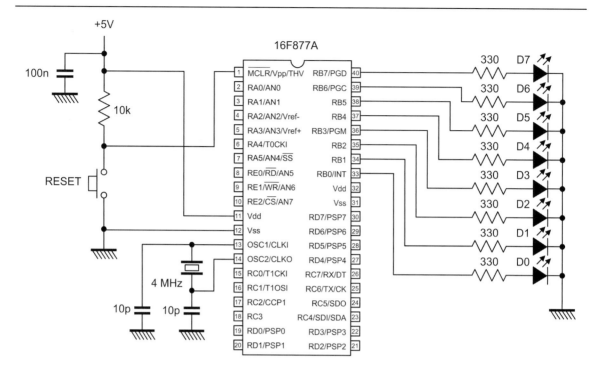

Figure 18.5   A simple microcontroller application based on the 16F877A device. This application uses eight LED outputs (D0 to D7). Note that the circuit must be manually reset at power-up.

- Two analogue comparators with programmable on-chip voltage reference, programmable input multiplexing, and externally accessible comparator output.
- 25mA current sink max. per pin
- 20mA current source per pin
- 8-bit timer/counter with prescaler
- Power-on reset (POR)
- Power-up timer (PWRT)
- Oscillator start-up timer (OST)
- Watchdog timer (WDT) with own on-chip RC oscillator
- Power saving 'sleep' function
- In-circuit serial programming (via two pins)
- Single supply 5V in-circuit serial programming
- Selectable oscillator options
- Self-programmable under software control
- 15 interrupts
- Operating frequency: DC to 20 MHz
- Supply voltage range: 2.0 V to 5.5 V
- Low power consumption.

PIC programming

PIC programming can be carried out using assembly language, C (see Table 18.2), BASIC, and Flowcode (see page 324) or a mixture of these languages. Software and a programmer (see Figs 18.6 and 18.7) will be required to do this. The two most commonly used methods of programming a PIC chip are shown in Fig. 18.8. In Fig. 18.8(a) a dedicated PIC programmer is used whilst in Fig. 18.8(b) the PIC is programmed whilst resident in the target system. In either case, the programming software and source code is resident on a PC and downloaded as **hex code** into the PIC. The process of generating the source code, compiling and/or assembling it into hex, code (see Fig. 18.10) is invariably performed by software known as an **Integrated Development Environment** (IDE), as shown in Fig. 18.9. Software is also available that can be used to simulate a PIC (see Figs 18.11 to 18.13) and full development systems are also available (see Figs 18.14 and 18.15).

Table 18.2 Switch and LED PIC test routine in C

```
//Defines for the PIC microcontroller
char PORTC@0x07;
char TRISC@0x87;
char PORTD@0x08;
char TRISD@0x88;
char PORTE@0x09;
char TRISE@0x89;

//PIC Functions
#pragma CLOCK_FREQ 4000000
#define P16F877A
#include <system.h>
#define MX_EE
#define MX_EE_TYPE2
const char MX_EE_SIZE = 256;
#define MX_SPI
#define MX_SPI_C
#define MX_SPI_SDI 4
#define MX_SPI_SDO 5
#define MX_SPI_SCK 3
#define MX_UART
#define MX_UART_C
#define MX_UART_TX 6
#define MX_UART_RX 7

//Macro function declarations

//Variable declarations
char FCV_SWITCHSTATE;

//Macro implementations

void main()
{
//PIC Initialisation
adcon1 = 0x07;

 //Interrupt initialisation code
option_reg = 0xC0;

while(1)
   {
    TRISA = TRISA | 0xff;
    FCV_SWITCHSTATE = PORTA;
    TRISB = 0x00;
    PORTB = FCV_SWITCHSTATE;
  if (( FCV_SWITCHSTATE ) == 0)
    break;
   }
mainendloop: goto mainendloop;
}
```

Figure 18.6 The E-blocks low-cost PIC microcontroller 'lite' programmer. This device is suitable for programming 18-pin PIC chips and it connects to a PC via a USB port. The board has two I/O ports for connecting to a wide variety of external peripheral modules.

Figure 18.7 E-blocks PIC microcontroller/microprogrammer. This versatile device connects to a PC via USB and can be used to program most 8-, 14-, 28- and 40-pin PIC chips. The board has five I/O ports for connection to a wide variety of external peripheral modules

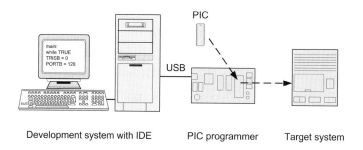

(a) Using a conventional PIC programmer

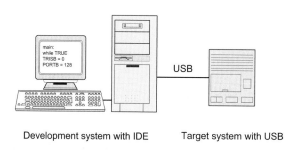

(b) Using a flash device in the target system

Figure 18.8   Two methods of programming a PIC device

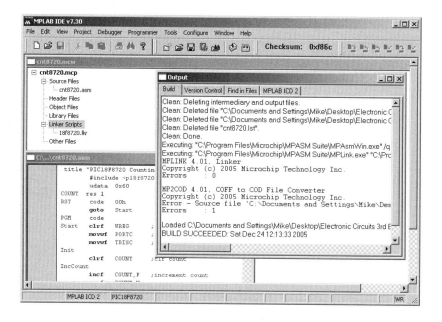

Figure 18.9   The MPLAB Integrated Development Environment (IDE)

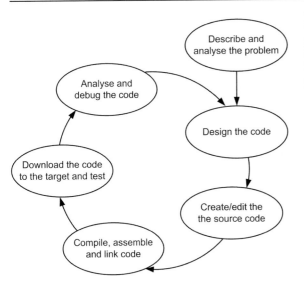

Figure 18.10 The PIC software development cycle

Figure 18.12 The switch and LED test program using the Oshonsoft PIC Simulator

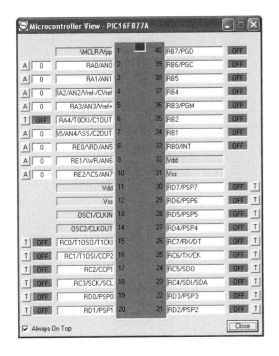

Figure 18.11 The Oshonsoft PIC Simulator showing the Microcontroller View window in which the current state of each pin is displayed

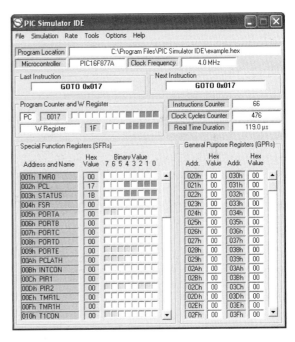

Figure 18.13 The main Oshonsoft PIC Simulator screen showing the state of the Special Function Registers (SFRs) and General Purpose Registers (GPRs). Note also the contents of the Program Counter and W Register.

Table 18.3 PIC16C/F84 instruction set

Command	Operation	Syntax	Status register bit(s) affected
ADDLW	ADD Literal to W	ADDLW K	C, DC, Z
ADDWF	ADD W to F	ADDWF F,D	C, DC, Z
ANDLW	AND Literal and W	ANDLW K	Z
ANDWF	AND W with F	ANDWF F,D	Z
BCF	Bit Clear F	BCF F,B	None
BSF	Bit Set F	BSF F,B	None
BTFSC	Bit Test, Skip if Clear	BTFSC F,B	None
BTFSS	Bit Test, Skip if Set	BTFSC F,B	None
CALL	Subroutine Call	CALL K	None
CLRF	Clear F	CLRF F	Z
CLRW	Clear W register	CLRW	Z
CLRWDT	Clear Watchdog Timer	CLRWDT	TO, PD
COMF	Complement F	COMF F,D	Z
DECF	Decrement F	DECF F,D	Z
DECFSZ	Decrement F, Skip if 0	DECFSZ F,D	None
GOTO	Unconditional branch	GOTO K	None
INCF	Increment F	INCF F,D	Z
INCFSZ	Increment F, Skip if 0	INCFSZ F,D	None
IORLW	Inclusive OR Literal with W	IORLW K	Z
IORWF	Inclusive OR W with F	IORWF F,D	Z
MOVF	Move F	MOVF F,D	Z
MOVLW	Move Literal to W	MOVLW K	None
MOVWF	Move W to F	MOVWF F	None
NOP	No Operation	NOP	None
RETFIE	Return from Interrupt	RETFIE	None
RETLW	Return with Literal in W	RETLW K	None
RETURN	Return from Subroutine	RETURN	None
RLF	Rotate Left through Carry	RLF F,D	C
RRF	Rotate Right through Carry	RRF F,D	C
SLEEP	Power-down	SLEEP	TO, PD
SUBLW	Subtract W from Literal	SUBLW K	C, DC, Z
SUBLW	Subtract W from F	SUBWF F,D	C, DC, Z
SWAPF	Swap F	SWAPF F,D	None
XORLW	Exclusive OR Literal with W	XORLW K	Z
XORWF	Exclusive OR W with F	XORWF F,D	Z

Abbreviations

C	Carry
b	Bit address
C	Carry
d	Destination
DC	Digit Carry
f	Register File Address
k	Literal or label
PD	Power-down
TO	Time-out
W	Working Register
Z	Zero

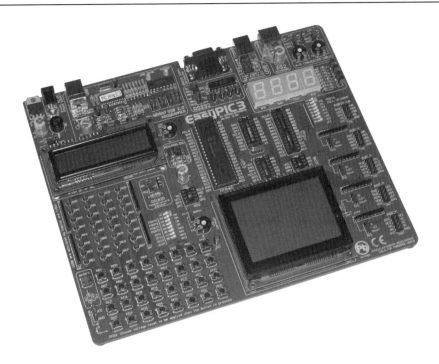

Figure 18.14   The MicroElektronika EasyPIC3 PIC development system (see Appendix 9). This system supports the PIC10F, 12F, 16F and 18F families and is connected to a PC via a USB port

Figure 18.15   The E-blocks system (see Appendix 9). The microcontroller system shown here uses a PIC programmer (left), keypad (bottom), LCD display (top) and serial communications interface (right).

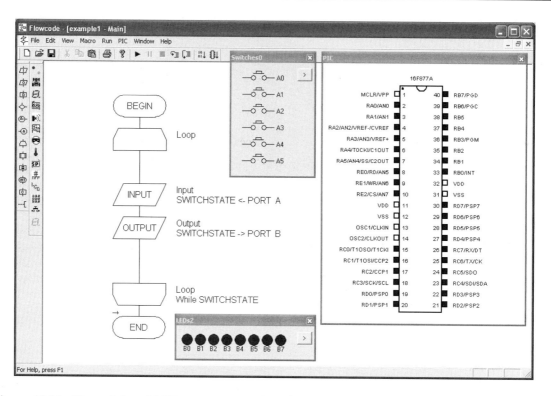

Figure 18.16 The switch and LED test program using Flowcode. Note how the flowchart symbols are simply dragged and dropped into the main program window from the toolbar on the left.

Flowcode

Flowcode (see Appendix 7) is a novel programming system that is designed to allow those new to PIC programming to develop complete applications with little or no knowledge of any higher level languages. However it is possible to embed code written in other languages into Flowcode programs.

Programs written in C and Assembly code can be easily embedded in Flowcode. This will allow you to incorporate more complex and sophisticated code in applications where this might become necessary. Note however that the embedded code cannot be checked or simulated by Flowcode but instead it will be passed on to the C compiler when the flowchart is compiled. It is therefore important to verify that the C code entered is correct, as syntax errors will cause the compilation of the whole flowchart to fail.

To access Flowcode variables, macro functions and connection points, it is necessary to colour the variable in your C code with the prefixes FCV_, FCM_ and FCC_MacroName_ respectively. For example, to use a Flowcode variable called DELAY in your C code, you must refer to it using FCV_DELAY. Note that all Flowcode defined variables must be entered in upper case.

To call a Flowcode macro called TEST in your C code, you must call FCM_TEST();. To jump to a connection point called A, defined in a Flowcode macro called TEST, your C code must be goto FCC_TEST_A;. Connection points defined in the main flowchart of a Flowcode file are prefixed FCC_Main_.

Assembly code (see Table 18.3) can be added into the code field in a C assembly code wrapper. For a single line of code you can use the asm operator in front of each instruction (e.g. asm movlw 5) or you can enclose several statements within an asm block. This makes adding assembly language routines very simple.

Practical investigation

Objective

To investigate the operation of a PIC16F877A microcontroller using a simulator to run a simple program.

Components and test equipment

Oshonsoft PIC Simulator or similar virtual Integrated Development Environment (IDE). The Oshonsoft IDE can be downloaded from www.Oshonsoft.com.

System configuration and required operation

The program is to simulate the operation of a PIC16F877A microcontroller and is to read an analogue voltage present on the AN0 analog input and display the 8-bit binary conversion result on PORTB.

The analogue-to-digital converter (ADC) BASIC source code used in the Practical investigation (shown in Table 18.4) is available for downloading from the companion website (see page 415).

Procedure

Start the PIC Simulator (see Fig. 18.17), click on Options and then Select Microcontroller. Select the 'PIC16F877' from the list and click on the Select button. Next click on File and Load Program. Select the file named adc.hex (generated as a result of compiling the BASIC source code shown in Table 18.4) and click on Open. This will load the program into the PIC's program memory.

Click on Tools and Microcontroller View in order to open the window shown in Fig. 18.18). If necessary, reposition the windows on the screen so that you can see both of them then Select the Rate and Extremely Fast before clicking on Simulation and Start. The simulation will start immediately.

To set an analogue input voltage, click on the analogue (A) button associated with the RA0/AN0 pin then use the slider to change the analogue value on this pin and click on Accept button to accept the value. The state of the PORTB pins should immediately reflect this value. Repeat with several different analogue input values and at reduced simulation speeds in order to examine the effect of each program instruction.

A further program

If you have been able to sun the simple ADC program successfully you might like to try your hand at entering, compiling and running a program that will loop continuously reading the state of the switches on Port A and output the result to Port B (see Figure 18.12).

Table 18.4 ADC program used in the Practical investigation

```
Symbol ad_action = ADCON0.GO_DONE 'set new name for A/D conversion start bit
Symbol display = PORTB 'set new name for PORTB used to display the conversion result

TRISB = %00000000 'set PORTB pins as outputs
TRISA = %111111 'set PORTA pins as inputs
ADCON0 = 0xc0 'set A/D conversion clock to internal source
ADCON1 = 0 'set PORTA pins as analog inputs
High ADCON0.ADON 'turn on A/D converter module

main:
Gosub getadresult 'go to conversion routine
display = ADRESH 'display the result of the conversion
Goto main 'repeat forever
End

getadresult: 'conversion routine
High ad_action 'start the conversion
While ad_action 'wait until conversion is completed
Wend
Return
```

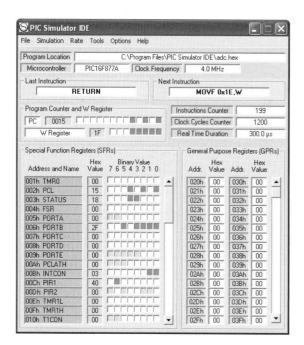

Figure 18.17   PIC Simulator IDE used in the Practical investigation. This screen shows the state of the PIC with the Program Counter at 0015.

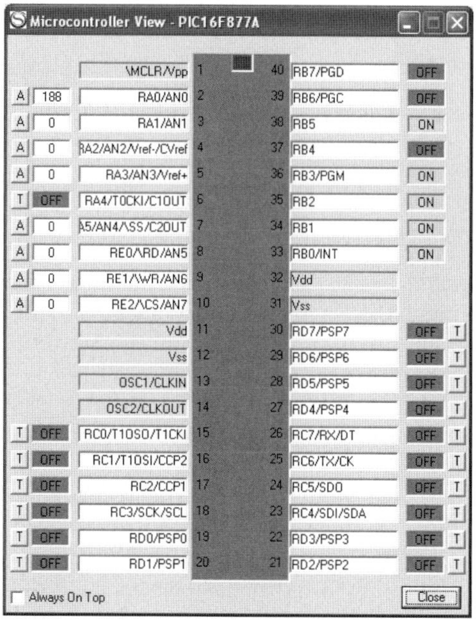

Figure 18.18   Microcontroller View showing the state of each of the PIC's pins during program execution

Problems

18.1   In relation to a PIC microcontroller, briefly explain the meaning of the following terms:
(a) Flash memory
(b) Harvard architecture
(c) RISC
(d) Sleep function
(e) IDE.

18.2   Explain, with the aid of a diagram, how a PIC can be programmed without having to remove it from the target system

18.3   Sketch a diagram to show the five stages in PIC software development cycle. Explain why it is usually necessary to follow this process several times before a fully functional program is obtained.

18.4   Explain how the PIC application circuit shown in Fig. 18.4 is automatically reset on power-up (Hint: the MCLR input needs to be taken low in order to reset the PIC).

18.5   Show how the PIC application shown in

Fig. 18.5 can be modified to support:
(a) automatic power-on reset.
(b) four switch inputs connected to Port D
(c) an analogue input connected to Port A.

18.6   List FOUR different types of peripheral I/O device that can be incorporated in a PIC microcontroller chip. Suggest a typical application for each device.

18.7   The following fragment of source code appears in a PIC BASIC program:

```
Dim inputstate As Byte
TRISA = %111111
TRISB = %00000000
While inputstate <> %111000
Wend
PORTB = %00000000
While inputstate <> %000111
Wend
PORTB = %11111111
```

Explain the function of the code by adding appropriate comments to each line.

Answers to these problems appear on page 376.

19

Circuit construction

This chapter deals with the techniques used to produce a working circuit from an initial 'paper design' to a working circuit ready for testing and performance measurement. Various methods are used for building electronic circuits. The method that's actually chosen for a particular application depends on a number of factors, including the available resources and the scale of production.

Techniques used for large-scale electronic manufacture generally involve fully automated assembly using equipment that can produce complex circuits quickly and accurately and at very low cost with minimal human intervention. On the other extreme, if only one circuit is to be built then a hand-built prototype is much more appropriate. It's also worth noting that, when a circuit is designed for a commercial application it will invariably be tested using computer simulation techniques before a prototype is manufactured (see Chapter 17). It's also possible for the computer simulation to output data that can be used by CNC equipment to manufacture a printed circuit board on which the real circuit will be assembled. These techniques have been instrumental in significantly reducing the time taken to get a prototype electronic circuit off the computer screen and into production.

We begin with a quick introduction to the different methods of circuit construction before looking in greater detail at the stages involved in the layout and manufacture of stripboards and printed circuit boards. Consideration is also given to the correct choice of enclosure and connectors as this is crucial in making the finished equipment both functional and attractive. The chapter also deals with soldering and desoldering techniques and the selection of heatsinks.

Circuit construction methods

There are several different ways of building an electronic circuit. The method that you choose depends on a number of factors, including the resources available to you and whether you are building a 'one-off' prototype or a large number of identical circuits. The methods that are available to you include:

Point-to-point wiring

With the advent of miniature components, printed circuit boards and integrated circuits, point-to-point wiring construction is a construction technique that is nowadays considered obsolete. The example shown in Fig. 19.1 is the underside of a valve amplifier chassis dating back to the early 1960s. Unless you are dealing with a very small number of components or have a particular desire to use **tag strips** and **group boards**, point-to-point wiring is not a particularly attractive construction method these days!

Breadboard construction

Breadboard construction is often used for assembling and testing simple circuit prior to production of a more permanent circuit using a stripboard or printed circuit board. The advantage of this technique is that changes can be quickly and easily made to a circuit and all of the components can be re-used. The obvious disadvantages of breadboard construction are that it is unsuitable for permanent use and also unsuitable for complex circuits (i.e. circuits with more than half a dozen, or so, active devices or integrated circuits). Figure 19.2 shows the simple bistable circuit that you met in Chapter 10 assembled and ready for testing.

Matrix board construction

Figure 19.3 shows matrix board construction. This low-cost technique avoids the need for a printed circuit but is generally only suitable for one-off prototypes. A matrix board consists of an insulated board into which a matrix of holes are drilled with copper tracks arranged as strips on the reverse side of the board. Component leads are inserted through the holes and soldered into place. Strips (or tracks) are linked together with short length of tinned copper wire (inserted through holes in the board and soldered into place on the underside of the board). Tracks can be broken at various points as appropriate. The advantage of this technique is that it avoids the need for a printed circuit board (which may be relatively expensive and may take some time to design). Disadvantages of matrix board construction are that it is usually only suitable for one-off production and the end result is invariably less compact than a printed circuit board. The matrix board shown in Fig. 19.3 is a simple oscilloscope calibrator based on a PIC16F84.

Printed circuit boards

Printed circuit board construction (see Fig. 19.4) technique is ideal for volume manufacture of electronic circuits where speed and repeatability of production are important. Depending on the complexity of a circuit, various types of printed circuit board are possible. The most basic form of printed circuit (and one which is suitable for home construction) has copper tracks on one side and components mounted on the other. More complex printed circuit boards have tracks on both sides (they are referred to as 'double-sided') whilst boards with up to four layers are used for some of the most sophisticated and densely packed electronic equipment (for example, computer motherboards). The double-sided printed circuit board shown in Fig. 19.4 is part of an instrument landing system (ILS) fitted to a passenger aircraft.

A further example of printed circuit construction is shown in Fig. 19.5. This shows 'zig-zag' inductive components fabricated using the copper track surface together with leadless 'chip' capacitors and a leadless transistor which is soldered directly to the surface of the copper track.

Surface mounting

Figure 19.6 shows an example of surface mounting construction. This technique is suitable for sub-miniature leadless components. These are designed for automated soldering directly to pads on the surface of a printed circuit board. This technique makes it possible to pack in the largest number of components into the smallest space but, since the components require specialized handling and soldering equipment, it is not suitable for home

construction nor is it suitable for hand-built prototypes. The example shown in Fig. 19.6 is part of the signal processing circuitry used in a computer monitor.

Using matrix boards and stripboards

Matrix boards and stripboards are ideal for simple prototype and one-off electronic circuit construction. The distinction between matrix boards and stripboards is simply that the former has no copper tracks and the user has to make extensive use of press-fit terminal pins which are used for component connection. Extensive inter-wiring is then necessary to link terminal pins together. This may be carried out using sleeved tinned copper wire (of appropriate gauge) or short lengths of PVC-insulated 'hook-up' or **equipment wire**.

Like their matrix board counterparts, stripboards are also pierced with a matrix of holes which, again, are almost invariably placed on a 0.1 in pitch. The important difference, however, is that stripboards have copper strips bounded to one surface which link together rows of holes along the complete length of the board. The result, therefore, is something of a compromise between a 'naked' matrix board and a true printed circuit. Compared with the matrix board, the stripboard has the advantage that relatively few wire links arc required and that components can be mounted and soldered directly to the copper strips without the need for terminal pins.

Conventional types of stripboard (those with parallel runs of strips throughout the entire board surface) are generally unsuitable for relatively complex circuitry of the type associated with microprocessor systems. Fortunately, several manufacturers have responded with special purpose stripboards. These have strips arranged in groups which not only permit the mounting of DIL integrated circuits (including the larger 28-pin and 40-pin types) but are also available in a standard range of 'card' sizes (both single and double-sided and with and without plated through holes). Board designed for prototyping using integrated circuits also tend to have sensibly arranged strips for supply distribution together with edge connectors (either direct or indirect types) for use with microprocessor bus systems.

Figure 19.1 Point-to-point wiring

Figure 19.4 Printed circuit construction

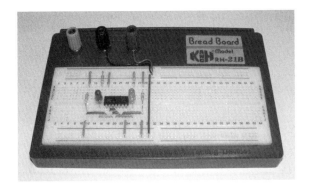

Figure 19.2 Breadboard construction

Figure 19.5 Printed circuit construction at UHF

Figure 19.3 Matrix board construction

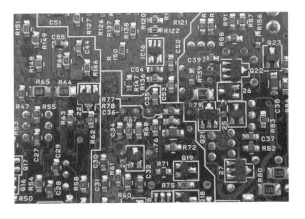

Figure 19.6 Surface mounted devices

Stripboard layout techniques

The following steps are required when laying out a circuit for stripboard construction:

1. Carefully examine a copy of the circuit diagram.
2. Mark all components to be mounted 'off-board' and identify (using appropriate letters and/or numbers, e.g. SK, pin-2) all points at which an 'off-board' connection is to be made.
3. Identify any multiple connections required between integrated circuits or between integrated circuits and connectors. Arrange such components in physical proximity and with such orientation that will effectively minimize the number of links required.
4. Identify components that require special attention (such as those which require heatsinks or have special screening requirements). Ensure that such components are positioned sensibly bearing in mind their particular needs.
5. Keep inputs and outputs at opposite ends of the stripboard. This not only helps maintain a logical circuit layout (progressing from input to output) but, in high gain circuits, it may also be instrumental in preventing instability due to unwanted feedback.
6. Use standard sizes of stripboard wherever possible. Where boards have to be cut to size, it is usually more efficient to align the strips along the major axis of the board.
7. Consider the means of mounting the stripboard. If it is to be secured using bolts and threaded spacers (or equivalent) it will be necessary to allow adequate clearance around the mounting holes.
8. Produce a rough layout for the stripboard first using paper ruled with squares, the corners of the squares representing the holes in the stripboard. This process can be carried out 'actual size' using 0.1 in. graph paper or suitably enlarged by means of an appropriate choice of paper. For preference, it is wise to choose paper with a feint blue or green grid as this will subsequently disappear after photocopying leaving you with a 'clean' layout.
9. Identify all conductors that will be handling high currents (i.e. those in excess of 1 A) and use adjacent strips connected in parallel at various points along the length of the board.
10. Identify the strips that will be used to convey the supply rails: as far as possible these should be continuous from one end of the board to the other. It is often convenient to use adjacent strips for supply and 0 V (or 'ground') since decoupling capacitors can easily be distributed at strategic points. Ideally, such capacitors should be positioned in close proximity to the positive supply input pin to all integrated circuits which are likely to demand sudden transient currents (e.g. 555 timers, comparators, IC power amplifiers).
11. In high-frequency circuits, link all unused strips to 0 V at regular points. This promotes stability by ensuring that the 0 V rail is effective as a common rail.
12. Minimize, as far as possible, the number of links required. These should be made on (he upper (component) side of the stripboard. Only in exceptional cases should links be made on the underside (foil side) of the board.
13. Experiment with positioning of integrated circuits (it is good practice, though not essential, to align them all in the same direction). In some cases, logic gates may be exchanged from package to package (or within the same package) in order to minimize strip usage and links. (If you have to resort to this dodge, do not forget to amend the circuit diagram!)

When the stripboard layout is complete, it is important to carefully check it against the circuit diagram. Not only can this save considerable frustration at a later stage but it can be instrumental in preventing some costly mistakes. In particular, one should follow the positive supply and 0 V (or 'ground') strips and check that all chips and other devices have supplies. The technique employed by the author involves the use of coloured pencils that are used to trace the circuit and stripboard layout; associating each line in the circuit diagram with a physical interconnection on the stripboard. Colours are used as follows:

Positive supply rails	Red
Negative supply rails	Black
Common 0 V rail	Green
Analogue signals	Yellow/Pink
Digital signals	White/Grey
Off-board connections	Orange/Violet
Mains wiring	Brown and blue

Assembly of the stripboard is happily a quite straightforward process. The sequence used for stripboard assembly will normally involve mounting IC sockets first followed by transistors, diodes, resistors, capacitors and other passive components. Finally, terminal pins and links should be fitted before making the track breaks. On completion, the board should be carefully checked, paying particular attention to all polarized components (e.g. diodes, transistors and electrolytic capacitors). Figure 19.7 shows a typical stripboard layout together with matching component overlay.

Using printed circuits

Printed circuit boards (PCB) comprise copper tracks bonded to an epoxy glass or synthetic resin bonded paper (SRBP) board. The result is a neat and professional looking circuit that is ideal for prototype as well as production quantities. Printed circuits can easily be duplicated or modified from original master artwork and the production techniques arc quite simple and should thus not deter the enthusiast working from home.

The following steps should be followed when laying out a printed circuit board:

1. Carefully examine a copy of the circuit diagram. Mark all components which are to be mounted 'off-board' and, using appropriate letters and/or numbers (e.g. SK1, pin-2), identify all points at which an 'off-board' connection is to be made.
2. Identify any multiple connections (e.g. bus lines) required between integrated circuits or between integrated circuits and edge connectors. Arrange such components in physical proximity and with such orientation that will effectively minimize the number of links required.
3. Identify components that require special attention (such as those that require heatsinks or have special screening requirements). Ensure that such components are positioned sensibly bearing in mind their particular needs.
4. Keep inputs and outputs at opposite ends of the PCB wherever possible. This not only helps maintain a logical circuit layout (progressing from input) but, in high gain circuits, it may also be instrumental in preventing instability due to unwanted feedback.
5. Use the minimum board area consistent with a layout which is uncramped. In practice, and to prevent wastage, you should aim to utilize as high a proportion of the PCB surface area as

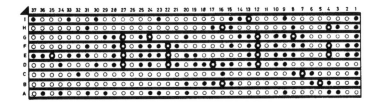

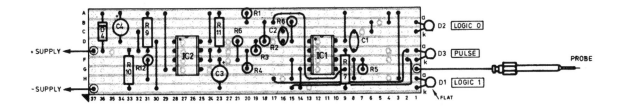

Figure 19.7 Stripboard layout and component overlay for the logic probe circuit shown in Figure 15.16 on page 283

possible. In the initial stages, however, it is wise to allow some room for manoeuvre as there will doubtless be subsequent modifications to the design.

6. Unless the design makes extensive use of PCB edge connectors, try to ensure that a common 0 V foil is run all round the periphery of the PCB. This has a number of advantages not the least of which is the fact that it will then be relatively simple to find a route to the 0 V rail from almost anywhere on the board.

7. Consider the means of mounting the PCB and, if it is to be secured using bolts and threaded spacers (or equivalent) you should ascertain the number and location of the holes required. You may also wish to ensure that the holes are coincident with the 0 V foil, alternatively where the 0 V rail is not to be taken to chassis ground, it will be necessary to ensure that the PCB mounting holes occur in an area of the PCB that is clear of foil.

8. Commence the PCB design in rough first using paper ruled with squares. This process can be carried out 'actual size' using 0.1 in. graph paper or suitably enlarged by means of an appropriate choice of paper. For preference, it is wise to choose paper with a feint blue or green grid as this will subsequently disappear after photocopying leaving you with a 'clean' layout.

9. Using the square grid as a guide, try to arrange all components so that they are mounted on the standard 0.1 in. matrix. This may complicate things a little but is important if you should subsequently wish to convert the design using computer-aided design (CAD) techniques (see later).

10. Arrange straight runs of track so that they align with one dimension of the board or another. Try to avoid haphazard track layout.

11. Identify all conductors that will be handling significant current (i.e. in excess of about 500 mA) and ensure that tracks have adequate widths. Table 19.1 will acts as a rough guide. Note that, as a general rule, the width of the 0 V track should be at least *twice* that used for any other track.

12. Identify all conductors that will be handling high voltages (i.e. those in excess of 150 V d.c. or 100 V r.m.s. a.c.) and ensure that these are adequately spaced from other tracks. Table 19.2 provides you with a rough guide.

Table 19.1 Current and track width

Current	Minimum track width
Less than 500 mA	0.6 mm
0.5 A to 1.5 A	1.6
1.5 A to 3 A	3
3 A to 6 A	6
6 A to 12 A	10

Table 19.2 Voltage and track spacing

Voltage	Minimum track spacing
Less than 50 V	0.6 mm
50 V to 150 V	1 mm
150 V to 300 V	1.6 mm
300 V to 600 V	3 mm
600 V to 1 kV	5 mm

13. Identify the point at which the principal supply rail is to be connected. Employ extra wide track widths (for both the 0 V and supply rail) in this area and check that suitably decoupling capacitors are placed as close as possible to the point of supply connection. Check that other decoupling capacitors are distributed at strategic points around the board. These should be positioned in close proximity to the positive supply input pin to all integrated circuits that are likely to demand sudden transient currents (e.g. 555 timers, comparators, IC power amplifiers). Ensure that there is adequate connection to the 0 V rail for each decoupling capacitor that you use. Additional decoupling may also be required for high-frequency devices and you should consult semiconductor manufacturers' recommendations for specific guidance.

14. Fill unused areas of PCB with 'land' (areas of foil which should be linked to 0 V). This helps

ensure that the 0 V rail is effective as a common rail, minimizes use of the etchant, helps to conduct heat away from heat producing components, and furthermore, is essential in promoting stability in high-frequency applications.

15. Lay out the 0 V and positive supply rails first. Then turn your attention to linking to the pads or edge connectors used for connecting the off-board components. Minimize, as far as possible, the number of links required. Do not use links in the 0 V rail and avoid using them in the positive supply rail.

16. Experiment with positioning of integrated circuits (it is good practice, though not essential, to align them all in the same direction). In some cases, logic gates may be exchanged from package to package (or within the same package) in order to minimize track runs and links. (If you have to resort to this dodge, do not forget to amend the circuit diagram.)

17. Be aware of the pin spacing used by components and try to keep this consistent throughout. With the exception of the larger wirewound resistors (which should be mounted on ceramic stand-off pillars) axial lead components should be mounted flat against the PCB (with their leads bent at right angles). Axial lead components should not be mounted vertically.

18. Do not forget that tracks may be conveniently routed beneath other components. Supply rails in particular can be routed between opposite rows of pads of DIL integrated circuits; this permits very effective supply distribution and decoupling.

19. Minimize track runs as far as possible and maintain constant spacing between parallel runs of track. Corners should be radiused and acute internal and external angles should be avoided. In exceptional circumstances, it may be necessary to run a track between adjacent pads of a DIL integrated circuit. In such cases, the track should not be a common 0 V path, neither should it be a supply rail.

Figure 19.8 shows examples of good and bad practice associated with PCB layout while Fig. 19.9 shows an example of a PCB layout and matching component overlay that embodies most of the techniques and principles discussed.

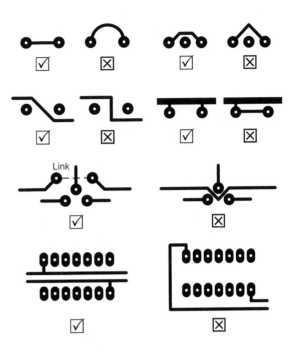

Figure 19.8 Some examples of good and bad practice in PCB layout

As with stripboard layouts, it is well worth devoting some time to checking the final draft PCB layout before starting on the master artwork. This can be instrumental in saving much agony and heartache at a later stage. The same procedure should be adopted as described on page 330 (i.e. simultaneously tracing the circuit diagram and PCB layout).

The next stage depends upon the actual PCB production. Four methods are commonly used for prototype and small-scale production. These may be summarized as follows:

(a) Drawing the track layout directly on the copper surface of the board using a special pen filled with etch resist ink. The track layout should, of course, conform as closely as possible with the draft layout.

(b) Laying down etch resist transfers of tracks and pads on to the copper surface of the PCB following the same layout as the draft but appropriately scaled.

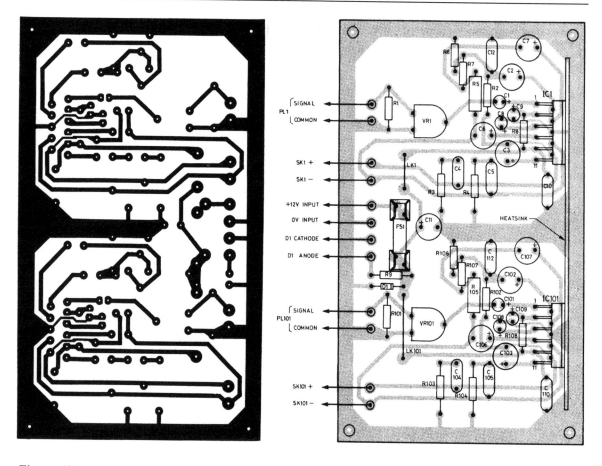

Figure 19.9 Example of a manually produced PCB layout for an automotive stereo amplifier. The copper foil layout is shown on the left and the matching component overlay is shown on the right

(c) Producing a transparency (using artwork transfers of tracks and pads) conforming to the draft layout and then applying photographic techniques.

(d) Using a PCB layout CAD package (see page 336).

Methods (a) and (b) have the obvious limitation that they are a strictly one-off process. Method (a) is also extremely crude and only applicable to very simple boards. Method (c) is by far the most superior and allows one to re-use or modify the master artwork transparency and produce as many further boards as are required. The disadvantage of the method is that it is slightly more expensive in terms of materials (specially coated copper board is

required) and requires some form of ultra-violet exposure unit. This device normally comprises a light-tight enclosure into the base of which one or more ultra-violet tubes are fitted. Smaller units are available which permit exposure of boards measuring 250 mm × 150 mm while the larger units are suitable for boards of up to 500 mm × 350 mm. The more expensive exposure units are fitted with timers which can be set to determine the actual exposure time. Low-cost units do not have such a facility and the operator has to refer to a clock or wristwatch in order to determine the exposure time. In use, the 1:1 master artwork (in the form of opaque transfers and tape on translucent polyester drafting film) is placed on the glass screen immediately in front of the ultra-violet tubes (taking care to ensure that it is placed so that the component side is uppermost). The opaque plastic film is then removed from the photo-resist board (previously cut roughly to size) and the board is then placed on top of the film (coated side down). The lid of the exposure unit is then closed and the timer set (usually for around four minutes but see individual manufacturer's recommendations. The inside of the lid is lined with foam which exerts an even pressure over the board such that it is held firmly in place during the exposure process.

It should be noted that, as with all photographic materials, sensitized copper board has a finite shelf-life. Furthermore, boards should ideally be stored in a cool place at a temperature of between 2°C and 10°C. Shelf-life at 20°C will only be around twelve months and thus boards should be used reasonably promptly after purchase.

Note that it is not absolutely essential to have access to an ultra-violet light-box as we all have at least occasional access to an entirely free source of ultra-violet light. Provided one is prepared to wait for a sunny day and prepared to experiment a little, the exposure process can be carried out in ordinary sunlight. As a guide, and with the full sun present overhead, exposure will take around fifteen minutes. Alternatively, one can make use of a sun-ray lamp. Again, some experimentation will be required in order to get the exposure right. With a lamp placed approximately 300 mm from the sun ray source (and arranged so that the whole board surface is evenly illuminated) an exposure time of around four minutes will be required. Note that, if you use this technique, it is important to follow the sun-ray lamp manufacturer's instructions concerning eye protection. A pair of goggles or dark sunglasses can be used to protect the eyes during the exposure process. However one should never look directly at the ultra-violet light source even when it is 'warming up'.

Finally, one can easily manufacture one's own light-box (using low-power ultra-violet tubes) or make use of a standard ultra-violet bulb (of between 150 W and 300 W) suspended above the work area. If this technique is used, the bulb should be hung approximately 400 mm above the table on which the sensitized board, artwork and glass sheet have been placed. A typical exposure time for a 300 W bulb is in the range of ten to fifteen minutes. A pair of dark sunglasses can again be used to protect the eyes during the process. Where one is

using an 'alternative' technique, a frame should be constructed along the lines shown in Figure 19.4. This can be used to hold the transparency in contact with the coated copper board during the exposure process.

Whichever method of exposure is used some experimentation may be required in order to determine the optimum exposure time. After this time has elapsed, the board should be removed and immersed in a solution of sodium hydroxide that acts as a developer. The solution should be freshly made and the normal concentration required is obtained by mixing approximately 500 ml of tap water (at 20°C) with one tablespoon of sodium hydroxide crystals. A photographic developing tray (or similar shallow plastic container) should be used to hold the developer. Note that care should be taken when handling the developer solution and the use of plastic or rubber gloves is strongly recommended. This process should be carried out immediately after exposure and care should be taken not to allow the board to be further exposed under room lights.

The board should be gently agitated while immersed in the developer and the ensuing process of development should be carefully watched. The board should be left for a sufficiently long period for the entire surface to be developed correctly but not so long that the tracks lift. Development times will depend upon the temperature and concentration of the developer and on the age of the sensitized board. Normal development times are in

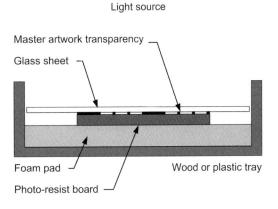

Figure 19.10 Exposure frame suitable for photo-etch PCB production

the region of 30 to 90 seconds and after this period the developed image of the track layout (an etch-resist positive) should be seen.

After developing the board it should be carefully washed under a running tap. It is advisable not to rub or touch the board (to avoid scratching the surface) and the jet of water should be sufficient to remove all traces of the developer. Finally, the board should be placed in the **etchant** which is a ferric chloride solution (FeCl3). For obvious reasons, ferric chloride is normally provided in crystalline form (though at least one major supplier is prepared to supply it on a 'mail order basis' in concentrated liquid form) and should be added to tap water (at 20°C) following the instructions provided by the supplier. If no instructions are given, the normal quantities involved are 750 ml of water to 500 g of ferric chloride crystals. Etching times will also be very much dependent upon temperature and concentration but, for a fresh solution warmed to around 40°C the time taken should typically be ten to fifteen minutes. During this time the board should be regularly agitated and checked to ascertain the state of etching. The board should be removed as soon as all areas not protected by resist have been cleared of copper; failure to observe this precaution will result in 'undercutting' of the resist and consequent thinning of tracks and pads. Where thermostatically controlled tanks are used, times of five minutes or less can be achieved when using fresh solution.

It should go without saying that great care should be exercised when handling ferric chloride. Plastic or rubber gloves should be worn and care must be taken to avoid spills and splashes. After cooling, the ferric chloride solution may be stored (using a sealed plastic container) for future use. In general, 750 ml of solution can be used to etch around six to ten boards of average size; the etching process taking longer as the solution nears the end of its working life. Finally, the exhausted solution must be disposed of with care (it should not be poured into an ordinary mains drainage system).

Having completed the etching process, the next stage involves thorough-ly washing, cleaning, and drying the printed circuit board. After this, the board will be ready for drilling. Drilling will normally involve the services of a 0.6 mm or 1 mm twist drill bit for standard component leads and IC pins. Larger drill bits may be required for the leads fitted to some larger components (e.g. power

diodes) and mounting holes. Drilling is greatly simplified if a special PCB drill and matching stand can be enlisted. Alternatively, provided it has a bench stand, a standard electric drill can be used. Problems sometimes arise when a standard drill or hand drill is unable to adequately grip a miniature twist drill bit. In such cases one should make use of a miniature pin chuck or a drill fitted with an enlarged shank (usually of 2.4 mm diameter).

PCB CAD packages

The task of laying out a PCB and producing master artwork (both for the copper foil side of a board and for the component overlay) is greatly simplified by the use of a dedicated PCB CAD package. Some examples of the use of these packages are shown in Figs 19.11 to 19.14.

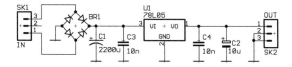

Figure 19.11   Circuit of a simple 5 V regulated power supply module based on a 78L05 three-terminal voltage regulator

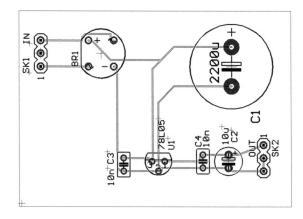

Figure 19.12   PCB track layout for the circuit shown in Fig. 19.5 produced by an autorouting CAD package (shown prior to optimization)

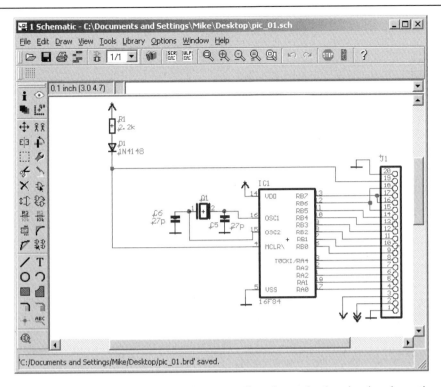

Figure 19.13 Circuit of a PIC16F84-based microcontroller shown in the circuit schematic editor of the Eagle CAD software (see Appendix 9)

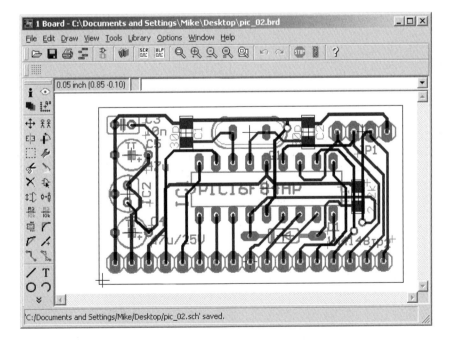

Figure 19.14 Double-sided PCB layout for the circuit shown in Fig. 19.13 generated by the Eagle CAD software (shown after optimization)

Connectors

Various forms of connectors may be required in any particular item of electronic equipment. These may be categorized in various ways but the following should serve as a guide:

(a) Mains connectors. These connectors are intended for use with an a.c. mains supply. Standard 3-pole IEC types should be employed wherever possible.

(b) Single-pole connectors. These are available in ranges having diameters of 4 mm, 3 mm, 2 mm and 1 mm and are ideal for use with test-leads (i.e. as input and output connectors in test equipment) and for low-voltage power supplies. Plugs and sockets are available in various colours to permit identification.

(c) Multi-pole connectors. A huge range of multi-pole connectors is currently available and the following types are worthy of special note:

 (i)   DIN standard connectors of the type commonly used in audio equipment
 (ii) DIN 41612 indirect stripboard and PCB edge connectors with 32, 64, or 96-ways.
 (iii) DIN 41617 low-cost indirect edge connectors
 (iv) D-connectors. These are available with 9, 15, 25, and 37-ways and are popularly used in microcomputer applications
 (v)   IEEE-488 connectors. These 14, 24, 36 and 50-way connectors are commonly used with equipment which uses the popular IEEE-488 instrument bus system.

 Note that **insulation displacement connector** (IDC) provide a means of terminating multi-way ribbon cables without the need for soldering. Simple tools are used to assemble the connector and strain relief clamp onto the cable, the insulation of which is pierced by the tines of the connector pins.)

(d) Coaxial connectors. These connectors are used for screened test leads and also for r.f. equipment. The following three types are most popular:

 (i)   BNC. These are available in both 50 Ω and 75 Ω series and are suitable for operation at frequencies up to 4 GHz.
 (ii) PL259/SO239 (popularly known as UHF). These 50 Ω types are suitable for use at frequencies up to 250 MHz.
 (iii) Belling-Lee (popularly known as TV). These low-cost connectors are suitable for use at frequencies up to 800 MHz with an impedance of 75 Ω.

The process of choosing a connector is usually very straightforward. It will first be necessary to ensure that an adequate number of ways are catered for and that the connector is suitably rated as far as current and voltage are concerned (one should consult individual manufacturer's ratings where any doubt arises). It is advisable to maintain compatibility with equipment of similar type and one should avoid using too wide a range of connectors and cables. In addition, a common pin-usage convention should be adopted and strictly adhered to. This will help to avoid problems later on and will make interwiring of equipment and exchange of modules a relatively straightforward process.

Enclosures

An appropriate choice of enclosure is vitally important to the 'packaging' of any electronic equipment. Not only will the enclosure provide protection for the equipment but it should also be attractive and add to the functionality of the equipment. Enclosures can be divided into five main types:

(a) Instrument cases. These are ideal for small items of test gear (e.g. meters and signal generators) and are available in a wide variety of styles and sizes. One of the most popular low-cost ranges of instrument cases is that manufactured by BICC-Vero and known as Veroboxes. This enclosure comprises plastic top and bottom sections with anodized aluminium front and back panels. Other ranges of instrument cases feature steel and aluminium construction and are thus eminently suited for larger projects or those fitted with larger mains transformers (such as heavy-duty power supplies).

(b) Plastic and diecast boxes. These low-cost enclosures comprise a box with removable lid secured by means of four or more screws. Boxes are available in a range of sizes and the diecast types (which are ideal for use in relatively hostile environments) are available both unpainted and with a textured paint finish.

(c) Rack systems. These are designed to accept standard cards and are ideal for modular projects. The outer case comprises an aluminium framework fitted with covers and a series of connectors at the rear from which the modules derive their power and exchange signals. Individual circuit cards (which may be stripboard or PCB) are fitted to a small supporting chassis and anodized aluminium front panel (available in various widths). The card assembly slides into the rack using appropriately positioned clip-in guides. Rack systems are expensive but inherently flexible.

(d) Desk consoles. These enclosures are ideal for desktop equipment and generally have sloping surfaces that are ideal for mounting key-boards and keypads.

(e) Special purpose enclosures. Apart from the general-purpose types of enclosure described earlier, a variety of special purpose enclosures are also available. These include such items as clock and calculator housings, enclosures for hand-held controllers, and cases for in-line mains power supplies (including types having integral 13 A plugs).

Having decided upon which of the basic types of enclosure is required, the following questions should be borne in mind when making a final selection of enclosure:

- Is the size adequate?
- Will the enclosure accommodate the stripboard or PCB circuitry together with the 'off-board' components (including, where appropriate, the mains transformer)?
- Will the enclosure be strong enough?
- Will other equipment be stacked on top of the enclosure?
- Can the mains transformer be supported adequately without deforming the case?
- Is the front panel of sufficient size to permit mounting all of the controls and displays?
- Is any protection (in the form of handles or a recessed front panel) necessary for the front panel mounted components?

- Is there any need for ventilation?
- How much heat is likely to be produced within the equipment?
- Can heat producing components be mounted on the rear panel?
- Will a cooling fan be required?
- Is the style of the equipment commensurate with other equipment of its type and within the same range?
- Is there a need to include screening or filtering in order to minimize incoming or outgoing noise or stray electric/magnetic fields?

When laying out the front panel of the equipment, it is important to bear in mind the basic principles of ergonomic design. All controls should be accessible. They should be logically arranged (grouping related functions together) and clearly labelled. Consideration should be given to the type of controls used (e.g. slider versus rotary potentiometers, push-button versus toggle switches).

It is also important to wire controls such that their action follows the expected outcome. Rotary 'gain' and 'volume' controls, for example, should produce an increase in output when turned in a clockwise direction. Indicators should operate with adequate brightness and should be viewable over an appropriate angle. Indicators and controls should be arranged so that it is possible to ascertain the status of the instrument at a glance. If necessary, a number of opinions should be sought before arriving at a final layout for the front panel; one's own personal preferences are unlikely to coincide exactly with those of the 'end user'.

Solder and soldering technique

Most solders are an alloy of two metals; tin (Sn) and lead (Pb). The proportion of tin and lead is usually expressed as a percentage or as a simple ratio. For example, the term **60/40 solder** refers to an alloy which comprises 60% tin and 40% lead by weight. The proportion of tin and lead is instrumental in determining the melting point of the solder alloy and the proportion of the constituents is usually clearly stated.

In the past, solder was often categorized by its intended application and the terms **radio solder**,

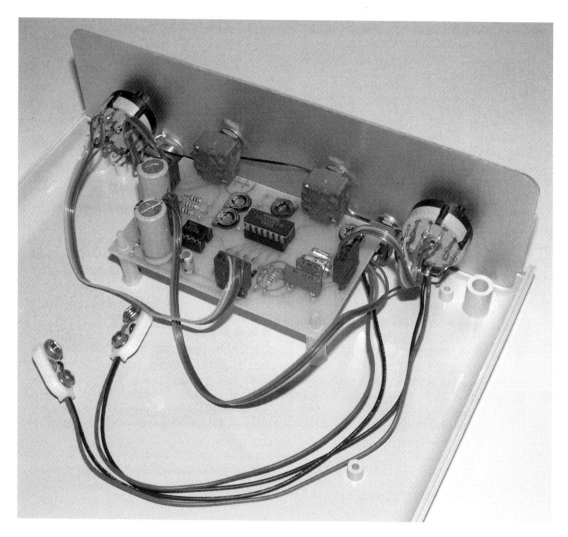

Figure 19.15 Internal wiring of a waveform generator based on PCB headers and coloured ribbon cable

electronic solder, and **general-purpose solder** and silver bearing solder. The difference between these types of solder can be briefly summarized as follows:

Radio solder

A medium melting point temperature (typically 230°C) solder used for making joints in point-to-point wiring as typically found in radios and early TV receivers. Typical joints would be quite substantial based on tag strips, group boards, valve

bases, etc., and each joint might involve several wires or component leads of substantial diameter.

Electronic solder

A low melting point temperature (typically 185°C) used for making joints on printed circuit boards. A joint would normally involve a single component lead and its associated copper pad and excessive temperature would have to be avoided in order to prevent damage both to the component (whose connecting lead might be of a very short length) and the PCB itself (where copper pads and tracks might lift due to the application of excessive heat).

General-purpose solder

A high melting point temperature (typically 255°C) designed for general electrical repair work (and unsuitable for use with electronic equipment).

Silver bearing solders

Silver bearing solders tend to flow more smoothly and form stronger joints. Unfortunately they generally require higher soldering and desoldering temperatures and this can make component removal and replacement difficult. In particular, the higher temperatures involved may cause damage to the bond between copper pads and tracks and the printed circuit substrate on which they are formed. In turn, this can result in the lifting of pads and tracks and permanent damage can result.

Flux

Flux has two roles in soldering. Firstly it helps with the removal of any oxide coating that may have formed on the surfaces that are to be soldered and secondly it helps to prevent the formation of oxides caused by the heat generated during the soldering process itself. Flux is available in both paste and liquid form. However, in the case of solder for electronic applications, it is also available incorporated within the solder itself. In this case, the flux is contained within a number of cores that permeate through the solder material rather like the letters that appear on a piece of souvenir seaside rock.

Safety when soldering

Whilst soldering is not a particularly hazardous operation, there are a number of essential safety and other precautions that should be observed. These are listed below:

Fumes

Solder fumes are an irritant and exposure, particularly if prolonged, can cause asthma attacks. Because of this it is essential to avoid the build up of fumes, particularly so when a soldering iron is in

continuous use. The best way to do this is to use a proprietary system for fume extraction. In addition, many professional quality irons can be fitted with fume extraction facilities which are designed to clear fumes from the proximity of the soldering iron bit.

Even when a soldering iron is only used intermittently it is essential to ensure that the equipment is used in a well-ventilated area where fumes cannot accumulate. Fume extraction should be fitted to all soldering equipment that is in constant use. Most good quality soldering irons that

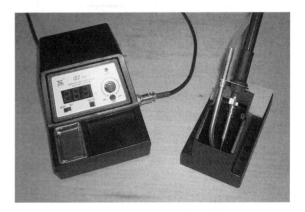

Figure 19.16   A modern temperature-controlled soldering station with ESD and fume extraction facilities

Figure 19.17   Fume extraction facility which can be added to a temperature controlled iron when necessary

Figure 19.18 Two different types of bit fitted to small soldering irons

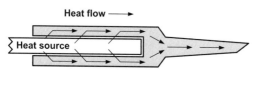

(a) Bit surrounds element

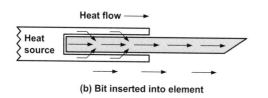

(b) Bit inserted into element

Figure 19.19 Heat transfer in the two types of soldering iron bit shown in Fig. 19.18

are designed for professional use can be fitted with fume extraction equipment. Figure 19.1 shows a modern soldering station equipped with fume extraction.

Damage to eyes

Molten solder or flux additives can cause permanent eye damage. Some means of eye protection is essential. This can take the form of safety glasses or the use of a bench magnifier. Ordinary reading or prescription glasses will also offer a measure of protection and may also be of benefit when undertaking close work.

Static hazards

Many of today's electronic devices are susceptible to damage from electric fields and stray static charges. Unfortunately, simply earthing the tip of a soldering iron is not sufficient to completely eliminate static damage and additional precautions may be required to avoid any risks of electro-static damage (**ESD**). These may involve using a low-voltage supply to feed the heating element fitted in the soldering iron, incorporating low/zero-voltage switching, and the use of conductive (antistatic) materials in the construction of the soldering iron body and the soldering station itself.

Burn, melt and fire hazards

The bit of a soldering iron is usually maintained at a temperature of between 250°C and 350°C. At this temperature, conventional plastic insulating materials will melt and many other materials (such as paper, cotton, etc.) will burn. It should also go without saying that personal contact should be avoided at all times and the use of a properly designed soldering iron stand is essential.

Shock hazards

Mains voltage is present in the supply lead to a mains operated soldering iron and also in the supply lead to any soldering station designed for use with a low-voltage iron. If the soldering iron bit comes into contact with a mains lead there can be a danger that the insulation will melt exposing the live conductors. In such situations, heatproof insulation is highly recommended both for the supply to the iron itself and also to any power unit.

Soldering guns

Soldering guns (where the bit is effectively a short-circuited turn) can become extremely hot and continuous operation of this type of iron can be extremely dangerous as the mains transformer fitted into the handle of the gun can easily become overheated. This can cause the insulation to reach a dangerously high temperature which can result in melting and short-circuits which will cause permanent damage to the gun. Because of their rather clumsy nature and high-power, soldering guns should be avoided for all general electronic work.

Soldering and desoldering techniques

Soldered joints effectively provide for both electrical and mechanical connection of components with pins, tags, stripboards, and PCB. Before a soldering operation is carried out, it is vitally important that all surfaces to be soldered are clean and completely free of grease and/or oxide films. It is also important to have an adequately rated soldering iron fitted with an appropriate bit. The soldering iron bit is the all-important point of contact between the soldering iron and the joint and it should be kept scrupulously clean (using a damp sponge) and free from oxide. To aid this process, and promote heat transfer generally, the bit should be regularly 'tinned' (i.e. given a surface coating of molten solder).

The selection of a soldering bit (see Figs 19.18 and 19.19) depends on the type of work undertaken. Figure 19.20 shows a number of popular bit profiles, the smaller bits being suitable for sub-miniature components and tightly packed boards. The procedure for making soldered joints to terminal pins and PCB pads are shown in Figs 19.21 and 19.22, respectively. In the case of terminal pins, the component lead or wire should be wrapped tightly around the pin using at least one turn of wire made using a small pair of long-nosed pliers. If necessary, the wire should be trimmed using a small pair of side cutters before soldering. Next, the pin and wire should then be simultaneously heated by suitable application of the soldering iron bit and then sufficient solder should be fed on to the pin and wire (not via the bit) for it to flow evenly around the joint thus forming an airtight 'seal'. The solder should then be left to cool (taking care not to disturb the component or wire during the process). The finished joint should be carefully inspected and re-made if it suffers from any of the following faults:

(a) Too little solder. The solder has failed to flow around the entire joint and some of the wire turn or pin remains exposed.
(b) Too much solder. The solder has formed into a large 'blob' the majority of which is not in direct contact with either the wire or the pin.
(c) The joint is 'dry'. This usually occurs if either the temperature of the joint was insufficient to permit the solder to flow adequately or if the joint was disturbed during cooling.

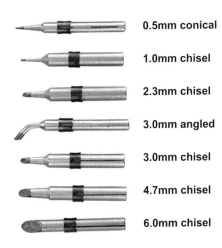

Figure 19.20 Typical soldering iron bit profiles

0.5mm conical
1.0mm chisel
2.3mm chisel
3.0mm angled
3.0mm chisel
4.7mm chisel
6.0mm chisel

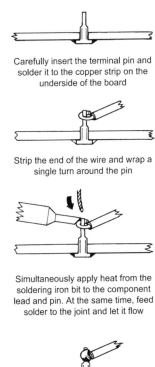

Carefully insert the terminal pin and solder it to the copper strip on the underside of the board

Strip the end of the wire and wrap a single turn around the pin

Simultaneously apply heat from the soldering iron bit to the component lead and pin. At the same time, feed solder to the joint and let it flow

Carefully inspect the completed joint for flaws

Figure 19.21 Procedure for making a soldered joint to a terminal pin

In the case of a joint to be made between a component and a PCB pad, a slightly different technique is used (though the requisites for cleanliness still apply). The component should be fitted to the PCB (bending its leads appropriately at right angles if it is an axial lead component) such that its leads protrude through the PCB to the copper foil side. The leads should be trimmed to within a few millimetres of the copper pad then bent slightly (so that the component does not fall out when the board is inverted) before soldering in place. Opinions differ concerning the angles through which the component leads should be bent. For easy removal, the leads should not be bent at all while, for the best mechanical joint, the leads should be bent through 90°. A good compromise, and that preferred by the author involves bending the leads through about 45°. Care should again be taken to use the minimum of solder consistent with making a sound electrical and mechanical joint.

Finally, it is important to realize that good soldering technique usually takes time to develop and the old adage 'practice makes perfect' is very apt in this respect. Do not despair if your first efforts fail to match with those of the professional!

Component removal and replacement

Care must be exercised when removing and replacing printed circuit mounted components (see Figs 19.23 to 19.28). It is first necessary to accurately locate the component to be removed on the upper (component) side of the PCB and then to correctly identify its solder pads on the underside of the PCB. Once located, the component pads should be gently heated using a soldering iron. The soldering iron bit should be regularly cleaned using a damp sponge (a small tin containing such an item is a useful adjunct to any soldering work station). The power rating of the iron should be the minimum consistent with effective removal of the components and should therefore not normally exceed 20 W for types which do not have temperature control since excessive heat can not only damage components but may also destroy the bond between the copper track and the board itself. This latter effect causes lifting of pads and tracks from the surface of the PCB and should be avoided at all costs.

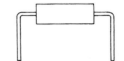

Prepare the component leads by
bending to size using pliers

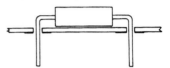

Insert the component into its correct
location on the PCB

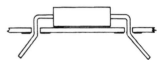

Bend the component leads through
approximately 45°

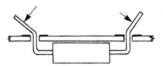

Trim the component leads using a
pair of side cutters

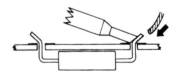

Simultaneously apply heat from
the soldering iron bit to the
component lead and copper pad.
At the same time, feed solder to
the joint and let it flow

Carefully inspect the completed
joint for any flaws and rework if
necessary

Figure 19.22   Procedure for making a soldered joint to a PCB pad

Figure 19.23 Making a soldered joint on a PCB. A fine point soldering bit is being used

Figure 19.26 Using a desoldering pump to remove excess solder

Figure 19.24 A PCB holder can be a useful soldering accessory when working on large boards

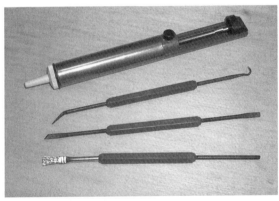

Figure 19.27 A variety of desoldering tools (including desoldering pump and wire brush)

Figure 19.25 A soldering iron bit cleaner. Regular cleaning is essential to ensure good soldered joints

Figure 19.28 A PCB bench magnifier is ideal for inspecting finished work for soldering defects

Once the solder in the vicinity of the pad has become molten (this usually only takes one or two seconds) and desoldering suction pump should be used to remove the solder. This will often require only one operation of the desoldering pump; however, where a large area of solder is present or where not all of the solder in the vicinity of the pad has become molten, a second application of the tool may be required. If the desoldering tool has to be repeatedly applied this usually indicates that either the tool is clogged with solder or that the soldering iron is not hot enough. With practice only one application of the soldering iron and soldering pump should be required. Once cleared of solder, the component lead should become free and a similar process should be used for any remaining leads or pins. Special tools arc available for the rapid removal of IC and SMD devices and these permit the simultaneous heating of all pins (see Figs 19.29 and 19.30).

When desoldering is complete, the component should be gently withdrawn from the upper (component) side of the board and the replacement component should be fitted, taking care to observe the correct polarity and orientation where appropriate. The leads should protrude through the PCB to the copper foil side and should be trimmed to within a few millimetres of the copper pad then bent slightly (so that the component do not fall out when the board is inverted) before soldering in place. Care should be taken to use the minimum of solder consistent with making a sound electrical and mechanical joint. As always, cleanliness of the soldering iron bit is extremely important if dry joints arc to be avoided.

Finally, a careful visual examination of the joint should be carried out. Any solder splashes or bridges between adjacent tracks should be removed and, if necessary, a sharp pointed instrument should be used to remove any surplus solder or flux which may be present.

In some cases it may be expedient (or essential when the PCB has been mounted in such a position that the copper foil side is not readily accessible) to remove a component by cutting its leads on the upper (component) side of the board. However, care should be taken to ensure that sufficient lead is left to which the replacement component (with its wires or pins suitably trimmed) may be soldered. During this operation, extra care must be taken when the soldering iron bit is placed in close

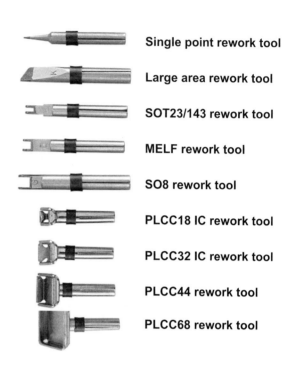

Single point rework tool

Large area rework tool

SOT23/143 rework tool

MELF rework tool

SO8 rework tool

PLCC18 IC rework tool

PLCC32 IC rework tool

PLCC44 rework tool

PLCC68 rework tool

Figure 19.29 Typical desoldering and rework tools

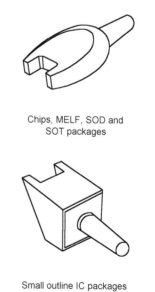

Chips, MELF, SOD and
SOT packages

Small outline IC packages

Figure 19.30 SMD desoldering tools for integrated circuits

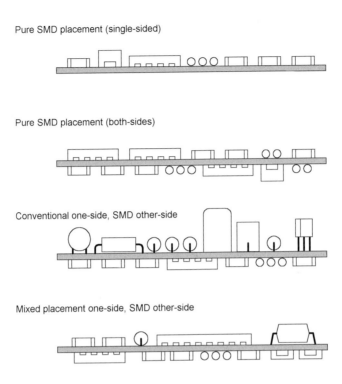

Figure 19.31 Various PCB layouts involving SMD and conventional components

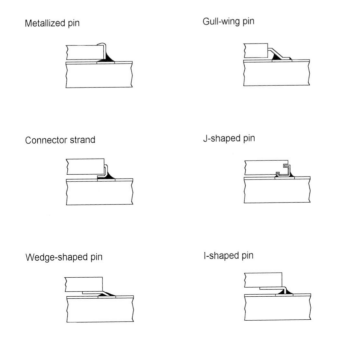

Figure 19.32 Different pin styles used with SMD devices

proximity to densely packed components on the upper side of the PCB. Polystyrene capacitors and other plastic encapsulated components are particularly vulnerable in this respect. Figures 19.31 and 19.32 show some typical arrangements used with SMD and conventional components.

Heatsinks

The electronic equipment designer is often faced with the task of designing circuits that will operate within a range of temperatures that will ensure safe and reliable operation. This task is normally carried out with the aid of a comprehensive set of manufacturers' data and a number of simple formulae. The symbols shown in Table 19.3 are commonly used in heatsink calculations:

Table 19.3 Symbols used in heatsink calculations

Symbol	Meaning	Units
T_A	Ambient temperature	°C
T_{Amax}	Maximum ambient temperature	°C
T_J	Junction temperature	°C
T_{Jmax}	Maximum junction temperature	°C
T_C	Case temperature	°C
T_S	Heatsink surface temperature (note 1)	°C
θ_{JA}	Thermal resistance (from junction to ambient)	°C/W
θ_{CS}	Thermal resistance (from case to surface)	°C/W
θ_{SA}	Thermal resistance (from surface to ambient)	°C/W
θ_T	Total thermal resistance (from junction to ambient)	°C/W
P_T	Total power dissipation (note 2)	W
P_{Tmax}	Maximum total power dissipation (note 2)	W

Notes:

1. At the point of contact with the semiconductor device
2. For one or more devices mounted on the heatsink

Determining temperature rise

The ability of a heatsink and transistor mounting arrangement to dissipate heat can be measured in terms of the amount by which the junction temperature exceeds the ambient (or surrounding air) temperature. A perfect heatsink and mounting arrangement would keep the junction temperature the same as that of the surrounding air. In practice, the junction temperature will rise under operational conditions.

In practice, a rise in junction temperature is unavoidable. What is important, however, is that the maximum junction temperature, T_{Jmax}, must never be exceeded. Furthermore, manufacturers provide a de-rating characteristic that must be used to determine device rating (percentage of full power rating at 25°C) when the junction temperature exceeds its normally specified value (see Fig. 19.33).

The temperature rise, ΔT, above ambient will be given by:

$$\Delta T = P_T \times \theta_T \tag{1}$$

where P_T is the total power dissipated by the semiconductor device(s) and θ_T is the total thermal resistance of the heatsink and mounting arrangement.

Determining the junction temperature

The temperature rise is simply the difference between the actual junction temperature and the surrounding (or ambient) temperature. Ultimately, the operating junction temperature is vitally important since exceeding the maximum operating junction temperature (T_{Jmax}) can result in the destruction of the semiconductor device!

Now $\Delta T = T_J - T_A \tag{2}$

Combining equations (1) and (2) gives:

$$T_J - T_A = P_T \times \theta_T$$

From which:

$$T_J = \left(P_T \times \theta_T \right) + T_A$$

It is worth using an example to explain the use of this formula in helping us to determine just how hot a semiconductor junction actually gets!

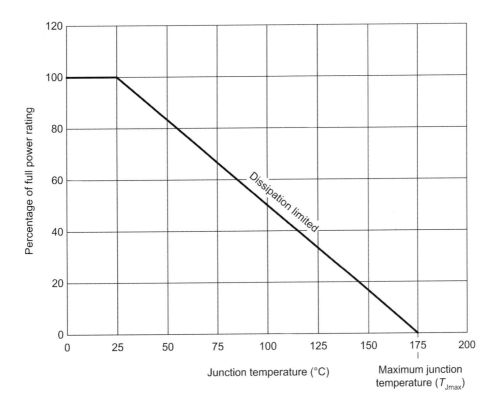

Figure 19.33 A typical semiconductor de-rating characteristic. Note how full power (100%) is available at temperatures of up to 20°C but above this the percentage must be reduced progressively

Example 19.1

A TO220 transistor dissipates a total power of 5W. If the total thermal resistance is 8°C/W and the ambient temperature is 30°C, determine the junction temperature.

Applying equation (3) gives:

$$T_J = \left(P_T \times \theta_T\right) + T_A = \left(5 \times 8\right) + 30 = 70°C$$

Determining thermal resistance

Figure 19.34 shows a typical TO3 case style transistor mounted on a finned **heat dissipator** (i.e. heatsink). Heat is conducted away from the semiconductor junction to the outer case of the TO3 package and then, via an insulating washer to the surface of the heatsink. From this point, heat is conducted to the extremities of the fins where it is radiated into the surrounding air space.

The total thermal resistance, θ_T, present in any heatsink arrangement is actually the sum of several individual thermal resistances. Consider the case of a transistor bolted directly to a metal heat radiator (see Fig. 19.35). The total thermal resistance, θ_T, is the sum of the following thermal resistances:

1. The thermal resistance that exists between the semiconductor junction and the case of the transistor (i.e. the thermal resistance *inside* the transistor package), θ_{JC}
2. The thermal resistance between the case of the transistor and the heat radiator (i.e. the thermal resistance of an insulating washer, where fitted), θ_{CS}

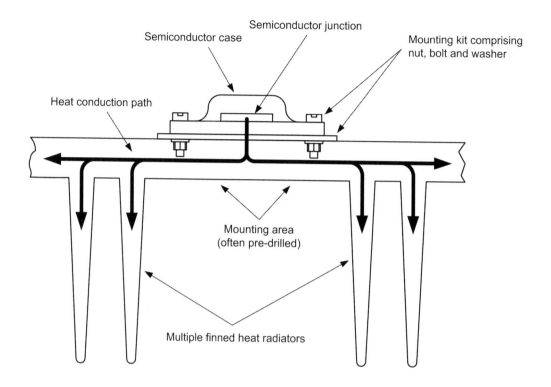

Figure 19.34   Typical mounting arrangement for a TO3 transistor

3. The thermal resistance between the surface of the heat radiator and the space surrounding it (i.e. the thermal resistance between the surface and ambient), θ_{SA}

Fig. 19.36 shows these three thermal resistances together with the temperatures that exist at each point in the arrangement shown in Fig. 19.35. The three thermal resistances shown in Fig. 19.36 actually appear 'in series' and we can use a simple electrical analogy to represent the thermal 'circuit' in electrical terms (see Fig. 19.37).

From Fig. 19.37 we can conclude that the total thermal resistance, θ_T, is given by:

$$\theta_T = \theta_{JC} + \theta_{CS} + \theta_{SA} \tag{4}$$

Note that the total thermal resistance is actually the same as the thermal resistance between the junction and ambient.

Thus:

$$\theta_T = \theta_{JC} + \theta_{CS} + \theta_{SA} = \theta_{JA} \tag{5}$$

The complete electrical equivalent circuit of the heatsink arrangement is shown in Fig. 19.37. Note that the source of power (P_{TOT}) is the semiconductor device and the 'potentials' at the two extreme ends of the series chain of thermal resistances are T_J (junction temperature) and T_A (ambient or surrounding air temperature).

Example 19.2

A transistor has a thermal resistance from junction to case of 1.5°C/W. If the transistor is fitted with a washer and mounting kit having a thermal resistance of 1.75°C/W, and a heatsink of

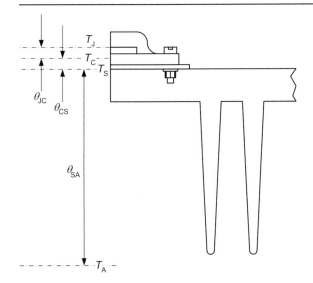

Figure 19.35 Typical mounting arrangement for a TO3 transistor

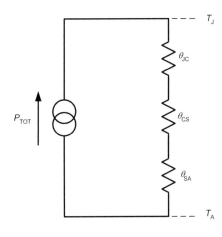

Figure 19.37 Typical mounting arrangement for a TO3 transistor

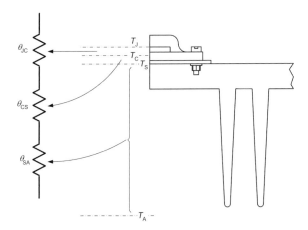

Figure 19.36 Typical mounting arrangement for a TO3 transistor

2.25°C/W, determine the total thermal resistance that exists between junction and ambient.

Applying equation (5) gives:

$$\theta_{JA} = \theta_{JC} + \theta_{CS} + \theta_{SA} = 1.5 + 1.75 + 2.25 = 5.5°C/W$$

Worst case conditions

In the design of electronic equipment we often need to plan for the **worst case** conditions, ensuring that the absolute maximum junction temperature, T_{Jmax}, is not exceeded when the total power dissipation and ambient temperature jointly reach their maximum working values.

From equation (3) we can infer that, under the worst case conditions:

$$T_{Jmax} = \left(P_{Tmax} \times \theta_T\right) + T_{Amax}$$

Combining this with equation (5) gives:

$$T_{Jmax} = \left(P_{Tmax} \times (\theta_{JC} + \theta_{CS} + \theta_{SA})\right) + T_{Amax} \qquad (6)$$

Example 19.3

A transistor has an absolute maximum junction temperature rating of 150° and a thermal resistance from junction to case of 1.0°C/W. If the device is fitted with a washer and mounting kit having a thermal resistance of 1.25°C/W and a heatsink of 2.75°C/W, determine whether the maximum ratings are exceeded when the total power dissipation reaches a maximum of 25 W at a maximum ambient temperature of 40°C.

Applying equation (6) gives:

$$T_{Jmax} = \left(25 \times (1.0 + 1.25 + 2.75) + 40 = 165°C\right)$$

The absolute maximum junction temperature rating (150°C) is exceeded and the designer should either reduce the power dissipation to a safe value or reduce the thermal resistance present (or both!).

Determining heatsink specifications

The electronic equipment designer often has to determine the required heatsink specifications given the absolute maximum junction temperature, thermal resistance from junction to case, maximum expected ambient temperature, etc. To do this, we need to rearrange equation (6) to make θ_{SA} the subject of the equation. Thus:

$$\theta_{SA} = \frac{T_{Jmax} - T_{Amax}}{P_{Tmax}} - \left(\theta_{JC} + \theta_{CS}\right) \qquad (7)$$

The value obtained for θ_{SA} will be the minimum acceptable rating for the required heatsink. In practice, we would choose a component with a higher rating to allow for a margin of safety.

Example 19.4

Determine the minimum acceptable thermal resistance rating for a heatsink for use with a transistor under the following conditions:

Maximum junction temperature:	175°
Thermal resistance of case to surface:	1.25°C/W
Thermal resistance of washer and mounting kit:	1.75°C/W
Maximum total power dissipation:	75W
Maximum ambient temperature:	45°C

Applying equation (7) gives:

$$\theta_{SA} = \frac{T_{Jmax} - T_{Amax}}{P_{Tmax}} - \left(\theta_{JC} + \theta_{CS}\right)$$

from which:

$$\theta_{SA} = \frac{175 - 45}{13} - \left(1.25 + 1.75\right) = 10 - 3 = 7°C/W$$

Example 19.5

A power FET operates with a maximum continuous drain current (I_D) of 0.5 A at a maximum continuous drain-source voltage (V_{DS}) of 40 V. Determine the minimum acceptable thermal resistance rating for a heatsink for use with the device under the following conditions:

Maximum junction temperature:	175°
Thermal resistance of case to surface:	1.35°C/W
Thermal resistance of washer and mounting kit:	1.45°C/W
Maximum ambient temperature:	55°C

First we need to determine the maximum total power dissipation. Since the device is a FET, the gate dissipation can be neglected. Thus the total power dissipation will be:

$$P_{Tmax} = V_{DS} \times I_D = 40 \times 0.5 = 20W$$

Applying equation (7) gives:

$$\theta_{SA} = \frac{T_{Jmax} - T_{Amax}}{P_{Tmax}} - \left(\theta_{JC} + \theta_{CS}\right)$$

$$\theta_{SA} = \frac{175 - 55}{20} - \left(1.35 + 1.45\right) = 6 - 2.8 = 3.2°C/W$$

All of the previous calculations assume that the heatsink is subject to natural convection. Increasing the airflow across the surface of the heatsink can dramatically increase the amount of radiated and consequently it will also significantly reduce the effective thermal resistance of the heatsink.

Practical heatsink arrangements

Having explained some of the theory behind the design of heat dissipators, we shall move on to describe some practical heatsink and mounting arrangements. Typical thermal resistance for several common case styles are listed in Table 19.4.

A selection of heatsink cross sections is shown in Fig. 19.41. These range from a simple folded U-section metal plate with a thermal resistance of

Table 19.4 Typical thermal resistances for various case styles

Case style	θ_{JA} (no heatsink) (°C/W)	θ_{JC} (°C/W)
TO92	200 to 350	40 to 60
TO126	83 to 100	3 to 10
TO220	60 to 70	1.5 to 4
TO202	62 to 75	6 to 13
TO218	30 to 45	1 to 1.56

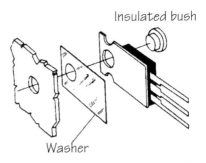

Figure 19.38 Typical mounting arrangement for a TO220 packaged device

20°C/W to a complex aluminium alloy extrusion with a thermal resistance of 1.2°C/W. Lower values of thermal resistance can be obtained with the use of forced air cooling.

Natural convection airflow can be enhanced by the proper placement of heatsinks and other heat producing components. Since warm air rises, vertical surfaces tend to transmit heat to the air better than comparable horizontal surfaces. The hottest devices should be located on the upper side of a horizontally mounted PCB or close to the upper edge of a vertically mounted PCB.

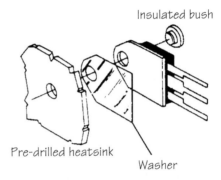

Figure 19.39 Typical mounting arrangement for a TO218 or TAB packaged device

Semiconductor mounting arrangements

A typical mounting arrangement for a TO220 semiconductor package is shown in Fig. 19.38. In many cases, the tab of the device is connected to one of three terminals (often the collector or drain) in which case a mica or thermally conductive plastic washer must be fitted. Note also that an insulated bush must be used in order to prevent the mounting bolt shorting the metal tab to the heatsink. Figure 19.39 shows a similar mounting arrangement used for a TO218 packaged device while Fig. 19.40 shows how a TO3 device is fitted.

The thermal resistance of the mounting kit used with a semiconductor device can have a major effect on the efficiency of the heat conduction from the surface of the case to the heat radiator. Special thermally impregnated washers have significantly lower thermal resistance than simple mica washers. Thermally conductive silicone grease should NOT be used with this type of washer.

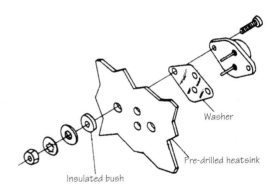

Figure 19.40 Typical mounting arrangement for a TO3 packaged device

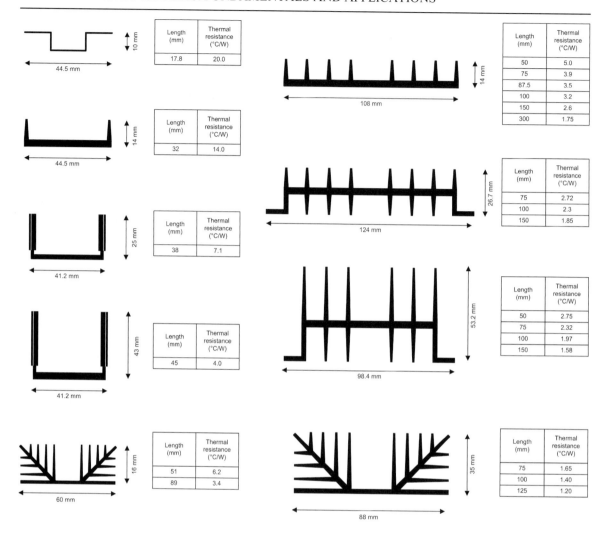

Figure 19.41   Some common heatsink types and their associated thermal resistances

A practical example

We conclude this chapter with an example of circuit construction based on a simple bench power supply. The requirement was for a small adjustable d.c. power supply that could be used for general-purpose bench testing of electronic circuits. After some preliminary investigation the following basic requirements were established:

1. It should be mechanically and electrically rugged
2. It should use proven and reliable technology
3. It should be low-cost and easy to maintain
4. It should comply with appropriate European legislation (e.g. the EMC and low-voltage directives)
5. It should have one variable (3 V to 15 V) output and one fixed (+5 V) accessory output and that both outputs should be protected against a short-circuit connected to the output.
6. With the exception of the accessory output and the a.c. mains input, all controls, connectors and switches should be made available on the front panel
7. It should have colour coded output terminals

that will accept standard 4 mm plugs. The terminals should also allow wires that have not been fitted with plugs to be clamped directly using a screw action

8. It should operate from a standard 220 V a.c. mains supply which should be connected using a standard 3-pin IEC connector

9. It should be 'tamper-proof' (it should not be possible to remove the knobs or the enclosure without having to resort to the use of special tools)

10. It should be lightweight and portable

11. It should have LED indicators to show that the power supply is switched on and that the outputs are present.

Design specification

Having determined a need for our product and obtained a detailed list of requirements, the next stage was that of firming up the design brief and producing a detailed specification for the power supply. This **design specification** was a detailed performance specification that included numerical values for relevant parameters (such as output voltage and output current). The detailed design specification was important because we returned when we needed to confirm that our prototype power supply met our requirements. We did this by comparing the measured performance specification with the original design specification. Taking into account the requirements listed above, we arrived at the following design specification:

Fixed output:
- Output voltage adjustable from 3 V to 15 V
- Output current adjustable from 50 mA to 1 A max.
- 4 mm binding post connectors (red and black)

Variable output:
- Output voltage fixed at +5V ±5%
- Output current 1 A max.
- 4 mm binding post connectors (yellow and black)

Input:
- 200 to 240 V a.c. via IEC connector, 1 A fuse and EMC filter

Prototype manufacture

The next stage in the development of the power

supply was that of examining a range of solutions before deciding on the particular solution that formed the basis of a prototype. Since the product has both electrical/electronic and mechanical aspects it was possible to consider these separately.

Research was needed to find a suitable electronic circuit and to identify the components that would be needed to build it. Various sources of information were available including books, magazines, data sheets (see Fig. 19.42) from manufacturers and suppliers as well as the Web. After carrying out some initial investigation, a suitable circuit design was located (see Fig. 19.43). This circuit was based on two low-cost readily available integrated circuit devices (see requirements 2 and 3) and the data sheets were obtained for each (see Fig. 19.42).

The initial prototype was constructed on stripboard and tested after fitting into the enclosure (as shown in Fig. 19.47). A further prototype (see Fig. 19.48) was developed using a printed circuit board (PCB), the track layout for which was designed using a PCB CAD package which had an autorouting facility.

In order to satisfy requirements 1 and 4, a fully-screened metal enclosure was used. This was based on a simple two-part construction (see requirements 3). An internal EMC filter was fitted in order to comply with the EMC directive (see requirement 4). Before manufacturing the enclosure it was necessary to produce some general arrangement and detail drawings showing how the sheet metal should be drilled, punched, cut and bent. These drawings had to be accurately dimensioned in order to accommodate the components used. The drawings were produced using a simple 2D CAD package. One drawing was produced to show the enclosure top and front panel detail (see Fig. 19.44) whilst a second drawing was used to show the enclosure bottom and rear panel detail.

The material selected for the enclosure was 1.22 mm (18 SWG) aluminium sheet. The reasons for the choice of this material were that it was reasonably lightweight (see requirement 10), low-cost (see requirement 3), and easy to process. It could also be attractively paint finished. Figures 19.49 and 19.50 show the assembled prototype.

Testing to specification

Having assembled a working prototype the next

Howard Associates

| DATA SHEET | L200 Adjustable Voltage Regulator |

MAIN FEATURES

- Adjustable output voltage down to 2.85 V
- Adjustable output current up to 2 A
- Input overload protection (up to 60 V for 10 ms)
- Thermal overload protection
- 5-pin Pentawatt® package
- Low bias current on regulation pin
- Low standby current drain
- Low cost

DESCRIPTION

The L200 is a monolithic integrated circuit voltage regulator which features variable voltage and variable current adjustment. The device is supplied in a 5-pin Pentawatt® package (a TO-3 packaged version is also available to special order). Current limiting, power limiting, thermal shutdown and input over-voltage protection (up to 60 V for 10 ms) make the L200 virtually blow-out proof. The L200 can be used in a wide range of applications wherever high-performance and adjustment of output voltage and current is required.

DIMENSIONS

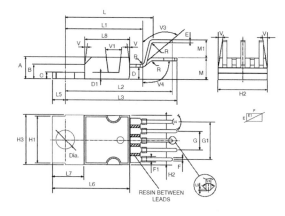

DIM.	mm			inch		
	MIN.	TYP.	MAX.	MIN.	TYP.	MAX.
A			4.8			0.189
C			1.37			0.054
D	2.4		2.8	0.094		0.110
D1	1.2		1.35	0.047		0.053
E	0.35		0.55	0.014		0.022
E1	0.76		1.19	0.030		0.047
F	0.8		1.05	0.031		0.041
F1	1		1.4	0.039		0.055
G	3.2	3.4	3.6	0.126	0.134	0.142
G1	6.6	6.8	7	0.260	0.268	0.276
H2			10.4			0.409
H3	10.05		10.4	0.396		0.409
L	17.55	17.85	18.15	0.691	0.703	0.715
L1	15.55	15.75	15.95	0.612	0.620	0.628
L2	21.2	21.4	21.6	0.831	0.843	0.850
L3	22.3	22.5	22.7	0.878	0.886	0.894
L4			1.29			0.051
L5	2.6		3	0.102		0.118
L6	15.1		15.8	0.594		0.622
L7	6		6.6	0.236		0.260
L9		0.2				0.008
M	4.23	4.5	4.75	0.167	0.177	0.187
M1	3.75	4	4.25	0.148	0.157	0.167
V4			40° (typ.)			

Page 1 of 8

Figure 19.42 Data sheet for the L200 integrated circuit variable voltage regulator. The data sheet provides all of the essential information required to produce a working circuit design, including maximum ratings, mounting details, and thermal characteristics

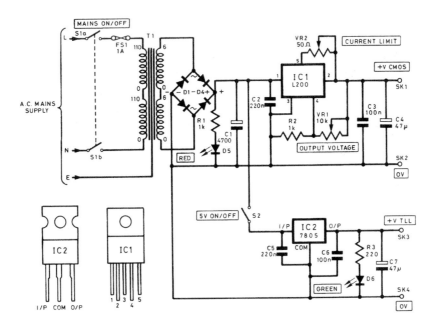

Figure 19.43 Circuit diagram for the variable bench power supply

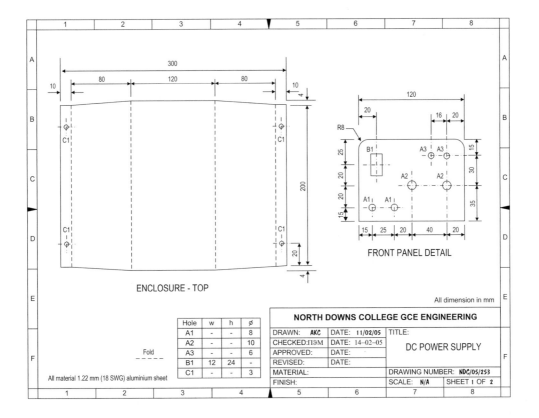

Figure 19.44 Mechanical details for the power supply enclosure

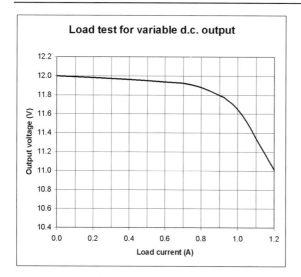

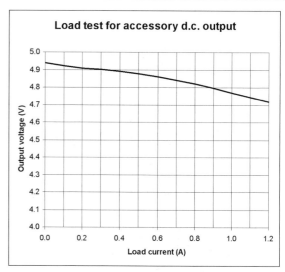

Figure 19.45 Load test graph for the variable d.c. output. Note how the output voltage begins to fall rapidly when the load current exceeds 0.9 A

Figure 19.46 Load test graph for the fixed accessory output. Note how the output voltage falls below 4.75 V at an output current of about 1.1 A

stage was that of measuring its performance in order to ensure that it fully conformed with the requirements set down in the design brief.

Prototype testing was carried out by applying a variable load to the output of the power supply and measuring corresponding values of output voltage and output current. To simplify the analysis, the test data was entered into a spreadsheet and graphs of each load test were generated (see Figs 19.45 and 19.46).

Performance measurement

The measured performance specifications for the power supply (obtained from the load tests) were as follows (figures from the original design specification are shown in square brackets):

Fixed output

- Minimum output voltage = 2.85V [3 V]
- Maximum output voltage = 15.23 V [15 V]
- Maximum output current = 1.2 A [1 A]

Variable output

- Output voltage (no load) = 4.94 V [5 V]
- Output voltage (1 A load) = 4.77 V [4.75 V]
- Output current = 1.2 A [1 A]

These results provided confirmation that the prototype conformed closely with the original design specification.

Evaluation and modification

Having carried detailed testing of the prototype power supply and having verified that the design specification had been met, the next stage was that of finalizing the prototype and carrying out any modifications prior to passing the design for production.

Feedback was obtained from a number of 'test users' who were asked to check that each of the original requirements had been satisfied. Several recommendations were made as a result of this feedback including:

- fitting a carrying handle to the enclosure
- fitting calibrated scales to the voltage and current controls
- relocating the +5 V accessory output switch to the front panel
- mounting both integrated circuit devices on the same heatsink (suitably up-rated) to reduce space and cost
- providing an over-current indicator that will become illuminated when/if the supply is overloaded.

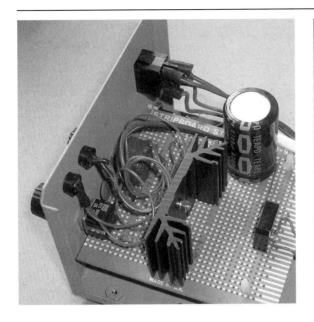

Figure 19.47   Internal arrangement showing the prototype stripboard and front panel wiring

Figure 19.49   Interior of the bench power supply showing transformer and stripboard mounting

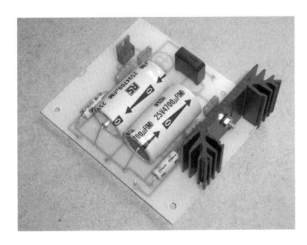

Figure 19.48   PCB layout developed for the final version of the bench power supply

Figure 19.50   External view of the bench power supply showing the front panel controls and enclosure

Practical investigation

Objective

To construct a simple circuit using (a) a stripboard and (b) a printed circuit board.

Components and test equipment

Stripboard and photo-resist copper laminate measuring approximately 6 cm square, 555 timer and 8-pin low-profile DIL socket (two required), resistors of 10 kΩ (four required), 560 kΩ (two

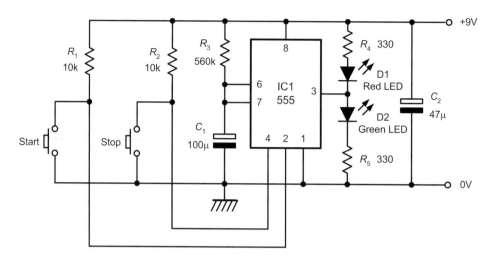

Figure 19.51 The one minute timer circuit used in the Practical investigation

required), 470 Ω (four required) 5% 0.25 W, capacitors of 100 μF and 47 μF 16 V (two of each required) , Red and Green LEDs (two of each required), four normally open (NO) printed circuit board mounting push-button switches, 9V PP3 battery connectors (two required).

Procedure

Following the procedures described earlier in this chapter, design and manufacture the one minute timer circuit using (a) stripboard and (b) printed circuit board construction techniques. Test each circuit when complete.

Table 19.5 See Problem 19.3

Maximum junction temperature:	155°
Thermal resistance of case to surface:	1.4°C/W
Thermal resistance of washer and mounting kit:	1.75°C/W
Maximum total power dissipation:	40W
Maximum ambient temperature:	45°C

Problems

19.1   Identify the type of circuit construction used for the low-pass L-C filter shown in Fig. 19.52.

19.2   Explain, with the aid of a diagram, how a PIC can be programmed without having to remove it from the target system

19.3   Determine the minimum acceptable thermal resistance rating for a heatsink for use with a power transistor operating under the conditions shown in Table 19.6.

Answers to these problems appear on page 376.

Figure 19.52 See Problem 19.1

Appendix 1

Student assignments

The student assignments provided here have been designed to support the topics introduced in this book. They can also be used to satisfy the assessment requirements of a taught course. Please note that the assignments are not exhaustive and may need modification to meet an individual awarding body's requirements as well as the locally available resources. The first 10 assignments satisfy the requirements for the level 2 courses while the remaining 12 are designed to meet the requirements of level 3 courses. Assignments can normally be carried out in three to five hours, including analysis, report writing and evaluation.

Level 2 assignments

Assignment 1 Electronic circuit construction

For each one of five simple electronic circuits shown in Figs A1.1 to A1.5:
(a) Identify and select the components required to build the circuit.
(b) Identify an appropriate method of construction selected from the following list:

- breadboard;
- tag board;
- stripboard;
- printed circuit board;
- wire wrapping.

Assemble and test each circuit according to its circuit diagram. Note that a different construction method must be selected for each circuit.

Assignment 2 Electronic circuit testing

For each of the circuits in Figs A1.1 to A1.5, describe the type and nature of the input and output signals (as appropriate). For each circuit select and use appropriate measuring instruments (e.g. multimeter and oscilloscope) to test and check the operation of each circuit. Write a short report to summarize your findings.

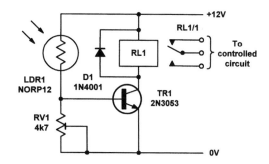

Figure A1.1   See Assignments 1 and 2

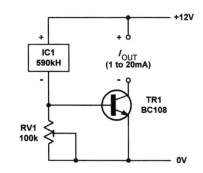

Figure A1.2   See Assignments 1 and 2

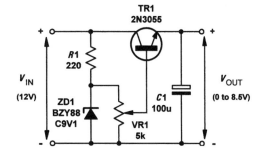

Figure A1.3   See Assignments 1 and 2

Assignment 3 Semiconductor investigation

Prepare a report describing the construction of (a) a junction diode and (b) a bipolar transistor. Describe, in your own words, the principle of

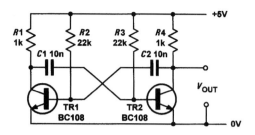

Figure A1.4   See Assignments 1 and 2

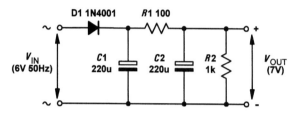

Figure A1.5   See Assignments 1 and 2

operation of each device. With the aid of a simple circuit diagram, describe a typical application for each type of semiconductor.

Assignment 4 Basic logic functions

Write a report identifying the types and symbols (both BS3939 and MIL/ANSI) used for all basic logic gates (AND, OR, NOT, NAND and NOR). Include in your report a description of the operation of each logic gate together with a truth table.

Assignment 5 Applications of logic circuits

With the aid of labelled diagrams, describe TWO applications of logic gates. One application should be based on combinational logic while the other should use sequential logic.

Assignment 6 Electronic measuring instruments

Write a report describing the operation and use of (a) a multimeter and (b) an oscilloscope. Illustrate your report with records of measurements carried out on three common electronic components and two simple electronic circuits.

Assignment 7 Microprocessors

Investigate TWO common types of microprocessor. Produce an A4-size data sheet for each device giving a brief specification of the chip, package and pin-connecting details, clock frequencies, details of the internal registers, descriptions of bus signals (including the number of address and address lines), etc.

Assignment 8 Monostable timer

Design, construct and test a variable timer based on a 555 device connected in monostable mode. The timer is to be adjustable over the range 10s to 90s and is to produce an output for driving a low-voltage relay.

Assignment 9 Square wave generator

Design, construct and test a 5 V 1 kHz square wave generator based on a 555 timer connected in astable mode.

Assignment 10 A dipole aerial

Design, construct and test a half-wave dipole aerial for DAB radio reception. Select a suitable feeder to connect the aerial to a DAB receiver and investigate the directional properties of the aerial. Compare these with what you would expect from radio theory.

Level 3 assignments

Assignment 11 Power supply investigation

With the aid of an electrical specification and operating manual for a typical low-voltage d.c. power supply, write a report explaining the char-acteristics of the unit. Also explain the meaning of each of the unit's specifications. Carry out a simple load test on the supply, plot a graph to illustrate your results and comment on your findings.

Assignment 12 Amplifier circuit investigation

Write a report that describes and explains one small-signal Class-A discrete amplifier circuit, one Class-B power amplifier circuit, and one amplifier circuit based on an operational amplifier. The report should identify and give typical specifications for each type of amplifier.

Carry out a simple gain and frequency response test on one of the amplifier circuits, plot a graph to illustrate your results and comment on your findings.

Assignment 13 Small-signal amplifier design and construction

Design, construct and test a single-stage transistor amplifier. Write a report describing the model used for the small-signal amplifier and include a detailed comparison of the predicted characteristics compared with the measured performance of the stage.

Assignment 14 Electronic counter investigation

Prepare a report describing the characteristics of J-K bistable elements when compared with D-type and R-S bistables. Using the data obtained, design, construct and test a four-stage binary counter. Modify the design using a standard logic gate to produce a decade counter. Include in your report timing diagrams for both the binary and decade counters.

Assignment 15 Pulse generator investigation

Prepare a report describing one oscillator circuit, one bistable circuit and one monostable circuit. The report should include a description of one industrial application for each of the circuits together with sample calculations of frequency and pulse rate for the two oscillator circuits, and pulse width for the monostable circuit.

Assignment 16 Performance testing

Carry out performance tests on two different amplifier circuits, two waveform generators and two digital circuits. Compare the measured performance of each circuit with the manufacturer's specification and present your findings in a written report. The report should include details of the calibration and operation of each test instrument in accordance with the manufacturer's handbooks as well as evidence of the adoption of safe working practice.

Assignment 17 Microprocessor clock

Design, construct and test (using an oscilloscope) a microprocessor clock based on two Schmitt inverting logic gates and a quartz crystal. The clock is to produce a square wave output at a frequency of between 1 MHz and 6 MHz (depending upon the quartz crystal used).

Assignment 18 Assembly language programming

Use a simple 8-bit microprocessor system to develop a simple assembly language program that will read the state of a set of eight input switches connected to an input port and illuminate a bank of eight light emitting diodes connected to an output port. The state of the input switches is to be indicated by the light emitting diodes.

Assignment 19 Variable pulse generator

Design, construct and test a variable pulse generator (along the lines of that shown on page 224). The pulse generator is to provide an output from 1 Hz to 10 kHz with a pulse width variable from 50 μs to 500 ms.

Assignment 20 AM radio tuner

Design, construct and test a simple AM radio tuner based on a diode demodulator and a variable tuned circuit. The radio tuner is to cover the medium wave band from 550 kHz to 1.5 MHz and its output is to be connected to an external audio amplifier. Investigate the performance of the tuner and suggest ways in which it could be improved.

Assignment 21 Analogue multimeter

Design, construct and test a simple analogue multimeter based on a 1 mA moving coil meter. The multimeter is to have three voltage ranges (10 V, 50 V and 100 V), two current ranges (10 mA and 50 mA) and one ohms range.

Assignment 22 PIC microcontroller

Design, construct and test a simple PIC microcontroller-based intruder alarm. The alarm is to be capable of identifying an intrusion into any one of four 'zones' using an LED panel and is to produce a master alarm signal using a piezoelectric transducer whenever any one of the alarm inputs is triggered. The alarm is to be 'set' and 'reset' using two further switch inputs.

Appendix 2

Revision problems

These problems provide you with a means of checking your understanding prior to an end-of-course assessment or formal examination. If you have difficulty with any of the questions you should refer to the page numbers indicated.

1. A 120 kΩ resistor is connected to a 6 V battery. Determine the current flowing. [Page 6]
2. A current of 45 mA flows in a resistor of 2.7 kΩ. Determine the voltage dropped across the resistor. [Page 6]
3. A 24 V d.c. supply delivers a current of 1.5 A. Determine the power supplied. [Page 8]
4. A 27 Ω resistor is rated at 3 W. Determine the maximum current that can safely be applied to the resistor. [Page 8]
5. A load resistor is required to dissipate a power of 50 W from a 12 V supply. Determine the value of resistance required. [Page 8]
6. An electrical conductor has a resistance of 0.05 Ω per metre. Determine the power wasted in a 175 m length of this conductor when a current of 8 A is flowing in it. [Page 8]
7. Figure A2.1 shows a node in a circuit. Determine the value of I_x. [Page 49]
8. Figure A2.2 shows part of a circuit. Determine the value of V_x. [Page 50]
9. A capacitor of 200 μF is charged to a potential of 50 V. Determine the amount of charge stored. [Page 33]
10. A sinusoidal a.c. supply has a frequency of 400 Hz and an r.m.s. value of 120 V. Determine

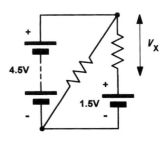

Figure A2.2   See Question 8

the periodic time and peak value of the supply. [Page 70]
11. Four complete cycles of a waveform occur in a time interval of 20 ms. Determine the frequency of the waveform. [Page 70]
12. Determine the periodic time, frequency and amplitude of each of the waveforms shown in Fig. A2.3. [Page 71]
13. Determine the effective resistance of each circuit shown in Fig. A2.4. [Page 26]
14. Determine the effective capacitance of each circuit shown in Fig. A2.5. [Page 36]
15. Determine the effective inductance of each circuit shown in Fig. A2.6. [Page 42]
16. A quantity of 100 nF capacitors is available, each rated at 100 V working. Determine how several of these capacitors can be connected to produce an equivalent capacitance of: (a) 50 nF rated at 200 V; (b) 250 nF rated at 100 V; and (c) 300 nF rated at 100 V. [Page 36]
17. Two 60 mH inductors and two 5mH inductors are available, each rated at 1 A. Determine how some or all of these can be connected to produce an equivalent inductance of: (a) 30 mH rated at 2 A; (b) 40 mH rated at 1A; and (c) 125 mH rated at 1 A. [Page 42]
18. Determine the resistance looking into the network shown in Fig. A2.7, (a) with C and D open-circuit and (b) with C and D shorted together. [Page 26]

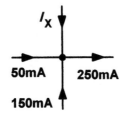

Figure A2.1   See Question 7

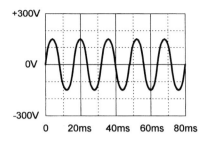

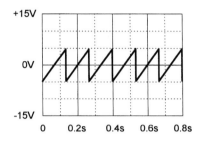

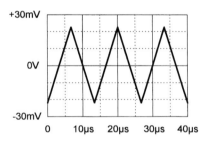

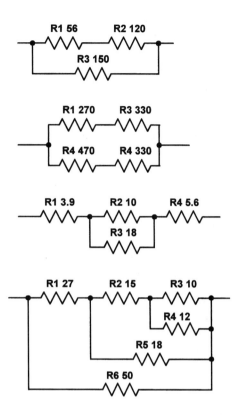

Figure A2.4 See Question 13

Figure A2.3 See Question 12

19. Determine the current flowing in each resistor and voltage dropped across each resistor in Fig. A2.8. [Page 49]
20. Determine the current flowing in the volt-meter movement shown in Fig. A2.9. [Page 54]
21. Assuming that the capacitor shown in Fig. A2.10 is initially fully discharged (by switching to position B), determine the current in $R1$ at the instant that S1 is switched to position A. Also determine the capacitor voltage 1 minute after operating the switch. [Page 58]
22. Determine the time taken for the output voltage in Fig. A2.11 to reach 4 V after the arrival of the pulse shown (assume that the capacitor is initially uncharged). [Page 58]

23. In Fig. A2.12 determine the current supplied to the inductor 100 ms after pressing the 'start' button. [Page 63]
24. Determine the reactance at 2 kHz of (a) a 60 mH inductor and (b) a 47 nF capacitor. [Page 72]
25. A 50 μF capacitor is connected to a 12 V, 50 Hz a.c. supply. Determine the current flowing. [Page 72]
26. An inductor of 2 H is connected to a 12 V, 50 Hz a.c. supply. If the inductor has a winding resistance of 40 Ω, determine the current flowing and the phase angle between the supply voltage and supply current. [Page 74]
27. An inductor of 100 μH is connected in series with a variable capacitor. If the capacitor is variable over the range 50 pF to 500 pF, determine the maximum and minimum values of resonant frequency for the circuit. [Page 77]

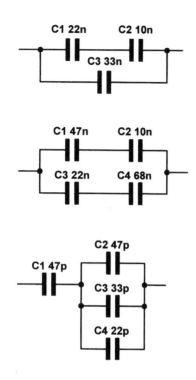

Figure A2.5   See Question 14

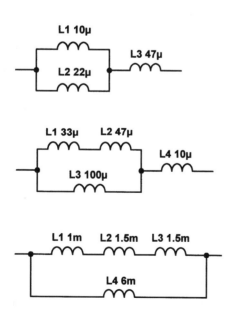

Figure A2.6   See Question 15

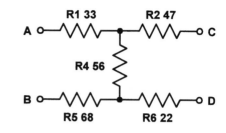

Figure A2.7   See Question 18

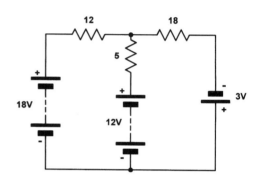

Figure A2.8   See Question 19

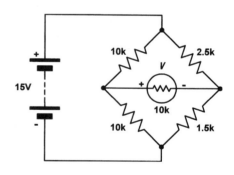

Figure A2.9   See Question 20

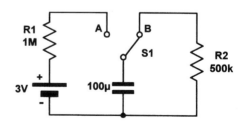

Figure A2.10   See Question 21

28. An audio amplifier delivers an output power of 40 W r.m.s. to an 8 Ω resistive load. What r.m.s. voltage will appear across the load terminals? [Page 8]

29. A transformer has 400 primary turns and 60 secondary turns. The primary is connected to a 220 V a.c. supply and the secondary is connected to a load resistance of 20 Ω. Assuming that the transformer is perfect, determine: (a) the secondary voltage; (b) the secondary current; and (c) the primary current. [Page 80]

30. Figure A2.13 shows the characteristic of a diode. Determine the resistance of the diode when (a) $V_F = 2$ V and (b) $I_F = 9$ mA. [Page 90]

31. A transistor operates with a collector current of 25 mA and a base current of 200 μA. Determine: (a) the value of emitter current; (b) the value of common-emitter current gain; and (c) the new value of collector current if the base current increases by 50%. [Page 96]

32. A zener diode rated at 5.6 V is connected to a 12 V d.c. supply via a fixed series resistor of 56 Ω. Determine the current flowing in the resistor, the power dissipated in the resistor and the power dissipated in the zener diode. [Page 122]

33. An amplifier has identical input and output resistances and provides a voltage gain of 26 dB. Determine the output voltage produced if an input of 50 mV is applied. [Page 388]

34. Figure A2.14 shows the frequency response of an amplifier. Determine the mid-band voltage gain and the upper and lower cut-off frequencies. [Page 135]

35. Figure A2.15 shows the frequency response of an amplifier. Determine the bandwidth of the amplifier. [Page 136]

36. The transfer characteristic of a transistor is shown in Fig. A2.16. Determine (a) the static value of common-emitter current gain at $I_C = 50$ mA and (b) the dynamic (small-signal) value of common-emitter current gain at $I_C = 50$ mA. [Page 145]

37. The output characteristics of a bipolar transistor are shown in Fig. A2.17. If the transistor operates with $V_{CC} = 15$ V, $R_L = 500$ and $I_B = 40$ A determine (a) the quiescent value of collector-emitter voltage; (b) the quiescent value of collector current; (c) the peak-peak output voltage produced by a base input current of 40 μA. [Page 146]

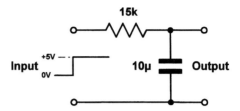

Figure A2.11   See Question 22

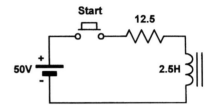

Figure A2.12   See Question 23

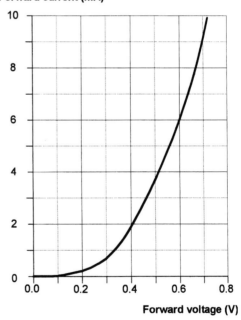

Figure A2.13   See Question 30

38. The output characteristics of a field effect transistor are shown in Fig. A2.18. If the transistor operates with $V_{DD} = 18$ V, $R_L = 3$ kΩ and $V_{GS} = -1.5$V determine: (a) the quiescent value of drain-source voltage; (b) the quiescent

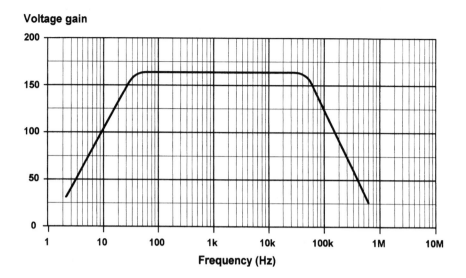

Figure A2.14   See Question 34

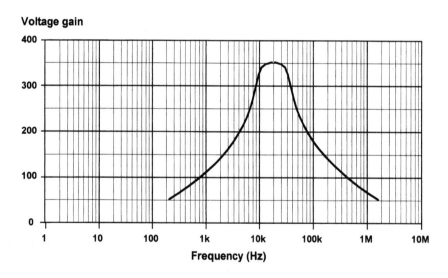

Figure A2.15   See Question 35

value of drain current; (c) the peak-peak output voltage produced by a gate input voltage of 1V pk-pk; (d) the voltage gain of the stage. [Page 146]

39. Figure A2.19 shows the circuit of a common-emitter amplifier stage. Determine the values of I_B, I_C, I_E and the voltage at the emitter. [Page 146]

40. A transistor having $h_{ie} = 2.5$ kΩ and $h_{fe} = 220$ is used in a common-emitter amplifier stage with $R_L = 3.3$ kΩ. Assuming that h_{oe} and h_{re} are both negligible, determine the voltage gain of the stage. [Page 143]

41. An astable multivibrator is based on coupling capacitors $C1 = C2 = 10$ nF and timing resistors $R1 = 10$ kΩ and $R2 = 4$ kΩ. Determine the frequency of the output signal. [Page 175]

42. A sine wave oscillator is based on a Wien bridge with $R = 5$ kΩ and $C = 15$ nF. Determine the frequency of the output signal. [Page 173]

43. The frequency response characteristic of an operational amplifier is shown in Fig. A2.20. If

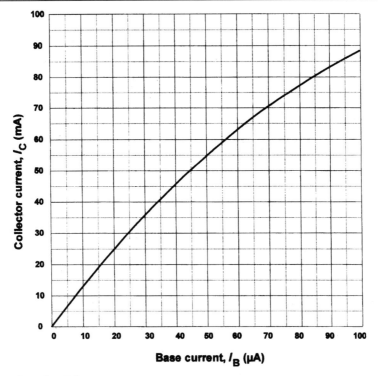

Figure A2.16   See Question 36

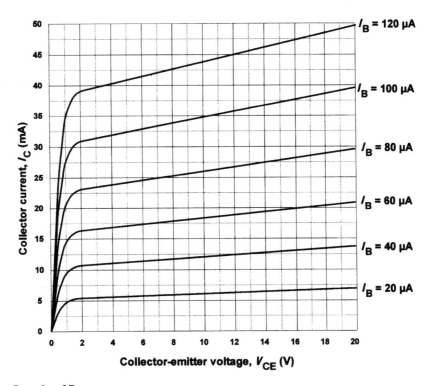

Figure A2.17   See Question 37

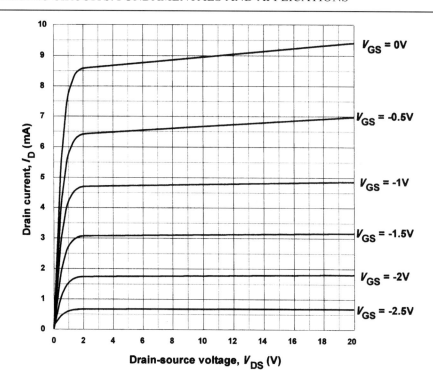

Figure A2.18 See Question 38

the device is configured for a closed-loop gain of 200, determine the resulting bandwidth. [Page 161]

44. Redraw Fig. A2.21 using American (MIL/ANSI) symbols. [Page 186]

45. Draw the truth table for the logic gate arrangement shown in Fig. A2.22. [Page 187]

46. Redraw Fig. A2.23 using BS symbols. [Page 186]

47. What single logic gate can be used to replace the logic circuit shown in Fig. A2.24? [Page 187]

48. What single logic gate can be used to replace the logic circuit shown in Fig. A2.25? [Page 187]

49. Devise arrangements of logic gates that will produce the truth tables shown in Fig. A2.26. Use the minimum number of logic gates in each case. [Page 187]

50. A 1 kHz square wave clock waveform is applied to the circuit shown in Fig. A2.27. Sketch the output waveform against a labelled time axis. [Page 188]

51. (a) Convert 7B hexadecimal to binary.
(b) Convert 11000011 binary to hexadecimal. [Page 201]

52. What is the largest value, expressed (a) in decimal, and (b) in binary, that can appear at any one time on a 16-bit data bus. [Page 201]

53. Sketch a diagram showing the basic arrangement of a microprocessor system. Label your drawing clearly. [Page 200]

54. (a) Explain the function of a microprocessor clock.
(b) Explain why a quartz crystal is used to determine the frequency of a microprocessor clock. [Page 206]

55. Sketch the circuit diagram of a typical microprocessor clock. Label your drawing clearly. [Page 206]

56. Explain, briefly, how a microprocessor fetches and executes instructions. Illustrate your answer with a timing diagram showing at least one fetch-execute cycle. [Page 207]

57. Sketch the circuit diagram of a monostable timer based on a 555 device. Explain, briefly,

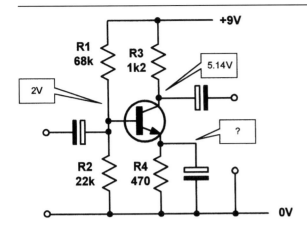

Figure A2.19   See Question 39

how the circuit operates. [Page 219]

58. A 555 timer is connected in monostable mode. If the values of the timing components used are C = 100 nF and R = 10 kΩ, determine the monostable pulse time. [Page 219]

59. Determine the frequency of a radio signal that has a wavelength of 1500 m. [Page 229]

60. Determine the wavelength of a radio signal that has a frequency of 40 MHz. [Page 229]

61. A superhet medium wave broadcast receiver with an intermediate frequency of 470 kHz is to cover the frequency range 560 kHz to 1.58 MHz. Over what frequency range should the local oscillator be tuned? [Page 232]

62. Explain, with the aid of waveforms, the operation of a simple AM demodulator. [Page 235]

63. Explain, with the aid of a labelled sketch, how the voltage and current are distributed in a half-wave dipole aerial. [Page 237]

64. Determine the length of a half-wave dipole for use at a frequency of 70 MHz. [Page 237]

65. Sketch the block schematic of a simple TRF radio receiver. Briefly explain the function of each stage. [Page 233]

66. A moving coil meter has a full-scale deflection current of 1 mA and a coil resistance of 400 Ω. Determine the value of the multiplier resistor if the meter is to be used as a voltmeter reading 0 to 15 V. [Page 245]

67. A moving coil meter has a full-scale deflection current of 5 mA and a coil resistance of 120 Ω. Determine the value of shunt resistor if the

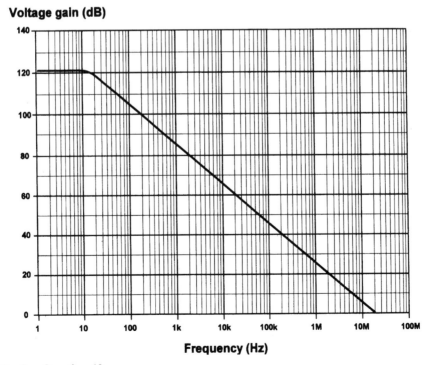

Figure A2.20   See Question 43

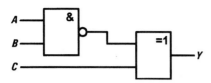

Figure A2.21   See Question 44

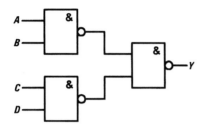

Figure A2.22   See Question 45

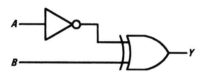

Figure A2.23   See Question 46

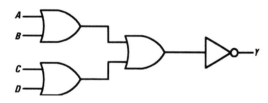

Figure A2.24   See Question 47

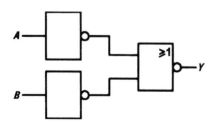

Figure A2.25   See Question 48

meter is to be used as an ammeter reading 0 to 20 mA. [Page 246]

68. Explain the term 'ohms-per-volt' as applied to an analogue voltmeter. [Page 248]

69. Sketch the circuit of a simple ohmmeter based on a moving coil meter. [Page 246]

70. Identify each of the forms of distortion shown in Fig. A2.28. [Page 267]

71. Give TWO reasons why a multimeter is generally considered to be unsuitable for checking the operation of logic circuits. [Page 282]

72. Explain how a logic probe can be used to check the operation of a combinational logic circuit. [Page 284]

73. Briefly explain the 'half-split' method of fault finding. [Page 274]

74. Define the terms 'sensor' and 'transducer'. Give one example of each. [Page 288]

75. Identify one type of sensor for use in each of the following applications:

(a) measuring the surface temperature of an integrated circuit package; (b) determining the pressure exerted on the walls of a gas storage vessel; (c) detecting the minimum level of fuel in a tank; (d) determining the flow rate of a liquid in a pipe. [Page 290]

76. Sketch interface circuits to show how a TTL logic signal can be used to control (a) a 24 V d.c. relay and (b) an a.c. mains operated motor. Identify all components used in each circuit. [Page 299]

77. Sketch the block schematic diagram of an electronic control system that incorporates feedback. Label your drawing clearly. [Page 287]

78. Sketch the block schematic diagram of an electronic instrumentation system that incorporates analogue input and digital display facilities. Label your drawing clearly. [Page 287]

79. Explain, with the aid of a circuit diagram, how a threshold light-level detector can be made with the aid of a comparator. [Page 296]

80. Explain the advantages of using a PIC microcontroller in simple control applications when compared with (a) logic gate arrangements and (b) conventional microprocessors. [Page 313]

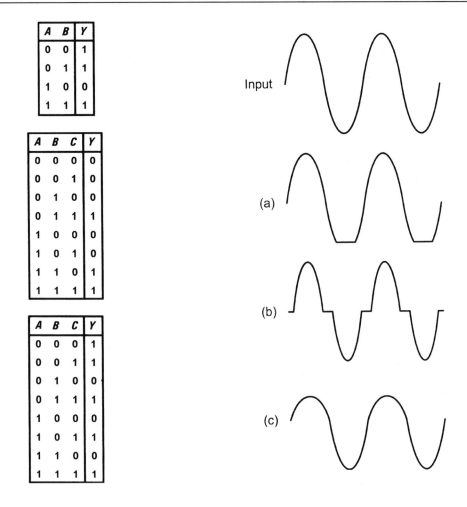

A	B	Y
0	0	1
0	1	1
1	0	0
1	1	1

A	B	C	Y
0	0	0	0
0	0	1	0
0	1	0	0
0	1	1	1
1	0	0	0
1	0	1	0
1	1	0	1
1	1	1	1

A	B	C	Y
0	0	0	1
0	0	1	1
0	1	0	0
0	1	1	1
1	0	0	0
1	0	1	1
1	1	0	0
1	1	1	1

Figure A2.26   See Question 49

Figure A2.28   See Question 70

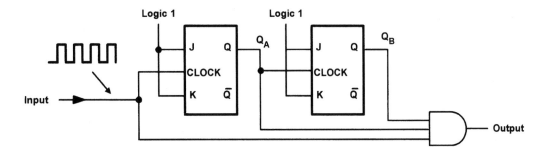

Figure A2.27   See Question 50

Appendix 3

Answers to problems

Chapter 1

1.1	Coulombs, Joules, Hertz
1.2	3.6 MJ
1.3	0.52 radian
1.4	11.46°
1.5	39.57 kΩ
1.6	0.68 H
1.7	2.45 nF
1.8	0.19 mA
1.9	4.75×10^{-4} V
1.10	16.5×10^{6} Ω
1.11	$4.8 \times 10^{6}, 7.2 \times 10^{3}, 4 \times 10^{3}, 0.5 \times 10^{-3}$
1.12	silver
1.13	33.3 mA
1.14	6.72 V
1.15	3.3 kΩ
1.16	150
1.17	0.436 Ω
1.18	0.029 W
1.19	0.675 W
1.20	57.7 mA
1.21	0.625×10^{6} V/m
1.22	12 A
1.23	6 μWb
1.24	1.103×10^{5}
1.25	0.5 A
1.26	1P 4W
1.27	1 kΩ, 3 kΩ, 10 kΩ, 30 kΩ

Chapter 2

2.1	60 Ω, 3.75 W, wirewound
2.2	270 kΩ 5%, 10 Ω 10%, 6.8 MΩ 5%, 0.39 Ω 5%, 2.2 kΩ 2%
2.3	44.65 Ω to 49.35 Ω
2.4	27 Ω and 33 Ω in series, 27 Ω and 33 Ω in parallel, 56 Ω and 68 Ω in series, 27 Ω, 33 Ω and 56 Ω in parallel, 27 Ω, 33 Ω and 68 Ω in series
2.5	66.67 Ω
2.6	10 Ω
2.7	102 Ω, 78.5 Ω
2.8	407.2 Ω
2.9	98.7 kΩ
2.10	3.21×10^{-4}
2.11	3.3 μF and 4.7 μF in parallel, 1 μF and 10 μF in parallel, 1 μF, 3.3 μF, 4.7 μF and 10 μF in parallel, 1 μF and 10 μF in series, 3.3 μF and 4.7 μF in series
2.12	60 pF, 360 pF
2.13	50 pF
2.14	20.79 mC
2.15	1.98 nF
2.16	69.4 μF
2.17	0.313 V
2.18	0.136 H
2.19	0.48 J
2.20	10 mH, 22 mH and 60 mH in parallel, 10 mH and 22 mH in parallel, 10 mH and 22 mH in series, 10 mH and 60 mH in series, 10mH 60mH and 100 mH in series

Chapter 3

3.1	275 mA
3.2	200 Ω
3.3	1.5 A away from the junction, 215 mA away from the junction
3.4	1.856 V, 6.6 V
3.5	0.1884 A, 0.1608 A, 0.0276 A, 5.09 V, 0.91V, 2.41V
3.6	1.8 V, 10.2 V
3.7	0.5 A, 1.5 A
3.8	1 V, 2 V, 3 V, 4 V, 5 V
3.9	40.6 ms
3.10	3.54 s
3.11	112.1 μF
3.12	0.128 s
3.13	2.625 V, 5 Ω
3.14	21 V, 7 Ω, 3 A, 7 Ω
3.15	50 mA, 10 V
3.16	50 V, 10 V

Chapter 4

4.1	4 ms, 35.35 V
4.2	59.88 Hz, 3394 V

4.3	50 Hz 30 V pk-pk, 15 Hz 10 V pk-pk, 150 kHz 0.1 Vpk-pk
4.4	19 V, −11.8 V
4.5	10.6 V
4.6	36.19 kΩ, 144.76 Ω
4.7	3.54 mA
4.8	10.362 Ω, 1.45 kΩ
4.9	4.71 V
4.10	592.5 Ω, 0.186 A
4.11	0.55, 0.487 A
4.12	157 nF
4.14	1.77 MHz to 7.58 MHz
4.15	7.5 mA, 2.71 V
4.16	281 kHz, 41.5, 6.77 kHz
4.20	18 V
4.21	245 V
4.22	4 turns per volt, 48 turns, 1.7 W, 0.36 A

Chapter 5

5.1	silicon (forward threshold appx. 0.6 V)
5.2	41 Ω, 150 Ω
5.4	9.1 V zener diode
5.5	250 Ω
5.6	germanium low-power high-frequency, silicon low-power low-frequency, silicon high-power low-frequency, silicon low-power high-frequency
5.7	2.625 A, 20
5.8	5 mA, 19.6
5.9	16.7
5.10	BC108
5.11	8 μA, 1.1 mA
5.12	47 mA, 94, 75
5.13	12.5mA, 12V, 60 μA
5.14	16 mA
5.16	dual-in-line, single-in-line, quad-in-line, large-scale integration
5.17	100 V, 500 mW, 50 nA, 0.7 V
5.18	signal or power diode, photo diode, zener diode, thyristor (or SCR)
5.19	NPN BJT, PNP Darlington BJT, P-channel enhancement MOSFET, N-channel JFET
5.20	PNP silicon general-purpose transistor, −40 V, −500 mA, −0.25 V, 625 mW, TO-92
5.21	The maximum power dissipation is not exceeded
5.22	N-channel JFET RF amplifier, 25 V, 50 mA, 2 nA, 2 mS, 8 pF, 350 mW, TO-92
5.23	9.6 mA, 5 V, 0 V

Chapter 6

6.1	80 mV
6.2	5 mV
6.3	200 Ω
6.4	12.74 V, 9.1 V, 8.4 V
6.5	36.4 mA, 0.33 W
6.6	0 V, 12.04 V, 0 V
6.7	14.35 V
6.8	0.5 Ω, 8.3 V
6.9	1%, 15.15 V
6.10	step-up (or 'boost'), pins 5 and 3, pin 7

Chapter 7

7.1	40, 160, 6400, 100 Ω
7.2	2 V
7.3	56, 560 kHz, 15 Hz
7.4	18.5
7.5	0.0144
7.6	2.25 V
7.7	13 μA, 4.487 mA, 3.39 V, 2.7 V, 4.51 V
7.8	5 V, 7 mA, 8.5 V
7.9	0.6 V, 18 μA, 3.9 V, 2 mA, 0 V, 4.5 V, 9 V
7.10	12.2 V, 6.1 mA, 5.5
7.11	2,500
7.12	40, 0.27 W, 133 mA
7.13	37%, 338 mW

Chapter 8

8.3	200
8.7	−7.5 V
8.8	82 dB

Chapter 9

9.1	4.44, 40
9.2	6.49 kΩ
9.3	18 kΩ
9.4	5.63 V pk-pk
9.5	23.32 ms, 42.9 Hz
9.6	14.3 kΩ, 42.9 kΩ
9.8	1.448 ms, 690 Hz
9.10	15.9 Hz
9.13	$R3 = 2R4$ (minimum gain = 2)

Chapter 10

10.9	Low-power Schottky (LS) TTL, 27th Month of 1998
10.10	0.6 V

Chapter 11

11.1   00111010
11.2   C2
11.3   (a) 111111, (b) 3F
11.4   (c)
11.5   1,048,576
11.6   1,024
11.7   −32,768
11.8   (a) Input port, FEH; output port, FFH
        (b) 01010000
11.10 See page 206
11.12 See page 200

Chapter 12

12.5   (a) *VR2*
        (b) *S2*
        (c) *RA*
        (d) *VR1*
        (e) *S1*
        (f) *R2*
        (g) *VR3*
        (h) *R5*
        (i) *C10*
        (j) *C5* and *R3*
12.6   50 Ω

Chapter 13

13.1   1.58 m
13.2   23.1 MHz
13.3   4 m
13.4   1.02 MHz to 2.07 MHz
13.5   10.7 MHz, 170 kHz, 63
13.6   (a) *L1/C2*
        (b) *R1/R2*
        (c) *L2*
        (d) *R4/C5*
        (e) *VR1*
        (f) *C6*
        (g) *C7*
        (h) *C1*
13.7   3m
13.8   24 W

Chapter 14

14.1   19.6 kΩ
14.2   11.11 Ω
14.3   16.67 kΩ
14.4   100 kΩ
14.5   (a) 19.99 V
        (b) 10 mV
14.6   25 kΩ
14.7   60 mA
14.8   35.3 μA
14.9   (a) 3.33 ms, 4 V
        (b) 125 ns, 150 mV
14.10 (a) 7.8 μs
        (b) 3.4 μs
        (c) 4.4 μs
        (d) 1.5 μs
        (e) 1 μs
        (f) 5 V

Chapter 15

15.1   Fault 1: *R2* open-circuit
        Fault 2: TR2 collector-emitter short-circuit
        Fault 3: *R3* open-circuit
        Fault 4: TR1 base-collector short-circuit
        Fault 5: *C2* short-circuit
        Fault 6: *R5* open-circuit

Chapter 16

16.1   1. Volume, float gauge, voltage or current
        2. Flow, flow sensor (rotating vane), voltage
        or pulse repetition frequency
        3. Semiconductor temperature sensor,
        threshold comparator, voltage (high or low)

Chapter 18

18.1   Analogue-to-digital converter;
        digital-to-analogue converter;
        bus interface device (RS-232, USB etc.);
        LED or LCD display driver

Chapter 19

19.1   point-to-point wiring
19.3   17.85 °C/W

Appendix 4

Semiconductor pin connections

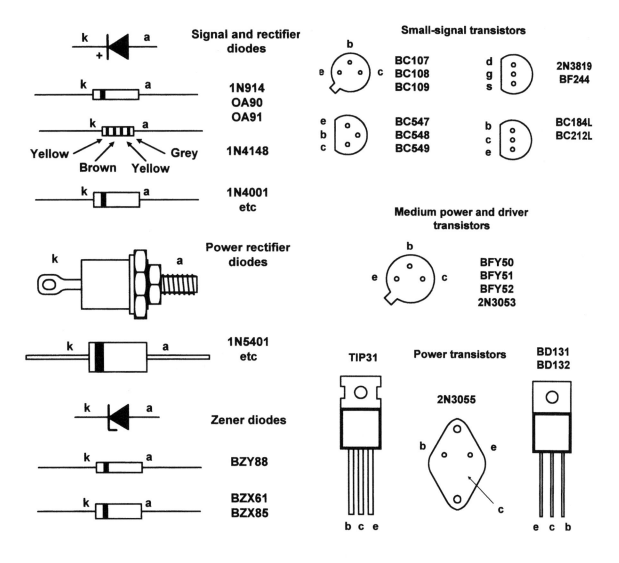

Figure A4.1   Diodes

Figure A4.2   Transistors

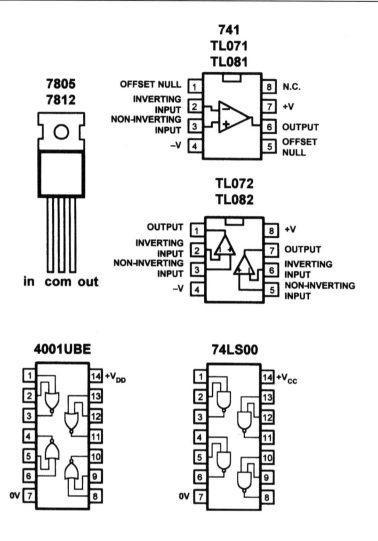

7805
7812

in com out

741
TL071
TL081

OFFSET NULL	1		8	N.C.
INVERTING INPUT	2		7	+V
NON-INVERTING INPUT	3		6	OUTPUT
–V	4		5	OFFSET NULL

TL072
TL082

OUTPUT	1		8	+V
INVERTING INPUT	2		7	OUTPUT
NON-INVERTING INPUT	3		6	INVERTING INPUT
–V	4		5	NON-INVERTING INPUT

4001UBE

74LS00

Figure A4.3   Integrated circuits

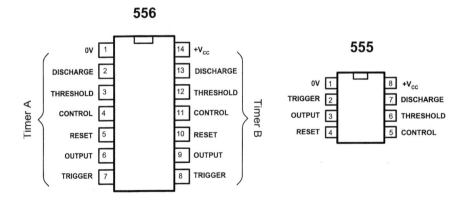

556

Timer A	0V	1		14	+V_cc	Timer B
	DISCHARGE	2		13	DISCHARGE	
	THRESHOLD	3		12	THRESHOLD	
	CONTROL	4		11	CONTROL	
	RESET	5		10	RESET	
	OUTPUT	6		9	OUTPUT	
	TRIGGER	7		8	TRIGGER	

555

0V	1		8	+V_cc
TRIGGER	2		7	DISCHARGE
OUTPUT	3		6	THRESHOLD
RESET	4		5	CONTROL

Figure A4.4   Integrated circuit timers

Appendix 5

1N4148 data sheet

FAIRCHILD
SEMICONDUCTOR®

1N/FDLL 914/A/B / 916/A/B / 4148 / 4448

DO-35

LL-34

THE PLACEMENT OF THE EXPANSION GAP
HAS NO RELATIONSHIP TO THE LOCATION
OF THE CATHODE TERMINAL

COLOR BAND MARKING		
DEVICE	**1ST BAND**	**2ND BAND**
FDLL914	BLACK	BROWN
FDLL914A	BLACK	GRAY
FDLL914B	BROWN	BLACK
FDLL916	BLACK	RED
FDLL916A	BLACK	WHITE
FDLL916B	BROWN	BROWN
FDLL4148	BLACK	BROWN
FDLL4448	BROWN	BLACK

Small Signal Diode

Absolute Maximum Ratings* $T_A = 25°C$ unless otherwise noted

Symbol	Parameter	Value	Units
V_{RRM}	Maximum Repetitive Reverse Voltage	100	V
$I_{F(AV)}$	Average Rectified Forward Current	200	mA
I_{FSM}	Non-repetitive Peak Forward Surge Current Pulse Width = 1.0 second Pulse Width = 1.0 microsecond	 1.0 4.0	 A A
T_{stg}	Storage Temperature Range	-65 to +200	°C
T_J	Operating Junction Temperature	175	°C

*These ratings are limiting values above which the serviceability of any semiconductor device may be impaired.

NOTES:
1) These ratings are based on a maximum junction temperature of 200 degrees C.
2) These are steady state limits. The factory should be consulted on applications involving pulsed or low duty cycle operations.

Thermal Characteristics

Symbol	Characteristic	Max	Units
		1N/FDLL 914/A/B / 4148 / 4448	
P_D	Power Dissipation	500	mW
$R_{\theta JA}$	Thermal Resistance, Junction to Ambient	300	°C/W

Small Signal Diode
(continued)

1N/FDLL 914/A/B / 916/A/B / 4148 / 4448

Electrical Characteristics $T_A = 25°C$ unless otherwise noted

Symbol	Parameter		Test Conditions	Min	Max	Units
V_R	Breakdown Voltage		$I_R = 100\ \mu A$	100		V
			$I_R = 5.0\ \mu A$	75		V
V_F	Forward Voltage	1N914B/4448	$I_F = 5.0$ mA	620	720	mV
		1N916B	$I_F = 5.0$ mA	630	730	mV
		1N914/916/4148	$I_F = 10$ mA		1.0	V
		1N914A/916A	$I_F = 20$ mA		1.0	V
		1N916B	$I_F = 20$ mA		1.0	V
		1N914B/4448	$I_F = 100$ mA		1.0	V
I_R	Reverse Current		$V_R = 20$ V		25	nA
			$V_R = 20$ V, $T_A = 150°C$		50	μA
			$V_R = 75$ V		5.0	μA
C_T	Total Capacitance					
		1N916A/B/4448	$V_R = 0$, $f = 1.0$ MHz		2.0	pF
		1N914A/B/4148	$V_R = 0$, $f = 1.0$ MHz		4.0	pF
t_{rr}	Reverse Recovery Time		$I_F = 10$ mA, $V_R = 6.0$ V (60mA), $I_{rr} = 1.0$ mA, $R_L = 100\Omega$		4.0	ns

Typical Characteristics

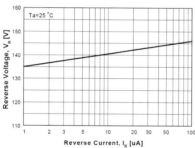

Figure 1. Reverse Voltage vs Reverse Current BV - 1.0 to 100 uA

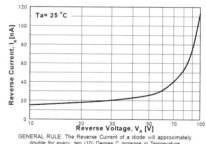

GENERAL RULE: The Reverse Current of a diode will approximately double for every ten (10) Degree C increase in Temperature

Figure 2. Reverse Current vs Reverse Voltage IR - 10 to 100 V

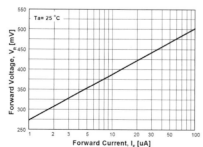

Figure 3. Forward Voltage vs Forward Current VF - 1 to 100 uA

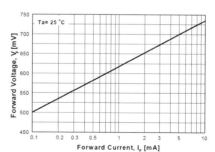

Figure 4. Forward Voltage vs Forward Current VF - 0.1 to 10 mA

Small Signal Diode
(continued)

1N/FDLL 914/A/B / 916/A/B / 4148 / 4448

Typical Characteristics (continued)

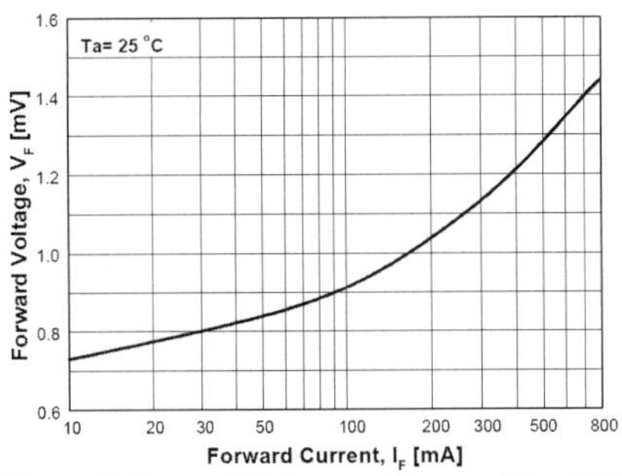

Figure 5. Forward Voltage vs Forward Current
VF - 10 to 800 mA

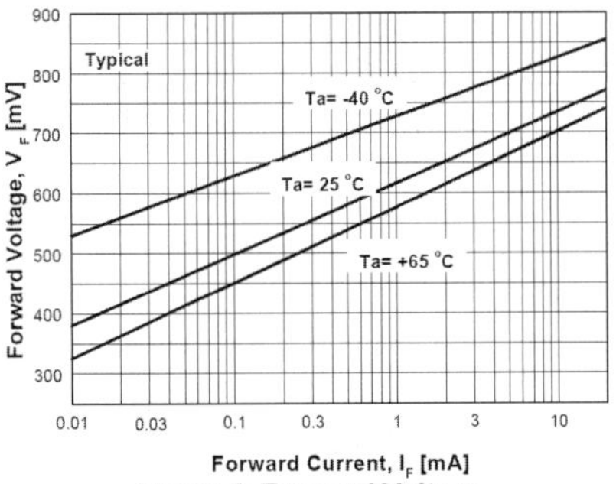

Figure 6. Forward Voltage
vs Ambient Temperature
VF - 0.01 - 20 mA (-40 to +65 Deg C)

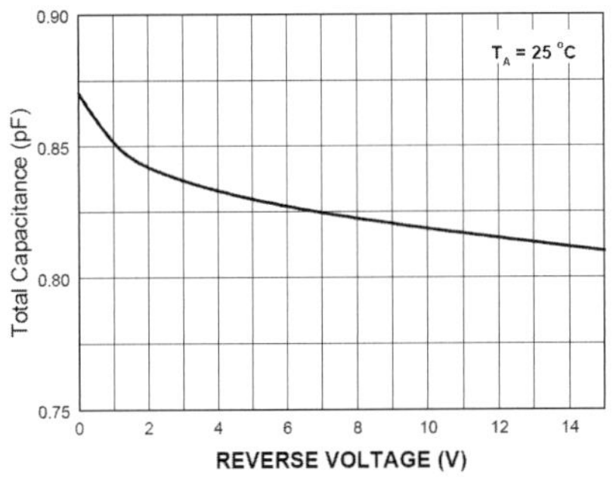

Figure 7. Total Capacitance

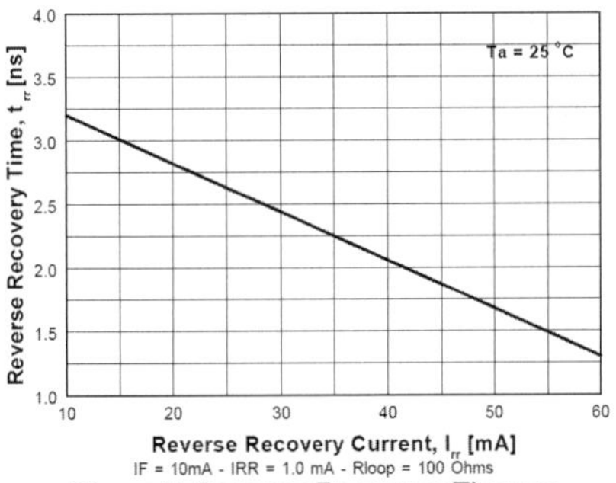

Figure 8. Reverse Recovery Time vs
Reverse Recovery Current

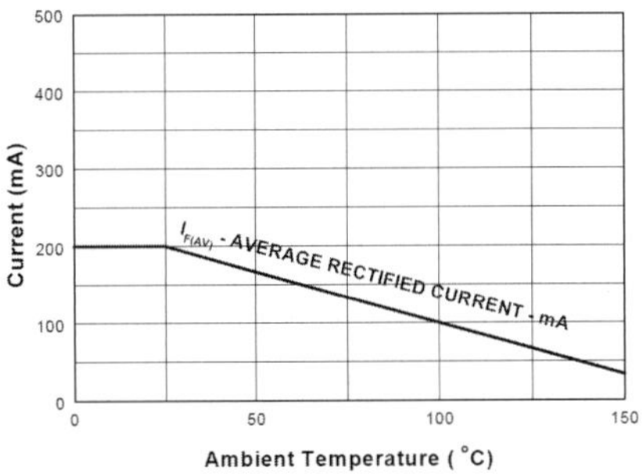

Figure 9. Average Rectified Current (I_F(AV))
versus Ambient Temperature (T_A)

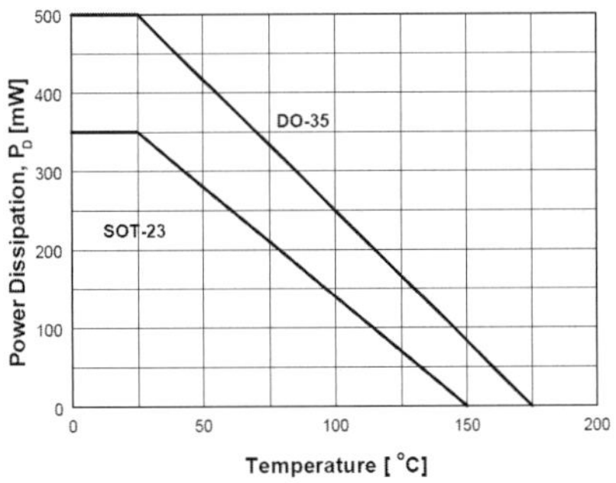

Figure 10. Power Derating Curve

Datasheet reproduced by kind permission of Fairchild Semiconductor

Appendix 6

2N3904 data sheet

FAIRCHILD
SEMICONDUCTOR ™

2N3904

C
B E   TO-92

MMBT3904

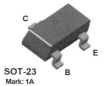

C
E
SOT-23   B
Mark: 1A

PZT3904

C
E
C
SOT-223   B

NPN General Purpose Amplifier

This device is designed as a general purpose amplifier and switch.
The useful dynamic range extends to 100 mA as a switch and to
100 MHz as an amplifier.

Absolute Maximum Ratings* T_A = 25°C unless otherwise noted

Symbol	Parameter	Value	Units
V_{CEO}	Collector-Emitter Voltage	40	V
V_{CBO}	Collector-Base Voltage	60	V
V_{EBO}	Emitter-Base Voltage	6.0	V
I_C	Collector Current - Continuous	200	mA
T_J, T_{stg}	Operating and Storage Junction Temperature Range	-55 to +150	°C

*These ratings are limiting values above which the serviceability of any semiconductor device may be impaired.

NOTES:
1) These ratings are based on a maximum junction temperature of 150 degrees C.
2) These are steady state limits. The factory should be consulted on applications involving pulsed or low duty cycle operations.

Thermal Characteristics T_A = 25°C unless otherwise noted

Symbol	Characteristic	Max 2N3904	Max *MMBT3904	Max **PZT3904	Units
P_D	Total Device Dissipation	625	350	1,000	mW
	Derate above 25°C	5.0	2.8	8.0	mW/°C
$R_{\theta JC}$	Thermal Resistance, Junction to Case	83.3			°C/W
$R_{\theta JA}$	Thermal Resistance, Junction to Ambient	200	357	125	°C/W

*Device mounted on FR-4 PCB 1.6" X 1.6" X 0.06."

**Device mounted on FR-4 PCB 36 mm X 18 mm X 1.5 mm; mounting pad for the collector lead min. 6 cm².

NPN General Purpose Amplifier
(continued)

Electrical Characteristics
T_A = 25°C unless otherwise noted

Symbol	Parameter	Test Conditions	Min	Max	Units
OFF CHARACTERISTICS					
$V_{(BR)CEO}$	Collector-Emitter Breakdown Voltage	I_C = 1.0 mA, I_B = 0	40		V
$V_{(BR)CBO}$	Collector-Base Breakdown Voltage	I_C = 10 µA, I_E = 0	60		V
$V_{(BR)EBO}$	Emitter-Base Breakdown Voltage	I_E = 10 µA, I_C = 0	6.0		V
I_{BL}	Base Cutoff Current	V_{CE} = 30 V, V_{EB} = 3V		50	nA
I_{CEX}	Collector Cutoff Current	V_{CE} = 30 V, V_{EB} = 3V		50	nA
ON CHARACTERISTICS*					
h_{FE}	DC Current Gain	I_C = 0.1 mA, V_{CE} = 1.0 V	40		
		I_C = 1.0 mA, V_{CE} = 1.0 V	70		
		I_C = 10 mA, V_{CE} = 1.0 V	100	300	
		I_C = 50 mA, V_{CE} = 1.0 V	60		
		I_C = 100 mA, V_{CE} = 1.0 V	30		
$V_{CE(sat)}$	Collector-Emitter Saturation Voltage	I_C = 10 mA, I_B = 1.0 mA		0.2	V
		I_C = 50 mA, I_B = 5.0 mA		0.3	V
$V_{BE(sat)}$	Base-Emitter Saturation Voltage	I_C = 10 mA, I_B = 1.0 mA	0.65	0.85	V
		I_C = 50 mA, I_B = 5.0 mA		0.95	V
SMALL SIGNAL CHARACTERISTICS					
f_T	Current Gain - Bandwidth Product	I_C = 10 mA, V_{CE} = 20 V, f = 100 MHz	300		MHz
C_{obo}	Output Capacitance	V_{CB} = 5.0 V, I_E = 0, f = 1.0 MHz		4.0	pF
C_{ibo}	Input Capacitance	V_{EB} = 0.5 V, I_C = 0, f = 1.0 MHz		8.0	pF
NF	Noise Figure	I_C = 100 µA, V_{CE} = 5.0 V, R_S =1.0kΩ,f=10 Hz to 15.7kHz		5.0	dB
SWITCHING CHARACTERISTICS					
t_d	Delay Time	V_{CC} = 3.0 V, V_{BE} = 0.5 V,		35	ns
t_r	Rise Time	I_C = 10 mA, I_{B1} = 1.0 mA		35	ns
t_s	Storage Time	V_{CC} = 3.0 V, I_C = 10mA		200	ns
t_f	Fall Time	I_{B1} = I_{B2} = 1.0 mA		50	ns

*Pulse Test: Pulse Width ≤ 300 µs, Duty Cycle ≤ 2.0%

Spice Model

NPN (Is=6.734f Xti=3 Eg=1.11 Vaf=74.03 Bf=416.4 Ne=1.259 Ise=6.734 Ikf=66.78m Xtb=1.5 Br=.7371 Nc=2 Isc=0 Ikr=0 Rc=1 Cjc=3.638p Mjc=.3085 Vjc=.75 Fc=.5 Cje=4.493p Mje=.2593 Vje=.75 Tr=239.5n Tf=301.2p Itf=.4 Vtf=4 Xtf=2 Rb=10)

NPN General Purpose Amplifier
(continued)

Typical Characteristics

Typical Pulsed Current Gain vs Collector Current

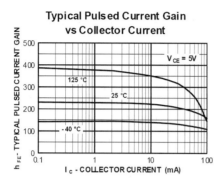

Collector-Emitter Saturation Voltage vs Collector Current

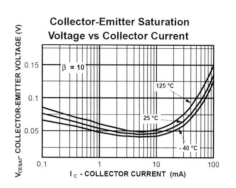

Base-Emitter Saturation Voltage vs Collector Current

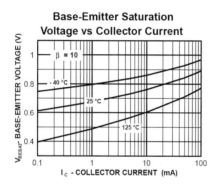

Base-Emitter ON Voltage vs Collector Current

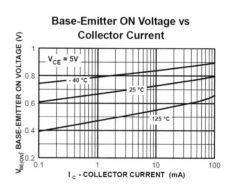

Collector-Cutoff Current vs Ambient Temperature

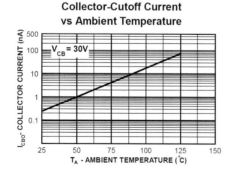

Capacitance vs Reverse Bias Voltage

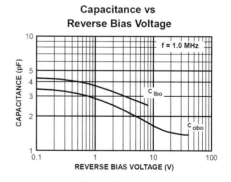

2N3904 / MMBT3904 / PZT3904

NPN General Purpose Amplifier
(continued)

Typical Characteristics (continued)

Noise Figure vs Frequency

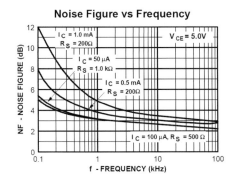

Noise Figure vs Source Resistance

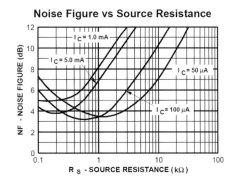

Current Gain and Phase Angle vs Frequency

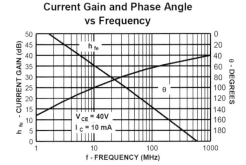

Power Dissipation vs Ambient Temperature

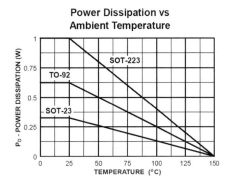

Turn-On Time vs Collector Current

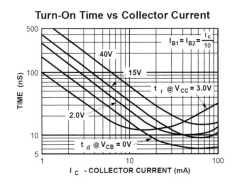

Rise Time vs Collector Current

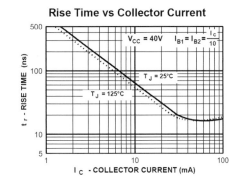

NPN General Purpose Amplifier
(continued)

Typical Characteristics (continued)

Storage Time vs Collector Current

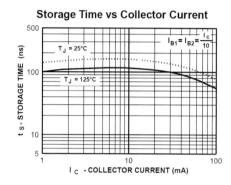

Fall Time vs Collector Current

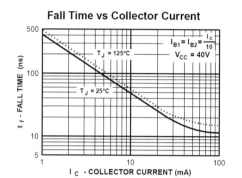

Current Gain

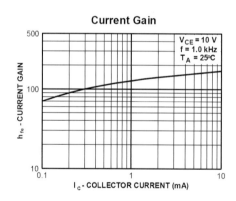

Output Admittance

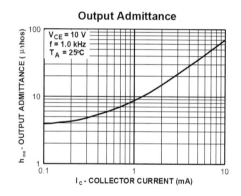

Input Impedance

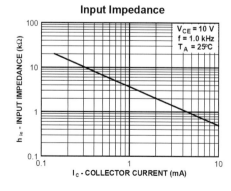

Voltage Feedback Ratio

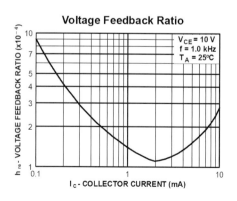

2N3904 / MMBT3904 / PZT3904

NPN General Purpose Amplifier
(continued)

Test Circuits

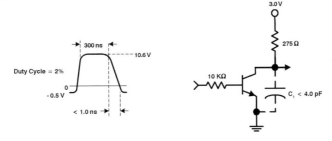

FIGURE 1: Delay and Rise Time Equivalent Test Circuit

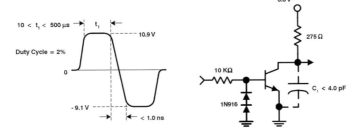

FIGURE 2: Storage and Fall Time Equivalent Test Circuit

Datasheet reproduced by kind permission of Fairchild Semiconductor

Appendix 7

Decibels

Decibels (dB) are a convenient means of expressing gain (amplification) and loss (attenuation) in electronic circuits. In this respect, they are used as a *relative* measure (i.e. comparing one voltage with another, one current with another, or one power with another). In conjunction with other units, decibels are sometimes also used as an *absolute* measure. Hence dBV are decibels relative to 1V, dBm are decibels relative to 1 mW, etc.

The decibel is one-tenth of a bel which, in turn, is defined as the logarithm, to the base 10, of the ratio of output power (P_{out}) to input power (P_{in}).

Gain and loss may be expressed in terms of power, voltage and current such that:

$$A_p = \frac{P_{out}}{P_{in}} \quad A_v = \frac{V_{out}}{V_{in}} \quad \text{and} \quad A_i = \frac{I_{out}}{I_{in}}$$

where A_p, A_v or A_i is the power, voltage or current gain (or loss) expressed as a ratio, P_{in} and P_{out} are the input and output powers, V_{in} and V_{out} are the input and output voltages, and I_{in} and I_{out} are the input and output currents. Note, however, that the powers, voltages or currents should be expressed in the same units/multiples (e.g. P_{in} and P_{out} should both be expressed in W, mW, µW or nW).

It is often more convenient to express gain in decibels (rather than as a simple ratio) using the following relationships:

$$A_p = 10\log_{10}\left(\frac{P_{out}}{P_{in}}\right) \quad A_v = 20\log_{10}\left(\frac{V_{out}}{V_{in}}\right)$$

$$\text{and} \quad A_i = 20\log_{10}\left(\frac{I_{out}}{I_{in}}\right)$$

Note that a positive result will be obtained whenever P_{out}, V_{out}, or I_{out} is greater than P_{in}, V_{out}, or I_{out}, respectively. A negative result will be obtained whenever P_{out}, V_{out}, or I_{out} is less than P_{in}, V_{in} or I_{in}. A negative result denotes attenuation rather than amplification. A negative gain is thus equivalent to an attenuation (or loss). If desired, the formulae may be adapted to produce a positive result for attenuation simply by inverting the ratios, as shown below:

$$A_p = 10\log_{10}\left(\frac{P_{in}}{P_{out}}\right) \quad A_v = 20\log_{10}\left(\frac{V_{in}}{V_{out}}\right)$$

$$\text{and} \quad A_i = 20\log_{10}\left(\frac{I_{in}}{I_{out}}\right)$$

where A_p, A_v or A_i is the power, voltage or current gain (or loss) expressed in decibels, P_{in} and P_{out} are the input and output powers, V_{in} and V_{out} are the input and output voltages, and I_{in} and I_{out} are the input and output currents. Note, again, that the powers, voltages or currents should be expressed in the same units/multiples (e.g. P_{in} and P_{out} should both be expressed in W, mW, µW or nW).

It is worth noting that, for identical decibel values, the values of voltage and current gain can be found by taking the square root of the corresponding value of power gain. As an example, a voltage gain of 20 dB results from a voltage ratio of 10 while a power gain of 20 dB corresponds to a power ratio of 100.

Finally, it is essential to note that the formulae for voltage and current gain are only meaningful when the input and output impedances (or resistances) are identical. Voltage and current gains expressed in decibels are thus only valid for matched (constant impedance) systems. Table A7.1 gives some useful decibel values.

Example A7.1

An amplifier with matched input and output resistances provides an output voltage of 1V for an input of 25 mV. Express the voltage gain of the amplifier in decibels.

Solution

The voltage gain can be determined from the formula:

Table A7.1   Decibels and ratios of power, voltage and current

Decibels (dB)	Power gain (ratio)	Voltage gain (ratio)	Current gain (ratio)
0	1	1	1
1	1.26	1.12	1.12
2	1.58	1.26	1.26
3	2	1.41	1.41
4	2.51	1.58	1.58
5	3.16	1.78	1.78
6	3.98	2	2
7	5.01	2.24	2.24
8	6.31	2.51	2.51
9	7.94	2.82	2.82
10	10	3.16	3.16
13	19.95	3.98	3.98
16	39.81	6.31	6.31
20	100	10	10
30	1,000	31.62	31.62
40	10,000	100	100
50	100,000	316.23	316.23
60	1,000,000	1,000	1,000
70	10,000,000	31,62.3	31,62.3

$A_v = 20 \, \log_{10}(V_{out}/V_{in})$

where $V_{in} = 25$ mV and $V_{out} = 1$ V.

Thus:

$A_v = 20 \, \log_{10}(1\text{V}/25 \text{ mV})$

$\quad = 20 \, \log_{10}(40) = 20 \times 1.6 = \textbf{32 dB}$

Example A7.2

A matched 600 Ω attenuator produces an output of 1 mV when an input of 20 mV is applied. Determine the attenuation in decibels.

Solution

The attenuation can be determined by applying the formula:

$A_v = 20 \, \log_{10}(V_{out}/V_{in})$

where $V_{in} = 20$ mV and $V_{out} = 1$ mV. Thus:

$A_v = 20 \, \log_{10}(20 \text{ mV}/1 \text{ mV})$

$\quad = 20 \, \log_{10}(20) = 20 \times 1.3 = \textbf{26 dB}$

Example A7.3

An amplifier provides a power gain of 33 dB. What output power will be produced if an input of 2 mW is applied?

Solution

Here we must re-arrange the formula to make P_{out} the subject, as follows:

$A_v = 10 \, \log_{10}(P_{out}/P_{in})$

thus

$A_p/10 = \log_{10}(P_{out}/P_{in})$

or

$\text{antilog}_{10}(A_p/10) = P_{out}/P_{in}$

Hence

$P_{out} = P_{in} \times \text{antilog}_{10}(A_p/10)$

Now $P_{in} = 2$ mW $= 20 \times 10^{-3}$ W and $A_v = 33$ dB, thus:

$P_{out} = 2 \times 10^{-3} \, \text{antilog}_{10}(33/10)$

$\quad = 2 \times 10^{-3} \times \text{antilog}_{10}(3.3)$

$\quad = 2 \times 10^{-3} \times 1.995 \times 10^{-3} = \textbf{3.99 W}$

Appendix 8

Mathematics for electronics

This section introduces the mathematical techniques that are essential to developing a good understanding of electronics. The content is divided into seven sections; notation, algebra, equations, graphs, trigonometry, Boolean algebra, logarithms and exponential growth and decay. We have included a number of examples and typical calculations.

Notation

The standard notation used in mathematics provides us with a shorthand that helps to simplify the writing of mathematical expressions. Notation is based on the use of symbols that you will already recognize. These include = (equals), + (addition), - (subtraction), × (multiplication), and , (division). Other symbols that you may not be so familiar with include < (less than), > (greater than), μ (proportional to) and Ö (square root).

Indices

The number 4 is the same as 2 × 2, that is, 2 multiplied by itself. We can write (2 × 2) as 2^2. In words, we would call this 'two raised to the power two' or simply 'two squared'. Thus:

$$2 \times 2 = \mathbf{2^2}$$

By similar reasoning we can say that:

$$2 \times 2 \times 2 = \mathbf{2^3}$$

and

$$2 \times 2 \times 2 \times 2 = \mathbf{2^4}$$

In these examples, the number that we have used (i.e. 2) is known as the *base* whilst the number that we have raised it to is known as an *index*. Thus, 2^4 is called 'two to the power of four', and it consists of a base of 2 and an index of 4. Similarly, 5^3 is called 'five to the power of 3' and has a base of 5

and an index of 3. Special names are used when the indices are 2 and 3, these being called 'squared' and 'cubed', respectively. Thus 7^2 is called 'seven squared' and 9^3 is called 'nine cubed'. When no index is shown, the power is 1, i.e. 2^1 means 2.

Reciprocals

The *reciprocal* of a number is when the index is −1 and its value is given by 1 divided by the base. Thus the reciprocal of 2 is 2^{-1} and its value is ½ or 0.5. Similarly, the reciprocal of 4 is 4^{-1} which means ¼ or 0.25.

Square roots

The *square root* of a number is when the index is ½. The square root of 2 is written as $2^{½}$ or Ö2. The value of a square root is the value of the base which when multiplied by itself gives the number. Since 3 ′ 3 = 9, then Ö9 = 3. However, (-3) ′ (-3) = 9, so we have a second possibility, i.e. Ö9 = ±3. There are always two answers when finding the square root of a number and we can indicate this is by placing a ± sign in front of the result meaning 'plus or minus'. Thus:

$$4^{\frac{1}{2}} = \sqrt{4} = \mathbf{\pm 2}$$

and

$$9^{\frac{1}{2}} = \sqrt{9} = \mathbf{\pm 3}$$

Variables and constants

Unfortunately, we don't always know the value of a particular quantity that we need to use in a calculation. In some cases the value might actually change, in which case we refer to it as a *variable*. In other cases, the value might be fixed but we might prefer not to actually quote it's value. In this case we refer to the value as a *constant*.

An example of a *variable quantity* is the output voltage produced by a power supply when the load current is changing. An example of a *constant quantity* might be the speed at which radio waves travel in space.

In either case, we use a symbol to represent the quantity. The symbol itself (often a single letter) is a form of shorthand notation. For example, in the case of the voltage from the power supply we would probably use v to represent *voltage*. Whereas, in the case of the speed of travel of a radio wave we normally use c to represent the speed of radio waves in space (i.e. the speed of light).

We can use letters to represent both variable and constant quantities in mathematical notation. For example, the statement:

voltage is equal to *current* multiplied by *resistance*

can be written in mathematical notation as follows:

$$V = I \times R$$

where V represents voltage, I represents current, and R represents resistance.

Similarly, the statement:

resistance is equal to *voltage* divided by *current*

can be written in mathematical notation as follows:

$$R = \frac{V}{I}$$

where R represents resistance, V represents voltage, and I represents current.

Proportionality

In electronic circuits, when one quantity changes it normally affects a number of other quantities. For example, if the output voltage produced by a power supply increases when the resistance of the load to which it is connected remains constant, the current supplied to the load will also increase. If the output voltage doubles, the current will also double in value. If, on the other hand, the output voltage falls by 50% the current will also fall by 50%. To put this in a mathematical way we can say that (provided that load resistance remains constant):

load current is *directly proportional* to output voltage

Using mathematical notation and symbols to represent the quantities, we would write this as:

$$I \propto V$$

where I represents load current, V represents output voltage, and α means 'is proportional to'.

In some cases, an increase in one quantity will produce a *reduction* in another quantity. For example, if the frequency of a wavelength of a radio wave increases its wavelength is reduced. To put this in a mathematical way we say that (provided the speed of travel remains constant):

wavelength is *inversely proportional* frequency

Using mathematical notation and symbols to represent the quantities, we would write this as follows:

$$T \propto \frac{1}{f}$$

where T represents periodic time and f represents frequency.

It's useful to illustrate these important concepts using some examples:

Example 1

The power in a loudspeaker is proportional to the square of the RMS output voltage and inversely proportional to the impedance of the speaker. Write down an equation for the power, P, in terms of the voltage, V, and impedance, Z. Obtain a formula for P in terms of V and Z.

From the information given, we can say that 'power is proportional to voltage squared' hence:

$$P \propto V^2 \tag{1}$$

and 'power is inversely proportional to impedance' thus:

$$P \propto \frac{1}{Z} \tag{2}$$

We can re-write this expression using a negative index, as follows:

$$P \propto Z^{-1}$$

We can go one stage further and combine these two relationships to obtain a formula for power that involves both variables, i.e. V and Z:

$$\frac{1}{Z}$$

Example 2

In Example 1, determine the power delivered to the loudspeaker if the RMS voltage, V, is 8 V and the loudspeaker has an impedance, Z, of 4 W.

$$P = V^2 Z^{-1} = \frac{V^2}{Z} = \frac{8^2}{4} = \frac{8 \times 8}{4} = \frac{64}{4} = \textbf{16 W}$$

Example 3

The equipment manufacturer in Example 2 recommends testing the amplifier at a power of 10 W. Determine the RMS voltage at the output that corresponds to this power level.

Here we must re-arrange the formula in order to make V the subject. If we multiply both sides of the equation by Z we will have P and Z on one side of the equation and V^2 on the other:

$$P \times Z = \frac{V^2}{Z} \times Z$$

from which

$$PZ = V^2 \frac{Z}{Z}$$

or

$$PZ = V^2 \tag{1}$$

Rearranging (1) gives:

$$V^2 = PZ \tag{2}$$

Taking square roots of *both* sides:

$$\sqrt{V^2} = \sqrt{PZ}$$

The square root of V^2 is V hence:

$$V = \sqrt{PZ} = \left(PZ \right)^{\frac{1}{2}}$$

We need to find V when $P = 10$ W and $Z = 4\,\Omega$:

$$V = \sqrt{10 \times 4} = \sqrt{40} = \textbf{6.32 V}$$

Laws of indices

When simplifying calculations involving indices, certain basic rules or laws can be applied, called the *laws of indices*. These are listed below:

(a) when multiplying two or more numbers having the same base, the indices are added. Thus $2^2 \times 2^4 = 2^{2+4} = \textbf{2}^\textbf{6}$.

(b) when a number is divided by a number having the same base, the indices are subtracted. Thus $2^5 / 2^2 = 2^{5-2} = \textbf{2}^\textbf{3}$.

(c) when a number which is raised to a power is raised to a further power, the indices are multiplied. Thus $(2^5)^2 = 2^{5 \times 2} = \textbf{2}^{\textbf{10}}$.

(d) when a number has an index of 0, its value is 1. Thus $2^0 = \textbf{1}$.

(e) when a number is raised to a negative power, the number is the reciprocal of that number raised to a positive power. Thus $2^{-4} = \textbf{1/2}^\textbf{4}$.

(f) when a number is raised to a fractional power the denominator of the fraction is the root of the number and the numerator is the power. Thus $25^{\frac{1}{2}} = \sqrt{25^1} = \textbf{±5}$

Standard form

A number written with one digit to the left of the decimal point and multiplied by 10 raised to some power is said to be written in standard form. Thus: 1234 is written as 1.234×10^3 in standard form, and 0.0456 is written as 4.56×10^{-2} in standard form.

When a number is written in standard form, the first factor is called the **mantissa** and the second factor is called the **exponent**. Thus the number 6.8 $\times 10^3$ has a mantissa of 6.8 and an exponent of 10^3.

Numbers having the same exponent can be added or subtracted in standard form by adding or subtracting the mantissae and keeping the exponent the same. Thus:

$$2.3 \times 10^4 + 3.7 \times 10^4 = (2.3 + 3.7) \times 10^4$$
$$= 6.0 \times 10^4,$$

and

$$5.7 \times 10^{-2} - 4.6 \times 10^{-2} = (5.7 - 4.6) \times 10^{-2}$$
$$= 1.1 \times 10^{-2}$$

When adding or subtracting numbers it is quite acceptable to express one of the numbers in non-standard form, so that both numbers have the same exponent. This makes things much easier as the following example shows:

$$2.3 \times 10^4 + 3.7 \times 10^3 = 2.3 \times 10^4 + 0.37 \times 10^4$$

$$= (2.3 + 0.37) \times 10^4$$

$$= 2.67 \times 10^4$$

Alternatively,

$$2.3 \times 10^4 + 3.7 \times 10^3 = 23000 + 3700$$

$$= 26700 = 2.67 \times 10^4$$

The laws of indices are used when multiplying or dividing numbers given in standard form.   For example,

$$(22.5 \times 10^3) \times (5 \times 10^2) = (2.5 \times 5) \times (10^{3+2})$$

$$= 12.5 \times 10^5 \text{ or } 1.25 \times 10^6$$

Example 4

Period, t, is the reciprocal of frequency, f. Thus

$$t = f^{-1} = \frac{1}{f}$$

Calculate the period of a radio frequency signal having a frequency of 2.5 MHz.

Now $f = 2.5$ MHz. Expressing this in standard form gives $f = 2.5 \times 10^6$ Hz.

Since

$$t = f^{-1} = \frac{1}{2.5 \times 10^6} = \frac{10^{-6}}{2.5} = \frac{1}{2.5} \times 10^{-6} = 0.4 \times 10^{-6}$$

Thus

$$t = 4 \times 10^{-7}\,\text{s}$$

Example 5

Resistors of 3.9 kΩ, 5.6 kΩ and 10 kΩ are connected in parallel as shown in Fig. A8.1. Calculate the effective resistance of the circuit.

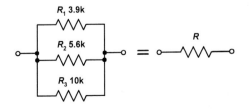

Figure A8.1   See Example 5

The resistance of a parallel circuit is given by the equation:

$$\frac{1}{R} = \frac{1}{R1} + \frac{1}{R2} + \frac{1}{R3}$$

Now we know that:

$$R1 = 3.9 \text{ k}\Omega = 3.9 \times 10^3 \,\Omega$$

$$R2 = 5.6 \text{ k}\Omega = 5.6 \times 10^3 \,\Omega$$

$$R3 = 10 \text{ k}\Omega = 1 \times 10^4 \,\Omega = 10 \times 10^3 \,\Omega$$

Hence:

$$\frac{1}{R} = \frac{1}{3.9 \times 10^3} + \frac{1}{5.6 \times 10^3} + \frac{1}{10 \times 10^3}$$

$$= \frac{10^{-3}}{3.9} + \frac{10^{-3}}{5.6} + \frac{10^{-3}}{10} = \left(\frac{1}{3.9} + \frac{1}{5.6} + \frac{1}{10}\right) \times 10^{-3}$$

$$= (0.256 + 0.179 + 0.1) \times 10^{-3} = 0.535 \times 10^{-3}$$

$$\frac{1}{R} = 0.535 \times 10^3$$

Now since we can invert both sides of the equation so that:

$$R = \frac{1}{0.535 \times 10^{-3}} = 1.87 \times 10^3$$

Hence

$$R = \textbf{1.87 k}\Omega$$

Equations

We frequently need to solve equations in order to find the value of an unknown quantity. Any arithmetic operation may be applied to an equation as long as the equality of the equation is maintained. In other words, the same operation must be applied to both the left-hand side (LHS) and the right-hand side (RHS) of the equation.

Example 6

A current of 0.5 A flows in the 56 Ω shown in Fig. A8.2. Given that $V = I R$ determine the voltage that appears across the resistor.

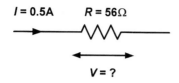

Figure A8.2   See Example 6

It's a good idea to get into the habit of writing down what you know before attempting to solve an equation. In this case:

$I = 0.5$ A

$R = 56$ Ω

$V = ?$

Now

$V = IR = 0.5 \times 56 = \textbf{28 V}$

Example 7

The current present at a junction in a circuit is shown in Fig. A8.3 where $I_1 = 0.1$ A, $I_2 = 0.25$ A and $I_3 = 0.3$ A. Determine the value of the unknown current, I_x.

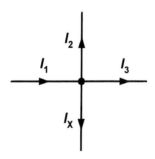

Figure A8.3   See Example 7

From Kirchhoff's Current Law, the algebraic sum of the current at a junction in a circuit is zero. Adopting the convention that current flowing towards the junction is positive and that flowing away from the junction is negative we can construct a formula along the following lines:

$0 = -I_x + I_1 - I_2 - I_3$

We can rearrange the equation to make I_x the subject by adding I_x to both sides:

$0 + I_x = -I_x + I_x + I_1 - I_2 - I_3$

From which:

$I_x = I_1 - I_2 - I_3$

Now $I_1 = 0.1$ A, $I_2 = 0.25$ A and $I_3 = 0.3$ A

Thus:

$I_x = 0.1 - 0.25 - 0.3 = \textbf{-0.45 A}$

Hence $I_x = \textbf{0.45 A flowing towards the junction}$

It is important to note here that our original 'guess' as to the direction of I_x as marked on Fig. A8.3 was incorrect and the current is actually flowing the other way!

Note that it is always a good idea to check the solution to an equation by substituting the solution back into the original equation. The equation should then balance such that the left hand side (LHS) can be shown to be equal to the right hand side (RHS).

In many cases it can be more convenient to change the subject of a formula *before* you insert values. The next few examples show how this is done:

Example 8

A copper wire has a length l of 1.5 km, a resistance R of 5 Ω and a resistivity ρ of 17.2×10^{-6} Ω mm. Find the cross-sectional area, a, of the wire, given the following relationship:

$$R = \frac{\rho l}{a}$$

Once again, it is worth getting into the habit of summarizing what you know from the question and what you need to find. Also, don't forget to include the units for each one of the quantities that you know.

$R = 5$ Ω

$\rho = 17.2 \times 10^{-6}$ Ω mm

$l = 1500 \times 10^3$ mm

$a = ?$

Now since

$$R = \frac{\rho l}{a}$$

then

$$5 = \frac{(17.2 \times 10^{-6})(1500 \times 10^{3})}{a}$$

Cross multiplying (i.e. exchanging the '5' for the 'a') gives:

$$a = \frac{(17.2 \times 10^{-6})(1500 \times 10^{3})}{5}$$

Now group the numbers and the powers of 10 as shown below:

$$a = \frac{17.2 \times 1500 \times 10^{-6} \times 10^{3}}{5}$$

Next simplify as far as possible:

$$a = \frac{17.2 \times 1500}{5} \times 10^{-6+3}$$

Finally, evaluate the result using your calculator:

$$a = 5160 \times 10^{-3} = 5.16$$

Don't forget the units! Since we have been working in mm, the result, a, will be in mm^2.

Hence $a = $ **5.16 mm^2**

It's worth noting from the previous example that we have used the laws of indices to simplify the powers of ten before attempting to use the calculator to determine the final result. The alternative to doing this is to make use of the exponent facility on your calculator. Whichever technique you use it's important to be confident that you are correctly using the exponent notation since it's not unknown for students to produce answers that are incorrect by a factor of 1,000 or even 1,000,000 and an undetected error of this magnitude could be totally disastrous!

Before attempting to substitute values into an equation, it's important to be clear about what you know (and what you don't know) and always make sure that you have the correct units. The values marked on components can sometimes be misleading and it's always worth checking that you

have interpreted the markings correctly before wasting time solving an equation that doesn't produce the right answer!

Example 9

The reactance, X, of a capacitor is given by the relationship:

$$X = \frac{1}{2\pi f C}$$

Find the value of capacitance that will exhibit a reactance of 10 kΩ at a frequency of 400 Hz.

First, let's summarize what we know:

$X = 10$ kΩ

$f = 400$ Hz

$\pi = 3.142$ (or use the 'π' button on your calculator)

$C = ?$

We need to re-arrange the formula to make C the subject. This is done as follows:

$$X = \frac{1}{2\pi f C}$$

Cross multiplying gives:

$$C = \frac{1}{2\pi f \times X}$$

(note that we have just 'swapped' the C and the X over).

Next, replacing π, C, and f by the values that we know gives:

$$C = \frac{1}{2 \times 3.142 \times 400 \times 10 \times 10^{3}} = \frac{1}{25136 \times 10^{3}}$$

$$= \frac{1}{2.5136 \times 10^{7}} = \frac{1}{2.5136} \times 10^{-7} = 0.398 \times 10^{-7} \text{ F}$$

Finally, it would be sensible to express the answer in nF (rather than F). To do this, we simply need to multiply the result by 10^9, as follows:

$$C = 0.398 \times 10^{-7} \times 10^{9} = 0.398 \times 10^{9-7}$$

$$= 0.398 \times 10^{2} = \textbf{39.8 nF}$$

Example 10

The frequency of resonance, f, of a tuned circuit (see Fig. A8.4) is given by the relationship:

$$f = \frac{1}{2\pi\sqrt{LC}}$$

If a tuned circuit is to be resonant at 6.25 MHz and $C = 100$ pF, determine the value of inductance, L.

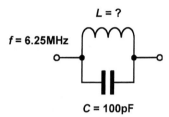

$$L = ?$$
$$f = 6.25\text{MHz}$$
$$C = 100\text{pF}$$

Figure A8.4   See Example 10

Here we know that:

$C = 100$ pF $= 100 \times 10^{-12}$ F

$f = 6.25$ MHz $= 6.25 \times 10^6$ Hz

$\pi = 3.142$ (use the 'π' button on your calculator!)

$L = ?$
First we will rearrange the formula in order to make L the subject:

$$f = \frac{1}{2\pi\sqrt{LC}}$$

Squaring both sides gives:

$$f^2 = \left(\frac{1}{2\pi\sqrt{LC}}\right)^2 = \frac{1^2}{2^2\pi^2\left(\sqrt{LC}\right)^2} = \frac{1}{4\pi^2 LC}$$

Re-arranging gives:

$$L = \frac{1}{4\pi^2 f^2 C}$$

We can now replace f, C and π by the values that we know:

$$L = \frac{1}{4 \times 3.142^2 \times \left(6.25 \times 10^6\right)^2 \times 100 \times 10^{-12}} \text{ H}$$

$$L = \frac{1}{39.49 \times 39.06 \times 10^{12} \times 100 \times 10^{-12}} \text{ H}$$

$$= \frac{1}{154248} = \frac{1}{1.54248 \times 10^5} = \frac{1}{1.54248} \times 10^{-5}$$

$$= 0.648 \times 10^{-5} = 6.48 \times 10^{-6} \text{ H} = \mathbf{6.8\ \mu H}$$

More complex equations

More complex equations are solved in essentially the same way as the simple equations that we have just met. Note that equations with square (or higher) laws may have more than one solution. Here is an example:

Example 11

The impedance of the AC circuit shown in Fig. A8.5 is given by:

$$Z = \sqrt{R^2 + X^2}$$

Determine the reactance, X, of the circuit when $R = 25\ \Omega$ and $Z = 50\ \Omega$.

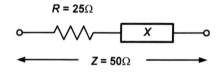

$$R = 25\Omega$$
$$X$$
$$Z = 50\Omega$$

Figure A8.5   See Example 11

Here we know that:

$R = 25\ \Omega$

$Z = 50\ \Omega$

$X = ?$

First we will re-arrange the formula in order to make X the subject:

$$Z = \sqrt{R^2 + X^2}$$

Squaring both sides gives:

$$Z^2 = \left(\sqrt{R^2 + X^2}\right)^2 = R^2 + X^2$$

Re-arranging gives:

$$X^2 = Z^2 - R^2$$

Taking the square root of both sides gives:

$$X = \sqrt{Z^2 - R^2}$$

We can now replace Z and R by the values that we know:

$$X = \sqrt{50^2 - 25^2} = \sqrt{2500 - 625} = \sqrt{1875}$$

thus

$$X = \mathbf{43.3\ \Omega}$$

In the previous example we arrived at a solution of 43.3 Ω for X. This is not the only solution as we will now show:

As before, we can check that we have arrived at the correct answer by substituting values back into the original equation, as follows:

$$50 = Z = \sqrt{R^2 + X^2} = \sqrt{25^2 + 43.3^2}$$

$$50 = \sqrt{625 + 1875} = \sqrt{2500} = 50$$

Thus LHS = RHS.

However, we could also have used –43.3 Ω (instead of +43.3 Ω) for X and produced the *same* result:

$$50 = Z = \sqrt{R^2 + X^2} = \sqrt{25^2 + (-43.3)^2}$$

$$50 = \sqrt{625 + 1875} = \sqrt{2500} = 50$$

Once again, LHS = RHS.

The reason for this apparent anomaly is simply that the result of squaring a negative number will be a positive number (i.e. $(-2)^2 = (-2) \times (-2) = +4$).

Hence, a correct answer to the problem in Example 11 would be:

$$X = \mathbf{\pm\ 43.3\ \Omega}$$

Don't panic if the ambiguity of this answer is worrying you! Reactance in an AC circuit *can* be either positive or negative depending upon whether the component in question is an inductor or a capacitor. In this case, the numerical value of the impedance is the same regardless of whether the reactance is inductive or capacitive.

Graphs

Graphs provide us with a visual way of representing data. They can also be used to show, in a simple pictorial way, how one variable affects another variable. Several different types of graph are used in electronics. We shall start by looking at the most basic type, the straight line graph.

Straight line graphs

Earlier we introduced the idea of *proportionality*. In particular, we showed that the current flowing in a circuit was directly proportional to the voltage applied to it. We expressed this using the following mathematical notation:

$$i \propto v$$

where i represents load current and v represents output voltage.

We can illustrate this relationship using a simple graph showing current, i, plotted against voltage, v. Let's assume that the voltage applied to the circuit varies over the range 1 V to 6 V and the circuit has a resistance of 3 W. By taking a set of measurements of v and i (see Fig. A8.6) we would obtain the following table of corresponding values shown below:

Voltage, v (V)	1	2	3	4	5	6
Current, i (A)	0.33	0.66	1.0	1.33	1.66	2.0

The resulting graph is shown in Fig. A8.7. To obtain the graph, a point is plotted for each pair of corresponding values for v and i. When all the points have been drawn they are connected together by drawing a line. Notice that, in this case, the line that connects the points together takes the form of a straight line. This is *always* the case when two variables are directly proportional to one another.

It is conventional to show the *dependent variable* (in this case it is current, *i*) plotted on the vertical axis and the *independent variable* (in this case it is voltage, *v*) plotted on the horizontal axis. If you find these terms a little confusing, just remember that, what you know is usually plotted on the horizontal scale whilst what you don't know (and may be trying to find) is usually plotted on the vertical scale. In fact, the graph contains the same information regardless of which way round it is drawn!

Now let's take another example. Assume that the following measurements are made on an electronic component:

Temperature, t (°C)	10	20	30	40	50	60
Resistance, R (W)		105	110	115	120	125	130

The results of the experiment are shown plotted in graphical form in Fig. A8.8. Note that the graph consists of a straight line but that it does not pass through the *origin* of the graph (i.e. the point at which *t* and *V* are 0°C and 0 V respectively). The second most important feature to note (after having noticed that the graph is a straight line) is that, when $t = 0$°C, $R = 100$ W.

By looking at the graph we could suggest a relationship (i.e. an *equation*) that will allow us to find the resistance, *R*, of the component at any given temperature, *t*. The relationship is simply:

$$R = 100 + \frac{t}{2} \ \Omega$$

If you need to check that this works, just try inserting a few pairs of values from those given in

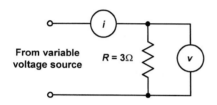

Figure A8.6   Corresponding readings of current, *i*, and voltage, *v*, are taken and used to construct the graph shown in Fig. A8.7

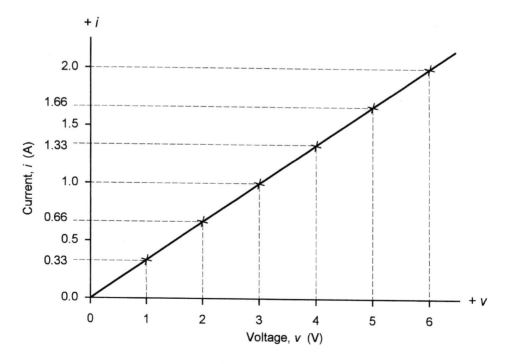

Figure A8.7   Graph of *i* plotted against *v*

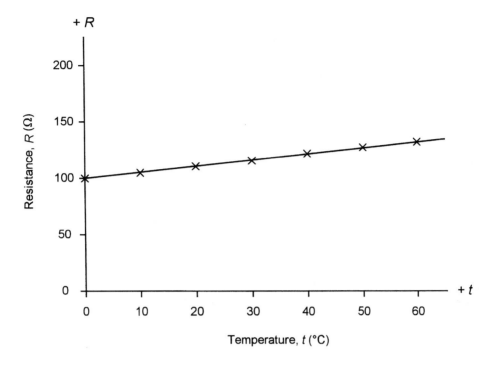

Figure A8.8   Graph of resistance, *R*, plotted against temperature, *t*, for a thermistor

the table. You should find that the equation balances every time!

Earlier we looked at equations. The shape of a graph is dictated by the equation that connects its two variables. For example, the general equation for a straight line takes the form:

$$y = m\,x + c$$

where *y* is the *dependent variable* (plotted on the vertical or *y-axis*), *x* is the *independent variable* (plotted on the horizontal or *x-axis*), *m* is the slope (or *gradient*) of the graph and *c* is the *intercept* on the *y*-axis. Fig. A8.9 shows this information plotted on a graph.

The values of *m* (the gradient) and *c* (the y-axis intercept) are useful when quoting the specifications for electronic components. In the previous example, the electronic component (in this case a *thermistor*) has:

(a)   a resistance of 100 W at 0°C (thus *c* = 100 Ω)

(a)   a characteristic that exhibits an increase in resistance of 0.5 W per °C (thus *m* = 0.5 W/°C).

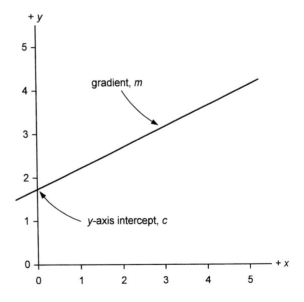

Figure A8.9   Definition of gradient (slope), *m*, and *y*-axis intercept, *c*

Example 12

The following data was obtained whilst making measurements on an N-channel field effect transistor:

V_{GS} (V)	−1	−2	−3	−4	−5	−6
I_D (mA)	8	6	4	2	0.4	0.1

Plot a graph showing how drain current (I_D) varies with gate-source voltage (V_{GS}) and use the graph to determine:

(a) the value of I_D when $V_{GS} = 0$ V, and
(b) the slope of the graph (expressed in mA/V) when $V_{GS} = -2$ V.

The data has been shown plotted in Fig. A8.10. Note that, since the values of V_{GS} that we have been given are negative, we have plotted these to the left of the vertical axis rather than to the right of it.

(a) We can use the graph to determine the crossing point (the *intercept*) on the drain current (I_D) axis. This occurs when $I_D = \mathbf{10\ mA}$.

(b) The slope (or *gradient*) of the graph is found by taking a small change in drain current and dividing it by a corresponding small change in gate-source voltage. In order to do this we have drawn a triangle on the graph at the point where $V_{GS} = -2$ V. The slope of the graph is found by dividing the vertical height of the triangle (expressed in mA) by the horizontal length of the triangle (expressed in V). From Fig. A8.10 we can see that the vertical height of triangle is 2 mA whilst its horizontal length is 1 V. The slope of the graph is thus given by 2 mA/1 V = **2 mA/V**.

Square law graphs

Of course, not all graphs have a straight line shape. In the previous example we saw a graph that, whilst

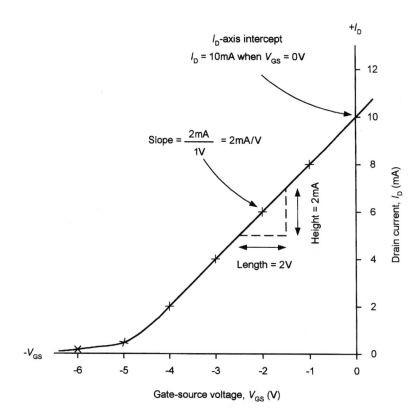

Figure A8.10   Drain current, I_D, plotted against gate-source voltage, V_{GS}, for a field effect transistor

substantially linear, became distinctly curved at one end. Many graphs are curved rather than linear. One common type of curve is the square law. To put this into context, consider the relationship between the power developed in a load resistor and the current applied to it. Assuming that the load has a resistance of 15 W we could easily construct a table showing corresponding values of power and current, as shown below:

Current, I (A)	0.25	0.5	0.75	1	1.25	1.5
Power, P (W)	0.94	3.75	8.44	15	23.44	33.75

We can plot this information on a graph showing power, P, on the vertical axis plotted against current, I, on the horizontal axis. In this case, P is the *dependent variable* and I is the *independent variable*. The graph is shown in Fig. A8.11.

It can be seen that the relationship between P and I in Fig. A8.11 is far from linear. The relationship is, in fact, a *square law relationship*. We can actually deduce this from what we know about the power dissipated in a circuit and the current flowing in the circuit. You may recall that:

$$P = I^2 R$$

where P represents power in Watts, I is current in Amps, and R is resistance in ohms.

Since R remains constant, we can deduce that:

$$P \propto I^2$$

In words, we would say that 'power is proportional to current squared'.

Many other examples of square law relationships are found in electronics.

More complex graphs

Many more complex graphs exist and Fig. A8.13 shows some of the most common types. Note that these graphs have all been plotted over the range $x = \pm 4$, $y = \pm 4$. Each graph consists of four quadrants. These are defined as follows (see Fig. A8.12):

First quadrant	Values of x and y are both positive
Second quadrant	Values of x are negative whilst those for y are positive
Third quadrant	Values of x and y are both negative
Fourth quadrant	Values of x are positive whilst those for y are negative.

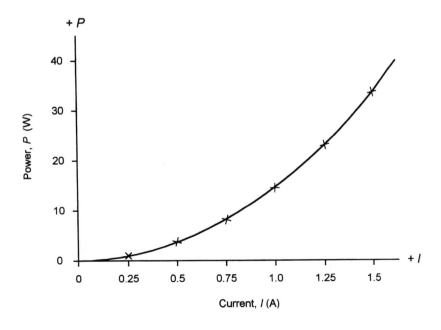

Figure A8.11 Power, P, plotted against current, I, showing a square law graph

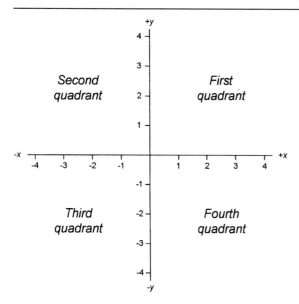

Figure A8.12   The first, second, third and fourth quadrants of a graph

The straight line relationship, $y = x$, is shown in Fig. A8.13(a). This graph consists of a straight line with a gradient of 1 that passes through the *origin* (i.e. the point where $x = 0$ and $y = 0$). The graph has values in the first and third quadrants.

The relationship $y = x^2$ is shown in Fig. A8.13(b). This graph also passes through the origin but its gradient changes, becoming steeper for larger values of x. As you can see, the graph has values in the first and second quadrants.

The graph of $y = x^3$ is shown in Fig. A8.13(c). This *cubic law* graph is steeper than the square law of Fig. A8.13(b) and it has values in the first and third quadrants.

Figure A8.13(d) shows the graph of $y = x^4$. This graph is even steeper than those in Figs. A8.13(b) and A8.13(c). Like the square law graph of Fig. A8.13(b), this graph has values in the first and second quadrants.

The graph of $y = x^5$ is shown in Fig. A8.13(e). Like the cubic law graph of Fig. A8.13(c), this graph has values in the first and third quadrants.

Finally, Fig. A8.13(f) shows the graph of $y = 1/x$ (or $y = x^{-1}$). Note how the y values are very large for small values of x and very small for very large values of x. This graph has values in the first and third quadrants.

If you take a careful look at Fig. A8.13 you should notice that, for odd powers of x (i.e. x^1, x^3, x^5, and x^{-1}) the graph will have values in the first and third quadrant whilst for even powers of x (i.e. x^2 and x^4) the graph will have values in the first and second quadrants.

Trigonometry

You might be wondering what trigonometry has to do with electronics. It's just that a familiarity with some basic trigonometry is fundamental to developing an understanding of signal, waveforms and AC circuits. Indeed, the most fundamental waveform used in AC circuits is the sine wave and this wave is a trigonometric function.

Basic trigonometrical ratios

Trigonometrical ratios are to do with the way in which we measure angles. Take a look at the right-angled triangle shown in Fig. A8.14. This triangle has three sides; a, b, and c. The angle that we are interested in (we have used the Greek symbol θ to denote this angle) is adjacent to side a and is opposite to side b. The third side of the triangle (the *hypotenuse*) is the longest side of the triangle.

In Fig. A8.14, the theorem of Pythagoras states that 'the square on the hypotenuse is equal to the sum of the squares on the other two sides'. Writing this as an equation we arrive at:

$$c^2 = a^2 + b^2$$

where c is the hypotenuse and a and b are the other two sides.

Taking square roots of both sides of the equation we can see that:

$$c = \sqrt{a^2 + b^2}$$

Thus if we know two of the sides (for example, a and b) of a right-angled triangle we can easily find the third side (c).

The ratios a/c, b/c, and a/b are known as the basic *trigonometric ratios*. They are known as sine (*sin*), cosine (*cos*) and tangent (*tan*) of angle θ respectively. Thus:

$$\sin \theta = \frac{opposite}{hypotenuse} = \frac{a}{c}$$

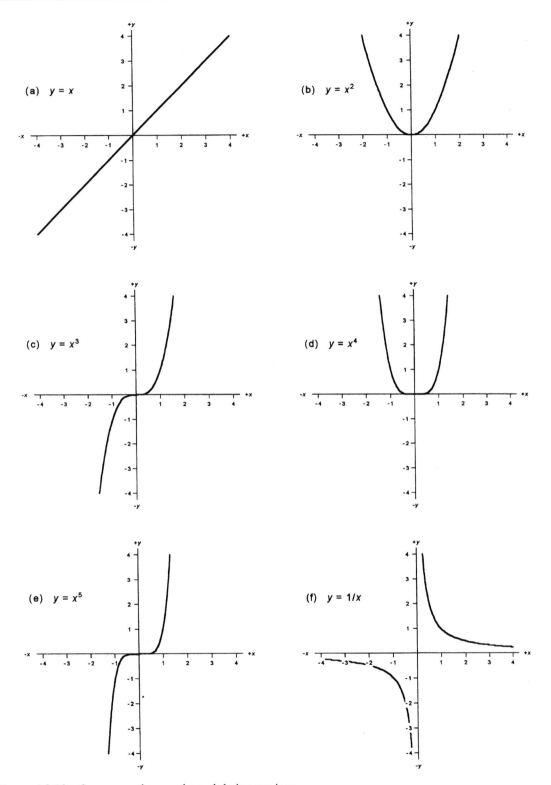

Figure A8.13   Some complex graphs and their equations

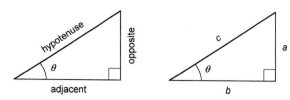

Figure A8.14   Right-angled triangles

and

$$\cos \theta = \frac{adjacent}{hypotenuse} = \frac{b}{c}$$

and

$$\tan \theta = \frac{opposite}{adjacent} = \frac{a}{b}$$

Trigonometrical equations

Equations that involve trigonometrical expressions are known as trigonometrical equations. Fortunately they are not quite so difficult to understand as they sound! Consider the equation:

$$\sin \theta = 0.5$$

This equation can be solved quite easily using a calculator. However, before doing so, you need to be sure to select the correct mode for expressing angles on your calculator. If you are using a 'scientific calculator' you will find that you can set the angular mode to either **radian** measure or **degrees**. A little later we will explain the difference between these two angular measures but for the time being we shall use degrees.

If you solve the equation (by keying in 0.5 and pressing the **inverse sine** function keys) you should see the result 30° displayed on your calculator. Hence we can conclude that:

$$\sin 30° = 0.5$$

Actually, a number of other angles will give the same result! Try pressing the sine function key and entering the following angles in turn:

30°, 210°, 390° and 570°

They should all produce the same result, 0.5! This should suggest to you that the graph of the sine function repeats itself (i.e. the shape of the graph is *periodic*). In the next section we shall plot the sine function but, before we do we shall take a look at using radian measure to specify angles.

The **radian** is defined as the angle subtended by an arc of a circle equal in length to the radius of the circle. This relationship is illustrated in Fig. A8.15.

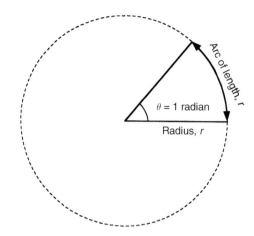

Figure A8.15   Definition of the radian

The circumference, *l*, of a circle is related to its radius, *r*, according to the formula:

$$l = 2\pi r$$

Thus

$$r = \frac{l}{2\pi}$$

Now, since there are 360° in one complete revolution we can deduce that one radian is the same as 360°/2π = 57.3°. On other words, to convert:

(a)   degrees to radians multiply by 57.3
(b)   radians to degrees divide by 57.3

It is important to note that one complete cycle of a periodic function (i.e. a waveform) occurs in a time, *T*. This is known as the *periodic time* or just the *period*. In a time interval equal to *T*, the angle will have changed by 360°. The relationship between time and angle expressed in degrees is thus:

$$\theta = \frac{T}{t} \times 360° \quad \text{and} \quad t = \frac{T}{\theta} \times 360°$$

Thus, if one complete cycle (360°) is completed in 0.02 s (i.e. $T = 20$ ms) an angle of 180° will correspond to a time of 0.01 s (i.e. $t = 10$ ms).

Conversely, if we wish to express angles in radians:

$$\theta = \frac{T}{t} \times 2\pi \quad \text{and} \quad t = \frac{T}{\theta} \times 2\pi$$

Thus, if one complete cycle (2π radians) is completed in 0.02 s (i.e. $T = 20$ ms) an angle of p radians will correspond to a time of 0.01 (i.e. $t = 10$ ms). It should be apparent from this that, when considering waveforms, time and angle (whether expressed in degrees or radians) are interchangeable. Note that the general equation for a sine wave *voltage* is:

$$v = V_{max} \sin(2\pi ft) = V_{max} \sin(6.28 \times \frac{t}{T})$$

where V_{max} is the maximum value of the voltage and T is the periodic time.

Graphs of trigonometrical functions

To plot a graph of $y = \sin \theta$ we can construct a table of values of $\sin \theta$ as θ is varied from 0° to 360° in suitable steps. This exercise (carried out using a scientific calculator) will produce a table that looks something like this:

Angle, θ	0°	30°	60°	90°	120°	150°
$\sin \theta$	0	0.5	0.87	1	0.87	0.5

Angle, θ	180°	210°	240°	270°	300°	330°
$\sin \theta$	0	−05	−0.87	−1	−0.87	−0.5

Plotting a graph of the values in the table reveals the graph shown in Fig. A8.16. We can use the same technique to produce graphs of $\cos \theta$ and $\tan \theta$, as shown in Figs A8.17 and A8.18.

Example 13

The voltage applied to a high voltage rectifier is given by the equation:

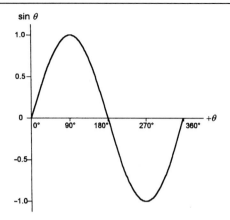

Figure A8.16   Graph of the sine function (the equation of the function is $y = \sin \theta$)

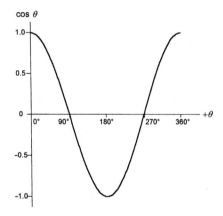

Figure A8.17   Graph of the cosine function (the equation of the function is $y = \cos \theta$)

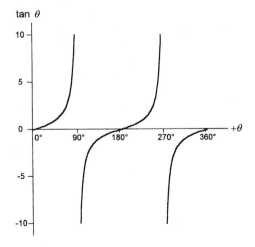

Figure A8.18   Graph of the tangent function (the equation of the function is $y = \tan \theta$)

$v = 800 \sin(314\ t)$

Determine the voltage when:

(a) $t = 1$ ms
(b) $t = 15$ ms.

We can easily solve this problem by substituting values into the expression.
   In case (a):

$v = 800 \sin(314 \times 0.001) = 800 \sin(0.314)$

$\quad = 800 \times 0.3088 = 247$ V

In case (b):

$v = 800 \sin(314 \times 0.015) = 800 \sin(4.71)$

$\quad = 800 \times -1 = \mathbf{-800\ V}$

Example 14

The voltage appearing at the secondary winding of a transformer is shown in Fig. A8.19. Determine:

(a) the period of the voltage
(b) the value of the voltage when (i) $t = 5$ ms and (ii) $t = 35$ ms
(c) the maximum and minimum values of voltage
(d) the average value of the voltage during the time interval 0 to 40 ms

(e) the equation for the voltage.

All of the above can be obtained from the graph of Fig. A8.19, as follows:

(a) period, $T = 20$ ms (this is the time for one complete cycle of the voltage)
(b) at $t = 5$ ms, $v = 20$ V; at $t = 35$ ms, $v = -20$ V
(c) $v_{max} = +20$ V and $v_{min} = -20$ V
(d) $v_{av.} = 0$ V (the positive and negative cycles of the voltage are equal)
(e) $v = 20 \sin 314\ t$ (the maximum value of the voltage, V_{max}, is 20 V and $T = 0.02$ s).

Boolean algebra

Boolean algebra is the algebra that we use to describe and simplify logic expressions. An understanding of Boolean algebra can be extremely useful when troubleshooting any circuit that employs digital logic. It is, for example, very useful to be able to predict the output of a logic circuit when a particular input condition is present.

Boolean notation

Just as with conventional algebra, we use letters to represent the variables in these expressions. The

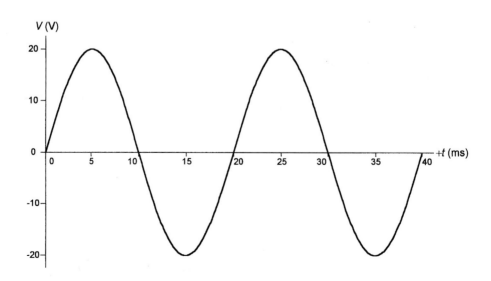

Figure A8.19   See Example 14

symbols used in Boolean algebra do not, however, have the same meanings as they have in conventional algebra:

Symbol	Meaning
$\cdot$	Logical AND
$+$	Logical OR
$\oplus$	Logical exclusive-OR
$\overline{}$	Logical NOT

Thus 'A AND B' can be written ($A \cdot B$) whereas 'A OR B' is written ($A + B$). The bar symbol, $\overline{}$, used above a variable (or over a part of an expression) indicates inversion (the NOT function). Thus 'NOT A' is written $\overline{A}$ and 'NOT (A AND B)' is written $\overline{A \cdot B}$. Note that $\overline{A \cdot B}$ is *not* the same as $\overline{A} \cdot \overline{B}$.

It is important to note that a logic variable (such as A or B) can exist in only two states variously (and interchangeably) described as true or false, asserted or non-asserted, 1 or 0, etc. The two states are mutually exclusive and there are no 'in between' conditions.

Truth tables

We often use tables to describe logic functions. These tables show all possible logical combinations and they are called *truth tables*. The columns in a truth table correspond to the input variables and an extra column is used to show the resulting output. The rows of the truth tables show all possible states of the input variables. Thus, if there are two input variables (A and B) there will be four possible input states, as follows:

$A = 0, B = 0$
$A = 0, B = 1$
$A = 1, B = 0$
$A = 1, B = 1$

Basic logic functions

The basic logic functions are: OR, AND, NOR, and NAND. We can describe each of these basic logic functions using truth tables as shown in Fig. A8.20 (for two input variables) and in Fig. A8.21 (for three input variables).

In general, a logic function with n input variables will have 2^n different possible states for those inputs. Thus, with two input variables there will be four possible input states, for three input variables there will be eight possible states, and for four input variables there will be 16 possible states.

Take a look at Fig. A8.20 and Fig. A8.21 and compare the output states for AND with those for NAND as well as OR with those for NOR. In every case, you should notice that the output states are complementary—in other words, all of 0's in the AND column have been inverted to become 1's in the NAND column, and vice versa. The same rule also applies to the OR and NOR columns.

A	B	A + B	A·B	$\overline{A+B}$	$\overline{A \cdot B}$
0	0	0	0	1	1
0	1	1	0	0	1
1	0	1	0	0	1
1	1	1	1	0	0

Figure A8.20 Truth tables for the basic logic functions using two variables: A and B

A	B	C	A+B+C	A·B·C	$\overline{A+B+C}$	$\overline{A \cdot B \cdot C}$
0	0	0	0	0	1	1
0	0	1	1	0	0	1
0	1	0	1	0	0	1
0	1	1	1	0	0	1
1	0	0	1	0	0	1
1	0	1	1	0	0	1
1	1	0	1	0	0	1
1	1	1	1	1	0	0

Figure A8.21 Truth tables for the basic logic functions using three variables: A, B and C

Venn diagrams

An alternative technique to that of using a truth table to describe a logic function is that of using a Venn diagram. This diagram consists of a number of overlapping areas that represent all of the possible logic states.

A simple Venn diagram with only one variable is shown in Fig. A8.22. The shaded area represents the condition when variable *A* is *true* (i.e. logic 1). The non-shaded area represents the condition when variable *A* is *false* (i.e. logic 0). Obviously, variable *A* cannot be both true and false at the same time hence the shaded and non-shaded areas are mutually exclusive.

Figure A8.23 shows Venn diagrams for the basic logic functions; AND, OR, NAND and NOR. You should note that these diagrams convey exactly the same information as the truth tables in Fig. A8.20 but presented in a different form.

Karnaugh maps

A Karnaugh map consists of a square or rectangular array of cells into which 0's and 1's may be placed to indicate false and true respectively. Two alternative representations of a Karnaugh map for two variables are shown in Fig. A8.24.

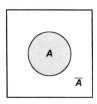

Figure A8.22   Venn diagram for a single logic variable, *A*

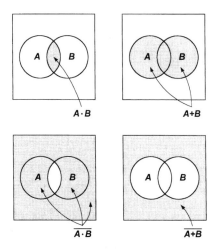

Figure A8.23   Venn diagrams for the basic logic functions

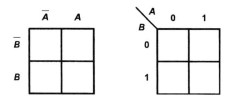

Figure A8.24   Alternative representations for a Karnaugh map showing two variables

The relationship between Boolean logic expressions for two variables and their Karnaugh maps is illustrated in Fig. A8.25 whilst Fig. A8.26 shows how the basic logic functions, AND, OR, NAND and NOR, can be plotted on Karnaugh maps.

Adjacent cells within a Karnaugh map may be grouped together in rectangles of two, four, eight, etc. cells in order to simplify a complex logic expression. Taking the NAND function, for example, the two groups of two adjacent cells in the Karnaugh map correspond to $\overline{A}$ or $\overline{B}$, as shown in Fig. A8.27. We thus conclude that:

$$\overline{A \cdot B} = \overline{A} + \overline{B}$$

Once again note how the overscore character is used to indicate the inverting function.

This important relationship is known as *De Morgan's theorem*.

Karnaugh maps can be drawn showing two, three, four, or even more variables. The Karnaugh map shown in Fig. A8.28 is for three variables—note that there are eight cells and that each cell corresponds to one of the eight possible combinations of the three variables; *A*, *B* and *C*.

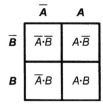

Figure A8.25   Karnaugh map for two variables showing the Boolean logic expression for each cell

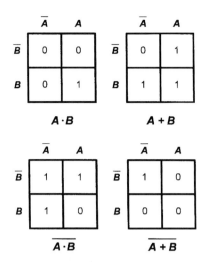

Figure A8.26   Karnaugh maps for the basic logic functions

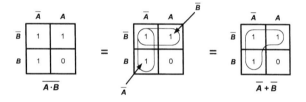

Figure A8.27   Grouping adjacent cells together to prove De Morgan's theorem

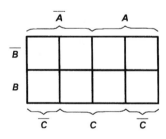

Figure A8.28   Karnaugh maps for three variables

The technique of grouping cells together is an extremely powerful one. On a Karnaugh map showing three variables, if two adjacent cells both contain 1's they can be grouped together to produce an expression containing just two variables. Similarly, if four adjacent cells can be grouped

together, they reduce to an expression containing only one variable.

It is also important to note that the map is a continuous surface which links edge to edge. This allows cells at opposite extremes of a row or column to be linked together. The four corner cells of a four (or more) variable map may likewise be grouped together (provided they all contain 1's).

As an example, consider the following Boolean function:

$$(A \cdot B \cdot C) + (A \cdot B \cdot \overline{C}) + (A \cdot \overline{B} \cdot C) + (A \cdot \overline{B} \cdot \overline{C})$$

Figure A8.29 shows how this function is plotted on the Karnaugh map.

As before, we have placed a 1 in each cell. Then we can begin to link together adjacent cells—notice how we have reduced the diagram to a group of four cells equivalent to the logic variable, A. In other words, the value of variables B and C, whether 0 or 1, will have no effect on the logic function. From Fig. A8.30 we can conclude that:

$$(A \cdot B \cdot C) + (A \cdot B \cdot \overline{C}) + (A \cdot \overline{B} \cdot C) + (A \cdot \overline{B} \cdot \overline{C}) = A$$

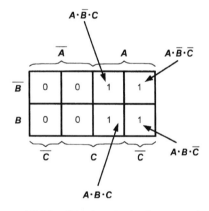

Figure A8.29   Plotting the logic expression on a Karnaugh map

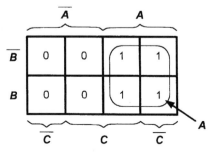

Figure A8.30   Grouping adjacent 1 cells together

Boolean algebra

The laws of Boolean algebra are quite different from those of ordinary algebra. For example:

$A \cdot 0 = 0$
$A \cdot 1 = A$
$A + 0 = A$
$A + 1 = 1$

The **Commutative Law** shows us that:

$A + B = B + A$

and also that:

$A \cdot B = B \cdot A$

The **Associative Law** shows us that:

$A + (B + C) = (A + B) + C = A + B + C$

and also that:

$A \cdot (B \cdot C) = (A \cdot B) \cdot C = A \cdot B \cdot C$

The **Distributive Law** shows us that:

$A \cdot (B + C) = A \cdot B + B \cdot C$

and also that:

$A + (B \cdot C) = (A + B) \cdot (A + C)$

All of the above laws can be proved using the various techniques that we met earlier.

Boolean simplification

Using the laws of Boolean algebra and De Morgan's theorem, $\overline{A \cdot B} = \overline{A} + \overline{B}$ (see earlier) we can reduce complex logical expressions in order to minimize the number of variables and the number of terms. For example, the expression that we met earlier can be simplified as follows:

$$\left(A \cdot B \cdot C\right) + \left(A \cdot B \cdot \overline{C}\right) + \left(A \cdot \overline{B} \cdot C\right) + \left(A \cdot \overline{B} \cdot \overline{C}\right) = \left(A \cdot B\right) + \left(A \cdot \overline{B}\right) = A$$

Here we have eliminated the variable C from the first and second pair of terms and then eliminated B from the result. You might like to compare this with the Karnaugh map that we drew earlier in Figure A8.30!

Example 15

Part of the control logic for an intruder alarm is shown in Fig. A8.31. Determine the logic required

to select the Y4 output and verify that the Y3 output is permanently selected and unaffected by the states of S1 and S2.

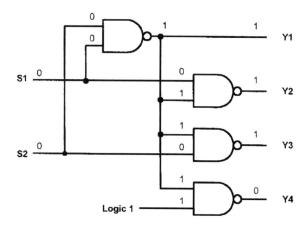

Figure A8.31   See Example 15

Each logic gate shown in Fig. A8.31 is a two input NAND gate. We can construct the truth table for the control circuit (see Fig. A8.32) by placing all four possible combinations of 0 and 1 on the S1 and S2 inputs and tracing the logic states through the circuit. We have shown the first stage of this process (the logic conditions that produce the first line of the truth table) in Fig. A8.32.

S1	S2	Y1	Y2	Y3	Y4
0	0	1	1	1	0
0	1	1	1	1	0
1	0	1	0	1	0
1	1	0	1	1	1

Logic states corresponding to this line are shown in Fig. **A8.31**

Figure A8.32   Truth table for Example 15

From Fig. A8.32 it can be seen that the Y4 output corresponds to the AND logic function:

$Y4 = S1 \cdot S2$

You can also see from Fig. A8.32 that the Y3 output remains at logic 1 regardless of the states of S1 and S2.

Logarithms

Many of the numbers that we have to deal with in electronics are extremely large whilst others can be extremely small. The problem of coping with a wide range of numbers is greatly simplified by using logarithms. Because of this, we frequently employ logarithmic scales, **Bels** (B) and **decibels** (dB), when comparing values. Note that a decibel (dB) is simply one tenth of a Bel (B). Hence:

10 dB = 1 B

It is important to note that the Bel is derived from the logarithm of a ratio—it's *not* a unit like the volt (V) or ampere (A).

The logarithm (to the base 10) of a number is defined as the power that 10 must be raised to in order to equal that number. This may sound difficult but it isn't, as a few examples will show:

Because $10 = 10^1$, the logarithm (to the base 10) of 10 is 1.

Because $100 = 10^2$, the logarithm (to the base 10) of 100 is 2.

Because $1,000 = 10^3$, the logarithm (to the base 10) of 1,000 is 3, and so on.

Writing this as a formula we can say that:

If $x = 10^n$, the logarithm (to the base 10) of x is n.

You might like to check that this works using the examples that we just looked at!

You may be wondering why we keep writing 'to the base 10'. Logarithms can be to any base but 10 just happens to be the most obvious and useful.

Logarithmic notation

The notation that we use with logarithms is quite straightforward:

If $x = 10^n$ we say that $\log_{10}(x) = n$.

The subscript that appears after the 'log' reminds us that we are using base 10 rather than any other (unspecified) number. Re-writing what we said earlier using our logarithmic notation gives:

$\log_{10}(10) = \log_{10}(10^1) = \mathbf{1}$

$\log_{10}(100) = \log_{10}(10^2) = \mathbf{2}$

$\log_{10}(1,000) = \log_{10}(10^3) = \mathbf{3}$

and so on.

So far we have used values (10, 100 and 1,000) in our examples that just happen to be exact powers of 10. So how do we cope with a number that isn't an exact power of 10? If you have a scientific calculator this is quite easy! Simply enter the number and press the 'log' button to find its logarithm (the instructions provided with your calculator will tell you how to do this).

You might like to practice with a few values, for example:

$\log_{10}(56) = \mathbf{1.748}$

$\log_{10}(1.35) = \mathbf{0.1303}$

$\log_{10}(4,028) = \mathbf{3.605}$

$\log_{10}(195,891) = \mathbf{5.292}$

$\log_{10}(0.3175) = \mathbf{-0.4983}$

Finally, there's another advantage of using logarithmic units (i.e. Bels and decibels) in electronics. If two values are multiplied together the result is equivalent to adding their values expressed in logarithmic units.

To put all of this into context, let's assume that the following items are connected in the signal path of an amplifier (see Fig. A8.33):

Pre-amplifier; input power = 1 mW, output power = 80 mW
Tone control; input power = 80 mW, output power = 20 mW
Power amplifier; input power = 20 mW, output power = 4 W

The power gain of each component, expressed as both a ratio and in dB is shown in the table below:

Component	Input power	Output power	Power gain (expressed as a ratio)	Power gain (expressed in decibels)
Pre-amplifier	1 mW	80 mW	80	19
Tone control	80 mW	20 mW	0.25	-6
Power amplifier	20 mW	4 W	200	23

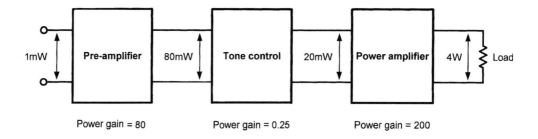

Figure A8.33 Amplifier system showing power gains and losses

The overall power gain can be calculated from the following:

Overall power gain (expressed as a ratio)

$$= \frac{\text{output power}}{\text{input power}} = \frac{4 \text{ W}}{1 \text{ mW}} = 4,000$$

Converting this gain to decibels gives:

$10 \log_{10} (4,000) =$ **36 dB**

Adding the power gains from the right-hand column gives: $19 - 6 + 23 =$ **36 dB**

Example 16

An amplifier has a power gain of 27,000. Express this in Bels (B) and decibels (dB).

In Bels, power gain $= \log_{10}(27,000) = 4.431$ B

In decibels, power gain

$= 10 \times \log_{10}(27,000) = 10 \times 4.431 =$ **44.31 dB**

Example 17

An amplifier has a power gain of 33 dB. What input power will be required to produce an output power of 11 W?

Now power gain expressed in decibels will be given by:

$$10 \log_{10} \left(\frac{P_{out}}{P_{in}} \right)$$

Thus $33 = 10 \log_{10} \left(\frac{11 \text{ W}}{P_{in}} \right)$

Dividing both sides by 10 gives:

$$3.3 = \log_{10} \left(\frac{11 \text{ W}}{P_{in}} \right)$$

Taking the antilog of both sides gives:

$$\text{antilog}(3.3) = \frac{11 \text{ W}}{P_{in}}$$

$$1,995 = \frac{11 \text{ W}}{P_{in}}$$

from which:

$$P_{in} = \frac{11 \text{ W}}{1,995} = 5.5 \text{ mW}$$

Note that the **antilog** or 'inverse log' function can be found using your calculator. This sometimes appears as a button marked '$10^{x'}$' or is obtained by pressing 'shift' (to enable the inverse function) and then 'log'. In any event, you should refer to your calculator's instruction book for more information on the keystrokes required!

Exponential growth and decay

In electronics we are often concerned with how quantities grow and decay in response to a sudden change. In such cases the laws that apply are the same as those that govern any natural system. The formula that relates to **exponential growth** takes the form:

$$y = Y_{max} \left(1 - e^x \right)$$

whilst that which relates to *exponential decay* takes the form:

$$y = Y_{max}e^{-x}$$

In the case of exponential growth, the value, Y_{max}, is simply the maximum value for y. In other words it is approximately the value that y will reach after a very long time (note that, even though y gets very close in value to it, Y_{max} is *never* quite reached).

In the case of exponential decay, the value, Y_{max}, is the initial value of y. In other words it is the value of y before the decay starts. After a very long period of time, the value of y will reach approximately zero (note that, even though y gets very close in value to zero, it is never quite reached).

Figure A8.34 shows a graph showing exponential growth whilst Fig. A8.35 shows a graph showing exponential decay. In both cases Y_{max} has been set to 1.

The name given to the constant, e, is the **exponential constant** and it has a value of approximately 2.713. The value of the constant is normally accessible from your calculator (in the same way that the value of π is stored as a constant in your calculator).

Finally, let's take a look at a practical example of the exponential function involving the discharge of a capacitor.

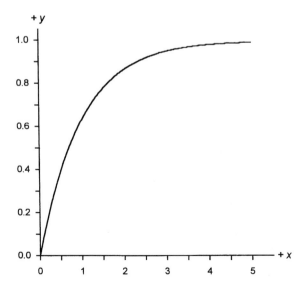

Figure A8.34   An exponential growth curve

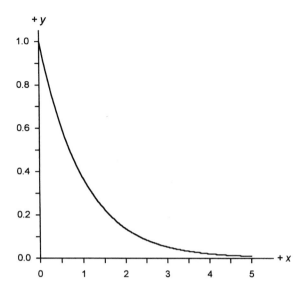

Figure A8.35   An exponential decay curve

Example 18

A capacitor is charged to a potential of 100 V. It is then disconnected from its charging source and left to discharge through a resistor. If it takes 10 s for the voltage to fall to 50 V, how long will it take to fall to 10 V?

Here we are dealing with exponential decay. If we use y to represent the capacitor voltage at any instant, and since the initial voltage is 100 V, we can determine the value of x (i.e. x_1) when the voltage reaches 50 V by substituting into the formula:

$$y = Y_{max}e^{-x}$$

hence:

$$50 = 100e^{-x_1}$$

from which:

$$e^{-x_1} = \frac{50}{100} = 0.5$$

taking logs (to the base e this time) of both sides:

$$\log_e\left(e^{-x_1}\right) = \log_e\left(0.5\right)$$

thus:

$-x_1 = -0.693$

hence:

$x_1 = 0.693$

We can use a similar process to find the value of x (i.e. x_2) that corresponds to a capacitor voltage of 10 V:

$y = Y_{max}e^{-x}$

hence:

$10 = 100e^{-x_2}$

from which:

$e^{-x_2} = \dfrac{10}{100} = 0.1$

taking logs (again to the base e) of both sides:

$\log_e\left(e^{-x_2}\right) = \log_e\left(0.1\right)$

thus:

$-x_2 = -2.3$

hence:

$x_2 = 2.3$

We can now apply direct proportionality to determine the time taken for the voltage to reach 10 V. In other words:

$\dfrac{t_{50}}{x_1} = \dfrac{t_{10}}{x_2}$

from which:

$t_{10} = t_{50} \times \dfrac{x_2}{x_1}$

thus:

$t_{10} = 10 \times \dfrac{2.3}{0.693} = 33.2\,\text{s}$

Figure A8.36 illustrates this relationship.

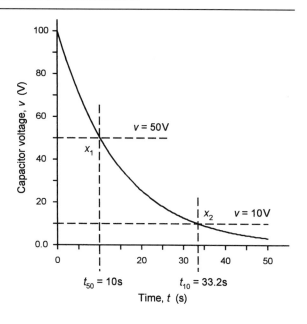

Figure A8.36   Capacitor voltage during discharge —see Example 18

Appendix 9

Useful web addresses

Author's website and additional resources to support this book

key2electronics   http://www.key2electronics.com/

Semiconductor manufacturers

Analog Devices   http://www.analog.com/
Dallas Semiconductor   http://www.dalsemi.com/
Fujitsu   http://www.fujitsu-fme.com/
Harris Semiconductor   http://www.harris.com/
Holtek Semiconductor   http://www.holtek.com
Intel   http://www.intel.com/
International Rectifier   http://www.irf.com/
Linear Technology   http://www.linear.com/
Maxim   http://www.maxim-ic.com/
Microchip Technology   http://www.microchip.com/
National Semiconductor   http://www.national.com/
Philips   http://www.semiconductors.philips.com/
Samsung   http://www.samsung.com/
Seiko   http://www.sii-ic.com/
SGS-Thompson   http://www.st.com/stonline/
Siemens   http://www.siemens.de/
Sony   http://www.sony.com/
Teridian Semiconductor   http://www.tsc.tdk.com/
Texas Instruments   http://www.ti.com/
Zilog   http://www.zilog.com/

Electronic component suppliers

Farnell Electronic Components   http://www.farnell.co.uk/
Greenweld   http://www.greenweld.co.uk/
Jaycar Electronics   http://www.jaycarelectronics.co.uk/
Magenta   http://www.magenta2000.co.uk/
Maplin Electronics   http://www.maplin.co.uk/
Quasar Electronics   http://www.quasarelectronics.com/
Rapid Electronics   http://www.rapidelectronics.co.uk/
RS Components   http://www.rswww.com/

Magazines

Everyday with Practical Electronics (EPE)	http://www.epemag.wimborne.co.uk/
EPE Online	http://www.epemag.com/
Elektor	http://www.elektor-electronics.co.uk/

Clubs and societies

American Radio Relay league	http://www.arrl.org/
British Amateur Electronics Club	http://members.tripod.com/baec/
Radio Society of Great Britain	http://www.rsgb.org.uk/

Miscellaneous

5Spice Analysis	http://www.5spice.com/
CadSoft Online (Eagle)	http://www.numberonesystems.com/
Electronics Workbench (Multisim)	http://www.electronicsworkbench.com/
Labcenter Electronics	http://www.labcenter.co.uk/
Matrix Multimedia (E-blocks)	http://www.matrixmultimedia.co.uk/
Microchip Technology (MPLab IDE)	http://www.microchip.com/
MicroElektronika (EasyPIC3)	http://www.breadboarding.co.uk/
Oshonsoft (Z80 and PIC simulators)	http://www.oshonsoft.com/
Pico Technology (Picoscope)	http://www.picotech.com/
Tina Design Suite	http://www.tina.com/
WebEE Electronic Engineering Homepage	http://www.web-ee.com/
WinSpice	http://www.winspice.com/

Index

Semiconductor devices